Environmental Footprints and Eco-design of Products and Processes

Series Editor

Subramanian Senthilkannan Muthu, Green Story Inc., Toronto, Canada

Indexed by Scopus

This series aims to broadly cover all the aspects related to environmental assessment of products, development of environmental and ecological indicators and eco-design of various products and processes. Below are the areas fall under the aims and scope of this series, but not limited to: Environmental Life Cycle Assessment; Social Life Cycle Assessment; Organizational and Product Carbon Footprints; Ecological, Energy and Water Footprints; Life cycle costing; Environmental and sustainable indicators; Environmental impact assessment methods and tools; Eco-design (sustainable design) aspects and tools; Biodegradation studies; Recycling; Solid waste management; Environmental and social audits; Green Purchasing and tools; Product environmental footprints; Environmental management standards and regulations; Eco-labels; Green Claims and green washing; Assessment of sustainability aspects.

Subramanian Senthilkannan Muthu
Editor

Carbon Footprint Assessments

Case Studies & Best Practices

Springer

Editor
Subramanian Senthilkannan Muthu
Chief Sustainability Officer
Green Story Inc.
Toronto, ON, Canada

ISSN 2345-7651 ISSN 2345-766X (electronic)
Environmental Footprints and Eco-design of Products and Processes
ISBN 978-3-031-70261-7 ISBN 978-3-031-70262-4 (eBook)
https://doi.org/10.1007/978-3-031-70262-4

This Springer imprint is published by the registered company Springer Nature Switzerland AG
The registered company address is: Gewerbestrasse 11, 6330 Cham, Switzerland

This book is dedicated to:
The lotus feet of my beloved Lord
Pazhaniandavar
My beloved late Father
My beloved Mother
My beloved Wife Karpagam and Daughters—
Anu and Karthika
My beloved Brother
Last but not least
To everyone working in various industrial
spaces to make the world "SUSTAINABLE"

Contents

Carbon Footprint Assessments: Case Studies and Best Practices

Subramanian Senthilkannan Muthu

In the face of escalating climate change, understanding and mitigating carbon footprints has become a critical priority for businesses, governments, and individuals alike. A carbon footprint represents the total greenhouse gas (GHG) emissions caused directly and indirectly by an entity, typically measured in carbon dioxide equivalents (CO_2e). This chapter delves into the methodologies, case studies, and best practices for carbon footprint assessments, providing a comprehensive overview of how various sectors are addressing their environmental impact.

This book, "Carbon Footprint Assessments: Case Studies and Best Practices," aims to provide a comprehensive overview of various methodologies for assessing carbon footprints, along with practical examples from diverse sectors. The following chapters will delve into specific case studies, offering insights into best practices and lessons learned from real-world applications. The book is arranged in a way that chapters. "Carbon Footrints in Agriculture: Challenges, Mitigation Strategies, and Future Prospects", "An In-depth Exploration of Carbon Footprint Analysis in the Construction Sector with Emphasis on the Dam Industry", "A Comprehensive Review of Carbon Footprints in the Wine Supply Chain in Italy", "Carbon Footprint of Fashion: Assessing and Addressing Carbon Emissions in Textile Production", and "Strategies to Reduce the Carbon Footprint of Protective Face Masks" deal with Carbon Footprint Assessment and Mitigation in Selected Industrial Spaces, while chapters. "Nature-Based Solutions to Reduce Carbon Footprint and Increase Resilience of Road Transportation Infrastructure", "Carbon Footprint as a Tool to Improve Sustainability: Evidence from the Agri-food Supply Chain", "Carbon Footprint Assessment of Agricultural Production Systems Under the Variable System Boundary Scenarios", "Offsetting Carbon Footprints: Exploring CCUS Challenges and Solutions Across Diverse Industries in India", "Climate Response

S. S. Muthu (✉)
Green Story Inc., Toronto, ON, Canada

S. S. Muthu (ed.), *Carbon Footprint Assessments*, Environmental Footprints and Eco-design of Products and Processes,
https://doi.org/10.1007/978-3-031-70262-4_1

and Nature-Based Solutions: Evaluating the Impact of the Belt and Road Initiative in Latin America", "Implementing Circular Economy Principles in Road Transport: Enhancing Efficiency to Mitigate Resource Depletion and Carbon Footprint", "Best Practices Applied in Logistic Intern: From the Perspective of Reducing the Carbon Footprint", "Sowing the Future: The Vital Role of Green Bonds in Sustainable Agribusiness to Reduce the Carbon Footprint", and "Green Bonds and the Incentive to Decrease the Carbon Footprint in the Brazilian Agribusiness Sector" explore broader perspectives and strategies.

The second chapter deals around the climate change impacts in Agriculture. Agriculture, being pivotal in global food production, has also an environmental impact, particularly in terms of carbon footprints, which in turn has become a subject of growing concern. This chapter will spotlight a comprehensive overview of the current state of research on carbon footprints in agriculture, focusing on the assessment of greenhouse gas (GHG) emissions, exploration of mitigation strategies, and the promotion of sustainable practices. The assessment of agricultural carbon involves quantifying emissions associated with cultivation practices, land use, livestock management, fertilizer application, and energy consumption, which in turn aims to identify key contributors and establish a legacy for effective mitigation strategies. Mitigation strategies encompass technological advancements, as well as policy interventions that incentivize sustainable practices. Thus, sustainable agricultural practices emerge as essential components in the quest to minimize carbon footprints. Mitigating the carbon footprint in agriculture is not only an environmental imperative but also an integral component of broader global efforts to achieve sustainable development goals. This chapter seeks to delve into the intricate challenges associated with the carbon footprint in agriculture, scrutinize existing mitigation strategies, and explore prospects that offer the potential for a more sustainable and resilient agricultural sector. The agricultural sector is not only a significant contributor to greenhouse gas emissions but also vulnerable to the impacts of climate change. As the global population continues to grow, the need for sustainable agricultural practices becomes increasingly urgent. Agriculture contributes to carbon footprints through various sources, including enteric fermentation, manure management, rice cultivation, and the use of synthetic fertilizers. Deforestation and changes in land use patterns further exacerbate carbon emissions, leading to a loss of biodiversity and ecosystem services. Conventional farming practices often rely heavily on fossil fuels, synthetic inputs, and water, contributing to resource depletion and environmental degradation. Implementing agroecological practices, such as organic farming, agroforestry, and crop rotation, can enhance soil health, reduce emissions, and promote biodiversity. Utilizing technology for precision agriculture helps optimize resource use, reduce waste, and improve overall efficiency. Integrating renewable energy sources on farms, such as solar and wind power, can offset emissions associated with energy-intensive agricultural activities. Adopting practices that enhance carbon sequestration in soils, such as cover cropping and conservation tillage, contributes to mitigating emissions. The authors of this chapter conclude that as we move forward, it is imperative that the agricultural sector transitions toward a more sustainable and climate-resilient model.

The third chapter presents the details pertaining to carbon footprint in construction sector. The construction industry is a significant contributor to the emission of greenhouse gases (GHG) emissions released during the production and transportation of construction materials, as well as energy consumption in construction and operation. According to the World Economic Forum, this whole buildings and construction business contributes a substantial 39% to CO_2 emissions from energy use. Here's the thing—construction activities also generate a lot of waste, from tearing things down and building them up, and when that waste ends up in landfills, it starts emitting methane, a pretty potent greenhouse gas. Dams play a crucial role in this sector, warranting an assessment of their carbon footprint. This chapter will explore the concept of the carbon footprint, exploring its introduction, pertinent standards and protocols, and the software instrumental in its precise calculation. The authors of this chapter first introduce the concept of the carbon footprint, emphasizing its significance in the construction industry. It then delves into the environmental concerns prevalent in construction, highlighting the need for sustainable practices. Following this, two types of dams are examined as case studies to illustrate the carbon footprint of construction projects. Subsequently, strategies for mitigating carbon emissions in the construction sector are explored. Finally, the chapter summarizes the most important findings derived from the analysis, offering insights into the implications for sustainable construction practices. This chapter employs the IMPACT2002+ methodology to evaluate the carbon footprint of two concrete dams: the Pine Flat dam (a concrete gravity dam) and the Morrow Point dam (a concrete arch dam). The first case study examines the complete life cycle, from construction to destruction, disposal, and recycling. It explores the impact of seismic retrofitting versus non-retrofitting, while the second case study focuses solely on the construction stage. Evaluation considers three stages: initial construction, seismic retrofitting, and decommissioning for the retrofitting and non-retrofitting part, assuming 20% concrete recycling for the latter. Results are presented for four life cycle scenarios, revealing a 32% reduction in pollution through concrete recycling during decommissioning. The study also highlights the positive environmental impact of retrofitting dams as compared to disposal, emphasizing considerations for carbon footprint and human health indicators. As per the results derived from this study reported in this chapter, the first case study involves the examination of four distinct life cycle scenarios for the Pine Flat dam, covering initial construction, retrofitting, and eventual decommissioning. Environmental assessments for the demolition and recycling phases, representing two scenarios for sections with and without retrofitting, were computed, incorporating a 20% recycling rate for demolished concrete. The second case study focused solely on the construction stage. Results revealed an estimated overall carbon footprint of approximately 495.1 kg/m^3 throughout the construction phase, with concrete usage contributing significantly at 71.36%. The carbon footprint during construction equated to around 120 kg/m^3 of concrete in GHG emissions. Assessment of the carbon footprint from decommissioning and retrofitting indicated values of approximately 346.4 kg/m^3 and 108.1 kg/m^3 of dam concreting, respectively, with landfilling of destroyed materials being the primary contributor. Introduction of recycling led to a noteworthy 20% reduction in emissions, demonstrating the potential of sustainable waste management practices. Landfilling accounted for over 50% of the impact. The construction stage notably influenced environmental impact, while retrofitting generated a 24% carbon footprint compared to construction. These findings offer decision-makers quantitative insights into the environmental benefits of retrofitting damaged dams versus constructing new ones, crucial given seismic risks. The life cycle assessment revealed a total carbon footprint of approximately 426.1 kg/m^3 for the arch dam construction.

The fourth chapter is dedicated to dealing around the climate change impacts around the wine sector. On the global scale, the agri-food sector represents one of most contributing to climate change, accounting for about one-fifth of the entire emissions. As per the statistics reported in the literature, nowadays, approx. 50 million km^2 of the habitable land is destined to agricultural, of which 80% to crop production and 20% to animal feed production. Coming to the point on wine sector, the wine industry has generated a revenue of about 172 billion USD in 2023 and its growth is expected to be about 4.5% per year, with an average per capita consumption estimated at 2.45 L. Italy ranks second in terms of wine consumption, soon after Portugal, while its production exceeds 4.91 billion liters, with over 1775 enterprises at the country level. Although the wine market presents good opportunities for economic profit, there is still an open debate on the sustainability of the wine sector, either from the economic, social, or environmental perspective. Carbon footprint represents a crucial indicator to be evaluated in order to support the decision-making process. It helps in assessing the environmental impacts of products and processes and managing supply chains, from farm to fork. By performing a systematic literature review, this chapter intends to collect and analyze all studies on the carbon footprint of the Italian wines. The study reported in this chapter presents all the research papers concerning the carbon footprint assessment of wines in Italy, discussing the environmental impacts along the entire supply chain. Based on the results of the content analysis, this work allows for the identification of the most impactful phase of the supply chain. This review contributes to the existing scientific literature by highlighting strengths and weaknesses of the current strategies on the assessment of carbon footprint and identifying further developments in this field. The authors of the chapter structured this chapter as follows: Section 2 presents the literature trends on the carbon footprint in the wine supply chain; Section 3 illustrates the main insights from the carbon footprints in the wine industry, distinguishing between viticulture, winemaking, bottling, distribution, and End-of-Life; last, Sect. 4 ends the chapter by discussing results and presenting future research directions. The chapter will present the main findings from a comprehensive literature review conducted on the carbon footprints of the wine industry in Italy, which ranks second in terms of wine consumption worldwide and whose production exceeds 4.91 billion liters per year. Specifically, the study will focus on five main supply chain stages, namely viticulture, winemaking, bottling, distribution, and end-of-life, with the purpose of identifying the most impactful phase of the supply chain. The following conclusions are reported by the authors of this chapter: In accordance with the selected studies, the winemaking contributes on average by 27% to the entire carbon emission of the wine supply chain, of which electricity used in the cellar stage significantly impacts. In terms of bottling, it represents a complex and delicate phase and was estimated as the most impactful stage to the environment, accounting for about 43% of the total carbon emissions from cradle-to-grave. Among several solutions, operators in the wine industry can substitute heavy bottles with lighter ones, considering that the adoption of ultralight bottles (ca. 0.36 kg of glass per bottle) can reduce emissions by 32%. Last, in the field of distribution and final consumption, the study revealed that distribution accounts for 13% of the carbon emissions, whereas the impacts of the consumption phase are approx. 2%. Considering end-of-life emissions, they vary according to recycling, incineration or landfilling.

The fifth chapter is earmarked to deal with the carbon footprints of apparel industry. The textile and apparel industry on global carbon emissions contributes approximately 10% to worldwide emissions. This sector, a major environmental polluter, involves energy-intensive processes from fiber production to textile care and disposal, including recycling and landfill. The focus of this chapter is on the importance of assessing textile products' CO_2 emissions throughout their lifecycle, addressing the role of carbon footprinting in understanding the relative importance of different greenhouse gases. This comprehensive book chapter focuses on the approaches for understanding and reducing the carbon footprint of apparel products, considering global production trends, consumer behavior, legislative impacts, and the implementation of sustainable practices and technologies. This chapter examines the global apparel industry and presents information on the industry's carbon impact today. Subsequently, it discusses different methodologies for measuring carbon footprints in industrial settings. The chapter concludes by exploring policy, regulation, and industry standards that aim to reduce the carbon footprint within the apparel industry, suggesting pathways for sustainable practices and compliance. Initially, the chapter reviews various studies that highlight the environmental impacts of apparel products. It then presents data on the carbon footprint of the modern apparel industry and explains the approaches to measuring carbon footprint in industrial settings. Finally, the chapter discusses policy, regulation, and industry standards for reducing the carbon footprint in the apparel industry. As concluded by the authors of this chapter, transitioning to low emission energy not only poses challenges but also offers significant opportunities for low- and middle-income countries, including social and environmental co-benefits. These opportunities often align with Sustainable Development Goals, showing more synergies than trade-offs. For instance, sustainable transport can improve air quality, health, and gender equality, while clean cooking options can enhance health and reduce deforestation. Achieving these benefits requires policies that maximize co-benefits. Additionally, structural economic changes, such as demand-side efficiency and advancement in the clean energy sector, can foster sustainable growth tailored to national contexts.

The sixth one is dedicated to deal around the strategies for reducing the carbon footprint of protective face masks. On the global scale, protective face masks played a fundamental role in limiting the spread of the COVID-19 pandemic. More than 380 and 402 billion units were sold in 2020 and 2021, respectively, compared to about 12 billion in 2019. Although sales returned gradually to the pre-pandemic levels, registering about 28 billion units in 2023, humans still face negative environmental effect mainly due to both production and disposal of protective face masks. The impacts largely depend on the production of polypropylene (PP) non-woven fabrics, during the manufacturing stage, followed by disposal. The production, consumption, and disposal of single use PP face masks impact on soil and marine pollution due to microplastics and better performances could be achieved by switching from single use to reusable face masks, either in environmental or economic terms. Sanitization techniques modelled for reuse foster reduction of carbon footprint. From the lifecycle thinking perspective, the current chapter illustrates the main tasks to improve the sustainability of protective face mask by evaluating their carbon footprint. Such an approach contributes (i) to enrich the scientific literature in the field of carbon footprint applied to the entire supply chain of protective face masks and (ii) to implement innovative materials, more suitable for recycling and with decreased environmental impacts, and to model the environmental impacts of the life cycle of protective face masks using sanitization techniques aimed to implement reuse strategies. The chapter aims to provide a useful tool for supporting policymakers to select the best climate change mitigation strategies in the face masks supply chain.

The seventh chapter brings a different dimension, which is, dealing around the Nature-Based Solutions for adaptation and mitigation for climate change impacts. Nature-Based Solutions (NbS) represent innovative approaches that use ecosystems and their services to tackle not only the challenges of climate change, but also to address a wide range of environmental issues. This pioneering approach encompasses strategies such as ecosystem restoration, the creation of green corridors and the implementation of sustainable soil and water management practices, aimed not only at minimizing the carbon footprint, but also at increasing the resilience of critical infrastructures. This chapter seeks to analyze the potential of NbS to synergistically integrate mitigation and adaptation, highlighting both their notable advantages and inherent challenges. The results reveal that NbS offers diverse benefits that transcend the mere mitigation of climate change. By incorporating ecosystems into coping strategies, NbS promotes biodiversity, protects water resources, and improves air and soil quality, providing resilience to critical infrastructure in the face of extreme weather events. They also play a crucial role in promoting human well-being, creating healthier environments for local communities, and fostering sustainable economic opportunities. However, some challenges need to be addressed, such as the time-consuming recovery of ecosystems and limitations of scale in dense urban areas, as well as the difficulty in obtaining funding due to the long-term nature of the benefits. The search for solutions to promote sustainability must involve the formation of solid partnerships between governments, non-governmental organizations (NGOs), and the private sector. This strategic collaboration is key to implementing and strengthening NbS practices. By joining forces, these entities can leverage their combined resources and expertise, creating synergies that boost the effectiveness of sustainable initiatives. Authors of the chapter reported the findings—Analyzing the evidence and experiences documented in the literature allowed for the identification of NbSs that have proven effective in enhancing the resilience of the road transportation sector. These measures range from incorporating natural elements into road infrastructure to the restoration of adjacent ecosystems, aiming to promote environmental sustainability and adaptation to changing climatic conditions. However, the research faced challenges in identifying specific studies on NbSs applied to road transportation. This lack of direct information indicates an opportunity to fill the existing literature gap in this area. By adopting a more comprehensive approach, relevant studies addressing critical infrastructure in general, including road transportation as an illustrative element, were identified. The inclusion of research from reputable institutions and well-regarded publications provided a solid foundation of scientific knowledge to support the study and contribute to the advancement of this emerging research area.

The eighth chapter brings an important topic of discussion, which is how Carbon footprint can be perceived as a tool to improve competitiveness of business, which is need of the hour. Concerns related to the environmental effects of economic activities have been addressed in an increasingly stringent European Union (EU) regulation, which aims at making Europe the first climate-neutral continent worldwide. Nowadays, companies are asked to adopt strategies to reduce their carbon emissions as to achieve the objectives and commitments of the European environmental policy. Moreover, it is essential for companies to prioritize sustainability and Environmental, Social, and Governance (ESG) objectives to meet regulatory requirements and build a reputation for Corporate Social Responsibility (CSR). So, in order to achieve either climate neutrality, stakeholders' expectations, or competitive advantage, companies must reduce their carbon emissions by implementing carbon accounting. This chapter describes the importance of applying the CF as an environmental sustainability indicator to enhance the performance of the agri-food supply chain. By describing the main international standards and guidelines, and by presenting the main findings from two case-study conducted in Italy, the chapter aims to highlight the advantages for the entire agri-food sector and to provide a useful tool for supporting the climate change mitigation strategies requested at the global level. The current chapter under discussion illustrates the main tasks to improve the sustainability of business by evaluating their carbon footprint and has a threefold goal. First, it introduces methods and data to enrich the scientific literature in the field of carbon footprint along the value and supply chain. Second, it presents some case-studies of carbon footprint related to food products. Third, it investigates the phenomenon of the so-called environmental dumping of goods imported from non-EU countries, which are characterized by an absence or a less severe environmental legislation. Such a practice brings an undue competitive advantage, due to the lack of costs linked to climate change mitigation strategies in the origin countries. Findings reported in this chapter highlight the need of switching from non-renewable to renewable energy, as well as the importance of replacing chemical with organic fertilizers. In addition, a suitable waste management represents a key toward sustainable development and economic growth.

The ninth chapter brings out an interesting case study to enumerate Carbon footprint assessment of agricultural production systems under the variable system boundary scenario. The widespread and scattered smallholders in China have a profound impact on agricultural production in response to climate change. The authors have used the data of peasant household obtained from the survey and the factor coefficients collected in Shaanxi Province of China to estimate the farmer's agricultural production (FAP) carbon footprint (CF) in a multi system boundary scenarios approach, and discussed and judged whether the agricultural production carbon footprint is the source or sink. The results showed that the agricultural activities' carbon footprint (ACF), farm's carbon footprint (FCF) and agricultural product's carbon footprint (PCF) of wheat, maize, rice, and apple crops in FAP exhibited significant differences (or even contradictory results) in four scenarios, whereas similar carbon effect characteristics in four crops. Distribution structure of ACF in subsystem of FAP for the four scenarios was consistent with the system boundary requirements set by different GHG accounting specifications. The authors integrated the theory of LCA and carbon nitrogen cycle in scenario 4 of the system boundary set in this study, and emphasized the duality of agricultural carbon effect, that top four carbon sources were heterotrophic soil respiration, raw material production, raw material utilization and directly N_2O emission (CH_4 emissions from cropland) in wheat (rice) production, and heterotrophic soil respiration, raw material production, raw material transportation, and directly N_2O emission in maize and apple production. Carbon sequestration from soil of the cropland was the main offset pathway in wheat, maize, and rice production, but the biomass of perennial woody crops in apple production. Therefore, the authors recommend considering not only the carbon source effect of agriculture but also the accompanying carbon sink effect in CF accounting progress, which can offset partial or total greenhouse gas emissions and provide significant climate benefits in FAP. In this chapter, the authors did strive to underscore the significant influence of system boundary uncertainty by presenting the results of various CF calculations that according to the different technical specifications, followed by the design of enhanced models aimed at enhancing the precision of the agricultural carbon footprint. To be specific, a systematic approach is utilized by the authors to establish a comprehensive CF accounting model at both the product and organizational levels in Sect. 2, which process includes rigorous formulation of functional units, system boundaries, accounting system formulas, factor coefficients, and activity data, among other aspects. The Sect. 3 of the chapter provides an overall perspective of the CF of agricultural production systems, and delves into the CF specificities of various agricultural products, system boundaries, and key categories. In Sects. 4 and 5 of this chapter, two crucial characteristics that need to be emphasized in the CF accounting of agricultural production systems are highlighted. Finally, Sect. 6 explores some potential limitations and suggests several promising directions for future research.

The tenth chapter will speak about Offsetting Carbon Footprints by Exploring Carbon Capture, Utilization, and Storage (CCUS) in terms of Challenges and Solutions Across Diverse Industries in India. With nations worldwide committing to net-zero targets, the spotlight intensifies on groundbreaking technologies, notably Carbon Capture, Utilization, and Storage (CCUS), as indispensable tools for achieving substantial reductions in CO_2 emissions on a large scale. This chapter specifically delves into the critical role that CCUS technologies play in mitigating carbon footprints across diverse industrial sectors. Focusing on India, currently the third-largest global CO_2 emitter, the study offers a comprehensive review of the initiatives undertaken by the Indian government, the policy frameworks in place, and the ongoing research efforts aimed at promoting the development of CCUS. This chapter undertakes an exhaustive review of the available literature to gauge the scope of research conducted on this topic both internationally and in India. The subsequent section of this chapter presents an analysis of key studies and their findings, serving as a comprehensive resource for gathering all pertinent information on CCUS and related subjects. As per the conclusion reported by the authors in this chapter, CCUS is a technology implemented across the globe to counteract climate change effects by mitigating carbon emissions. Developed Countries like the US, Germany, Canada, Norway, Europe, etc., are a few leading countries in adopting this technology. CCUS has been a topic of discussion in India for the past two decades, and there are some key initiatives taken to implement this technology. However, these are insufficient to achieve targets under the Paris climate agreement. To endorse CCUS in the global circular economy, numerous renewable energy sources will be required. CCUS Technology and its Contribution to Circular Economy is a topic of discussion for further research.

The eleventh chapter brings a new dimension to the discussions on climate change. It will revolve around the belt and road initiative, or new silk road. This chapter is dedicated to deal around the environmental and geo political impacts of the BRI in Latin America. The Belt and Road Initiative (BRI), commonly known as the "New Silk Road," represents a far-reaching strategy orchestrated by China to expand its influence significantly. This initiative entails the establishment of extensive networks of both land and maritime transportation, aiming to provide substantial opportunities for infrastructure development in various regions worldwide. Furthermore, the BRI seeks to stimulate international trade by fostering collaboration with Chinese enterprises on a global scale. Initially launched in 2013 to connect Asia, Eastern Europe, and Africa, the project has expanded its scope to include regions such as Latin America. The initiative has incorporated Nature-Based Solutions (NbS) into its environmental cooperation program after the Chinese Council for International Cooperation on Environment and Development (CCICED), a government-affiliated think tank, recommended its adoption to achieve domestic climate and biodiversity goals. Furthermore, China‘s 2021 climate action and environmental protection guidelines now require subnational governments to apply NbS "general principles" for biodiversity protection, ecosystem restoration, and climate change mitigation and adaptation. However, it is crucial to note that the growing global interdependence and the multifaceted economic, social, and environmental nature of this initiative are not exempt from challenges and tensions. This chapter aims to explore the opportunities arising from the BRI and its current expansion into Latin American territory, analyzing dilemmas that emerge within the context of shifts in global geopolitics. Considering the influence of the United States in the region and Sino- American rivalry, the analysis seeks to deepen the understanding of the underlying challenges in this constantly evolving dynamic, with a particular focus on the impacts on the global environment, as well as the principles grounded in the global agenda of the Sustainable Development Goals (SDGs) and the 2030 Agenda. The research methodology employed for this study primarily centered on a comprehensive literature review. This approach was selected to uphold the research's reliability and validity, as it drew from well-established scientific and academic sources. These sources encompassed a diverse range of materials, including books, articles, theses, and various scholarly works available in the academic literature.

The twelfth chapter interlinks the hot topic of the day, which is Circular Economy (CE) into climate change. One response to contemporary threats to the environment, mainly from human activities, is the Circular Economy (CE), which is a solution aimed at mitigating the adverse environmental impact of the production and consumption of goods and services. This is because the CE is an economic system that supports more sustainable management of resources throughout the life cycle of systems and is characterized by closed loops, promoting more sustainable practices such as reuse, remanufacturing, and recycling. In this sense, in view of the transport sector's high energy consumption and consequent greenhouse gas (GHG) and air pollutant emissions and waste generation, this chapter aims to identify articles in the scientific literature that seek to incorporate Circular Economy concepts into the road transport sector through a systematic review with a bibliometric approach. This chapter seeks to provide a more general overview of the application of CE in the urban road transport sector, which is still lacking in the literature, presenting a broader scope than the studies currently published, which direct their efforts to a specific field of transport engineering. This is necessary in order to understand which are the main areas of application on the subject, as well as to direct learning efforts on the subject. To achieve its objectives, this study is structured as follows. Section 1 presents a contextualization of the subject, the problem, and the research objectives. Section 2 deals with the importance of using CE concepts in the transportation sector. Section 3 presents and describes the methodological procedure. Section 4 presents and discusses the bibliometric and systematic results. Finally, Sect. 5 contains the concluding remarks. The findings of this research reveal that, although the topic of integrating the Circular Economy (CE) into the road transport sector is in its infancy, there is a notable process of growth and growing interest in academia. In this scenario, the current panorama highlights eight significant areas of study, highlighting the breadth and complexity of implementing CE principles in this specific context. These areas include Road Infrastructure, which addresses the sustainability of roads and structures, and Urban Planning, which seeks to integrate CE into decisions related to the development of cities. Life Cycle Analysis (LCA) has emerged as a crucial aspect, focusing on the holistic assessment of environmental impacts throughout the life cycle of road transport systems. In addition, Material Flow Analysis is emerging as an area of research aimed at optimizing the use of resources and minimizing waste. In the context of logistics and the supply chain, CE seeks to redefine practices to make transportation more efficient and environmentally responsible. End-of-Life Analysis focuses on the proper management of waste generated by the sector, while Resource Optimization aims to maximize the efficient use of inputs, thus reducing the ecological footprint. Finally, Urban Mobility emerges as a crucial area, exploring how CE can transform ways of getting around in cities, promoting sustainable and innovative alternatives. This diversity of areas suggests a promising field for future research and practice aimed at deepening the successful integration of CE into road transportation, aligning it with broader environmental and social objectives. As concluded by the authors of this chapter, these findings reported in this chapter not only consolidate the growing interest in CE integration in the urban road transport sector, but also provide a solid basis to guide future research. To advance this field, future studies could delve deeper into each specific area identified, conducting comprehensive analyses to gain a more nuanced understanding of the practices and challenges involved. Furthermore, expanding the scope of research to cover additional relevant databases such as ScienceDirect, Scielo, and Scopus could significantly increase knowledge on the application of the circular economy in this context. Furthermore, investigating the practical implementation of circular economy initiatives in real urban road transport contexts could provide valuable insights into their feasibility, effectiveness, and impact on sustainability metrics. Another possible front consisting of the socio-economic implications and stakeholder perspectives around the adoption of circular economy practices in road transport could offer a comprehensive understanding of the broader implications of such initiatives.

The sustainable logistics and the reduction of Greenhouse Gas (GHG) emissions related to its operations have been increasingly garnering attention from both the scientific and business communities. With this in mind, the thirteenth chapter presents the details of carbon footprint pertaining to logistics sector. Internal logistics constitutes a crucial aspect of supply chain management that encompasses the entire spectrum of activities within an organization. This involves the efficient handling of storage, transportation, and distribution of goods to meet domestic demands and support manufacturing processes, ensuring a seamless production and supply of products, materials, components, and semi-finished parts. To facilitate these internal logistics operations, various equipment and vehicles are employed, predominantly powered by fossil fuels such as diesel. Unfortunately, this reliance on traditional fuels contributes to the emission of atmospheric substances and greenhouse gas (GHG), both directly and indirectly, by a company. This study endeavors to present optimal practices aimed at mitigating greenhouse gas emissions and atmospheric pollutants in internal logistics. These practices encompass internal transport, movement, handling, and storage activities. Additionally, the study seeks to initiate a discussion on actionable measures, outlining steps necessary for implementation. The research underscores the importance of adopting alternative energy sources for equipment and emphasizes the utilization of innovative systems like Picker-to-Part, Part-to-Picker, Order Batching, and Picker Routing. These systems are instrumental in reducing overall energy consumption during logistics activities. Authors conclude that, reducing the carbon footprint is still a challenge for all sectors of the economy, including internal logistics, a crucial aspect of supply chain management that encompasses the entire spectrum of activities within an organization. Therefore, to meet the needs of this sector, this study aims to explore optimal practices for reducing activity, improving operation and mitigating greenhouse gas emissions in internal logistics, focusing on low-carbon practices within the operational boundaries of an organization and proposing a methodology to help evaluators and decision-makers define evaluation limits for the scopes, categories, activities, and elements of the internal logistics operation.

The fourteenth chapter will discuss around the green bonds and their implications on climate change. The growing concern about environmental issues has been directing investors worldwide who seek investments that offer expected returns and are linked to genuinely green projects. Thus, Green Bonds emerge as fixed-income securities with funds allocated exclusively to finance or refinance eligible new or existing Green Projects that must, necessarily, comply with a well-defined taxonomy. Additionally, in recent years, central banks have assimilated the potential risks to the financial market related to climate change and the abandonment of fossil fuels, as significant economic impacts arise with global warming, where the resulting carbon emissions risk affecting investment decisions. Sectors such as agribusiness, which contribute to biodiversity loss, deforestation, and the release of considerable amounts of carbon into the atmosphere, exacerbate the impacts of climate change and signal the urgency for the use of tools to make the sector sustainable. This study aims to discuss the issuance of Green Bonds directed at agribusiness, seeking to promote reflection on the essentiality of green agendas and the need for climate change mitigation and adaptation in agribusiness. Since climate change has exponentially become a significant challenge for the sustainable development of society as a whole, it has become equally essential to direct investments towards a low-carbon and climate-resilient economy. In addition to avoiding new environmental liabilities, improvements implemented through Green Bonds can control environmental impacts and benefit previously degraded areas with low productivity, making them fertile again. This chapter focuses on presenting a literature review of studies on the impacts of agribusiness on the environment and discussing possible strategies for reducing the carbon footprint through the issuance of green bonds, which, in line with the SDGs, continue with participation of private companies, public companies, and governments in financing and refinancing green projects. Furthermore, there is the objective of alerting agribusiness decision makers and scholars on the subject about the importance of more robust analyses, as the subject requires more in-depth studies. The relevance of this study lies in addressing a topic of global scope, directly linked to the 2030 Agenda and the SDGs, but which still lacks related studies. In addition to this introductory section, the rest of the document is organized as follows: Section 2 deals with the methodological procedure developed to efficiently conduct the research; Section 3 discusses the main impacts of agribusiness on the environment related to soil management; Section 4 presents an overview of the green bond market; Section 5 correlates the importance of the green bond market for agribusiness; and, finally, Sect. 6 concludes the research and proposes guidelines for future studies on the topic. Conclusion of the chapter as drawn by the authors—As the world faces increasingly urgent environmental challenges, the role of agribusiness in the search for sustainable solutions is undeniable. Green Bonds represent an innovative and effective approach to financing and accelerating the transition of agribusiness towards a more responsible and environmentally friendly model. It is hoped that this study has provided a comprehensive overview of the benefits of these financial instruments, while highlighting the continued importance of their responsible use and global collaboration in the pursuit of a truly green agricultural future. Furthermore, due to the need for new material on the subject, it is advisable that the study be used by other critical agribusiness sectors, as well as in studies of the financial sector.

In continuation to the last chapter, the fifteenth chapter will again talk about green bonds and climate change with specific to Brazilian agri sector. The agricultural activity in Brazil dates back to colonial times and highlights a multitude of product cycles that over time have been distributed in expanded regions with the advancement of agricultural frontiers, as well as in environmental preservation areas, in order to meet internal and external demands. As a global protagonist, having reached the position of the largest grain exporter by value in 2020, Brazil still finds an active participant in its trade balance in agribusiness. However, in addition to concerns about the sector's liabilities, such as the advancement of deforestation and direct involvement in greenhouse gas emissions, it is equally important to consider the growing market pressure for the adoption of sustainable practices in agribusiness. Brazilian agribusiness faces a series of challenges related to issues such as deforestation in the Amazon, soil degradation, and excessive use of pesticides, worrying environmentalists, consumers, and investors who seek the adoption of alternatives that mitigate the damage already caused and prevent new liabilities. Thus, this paper aims to discuss Green Bonds as positive tools with well-defined taxonomies, with the potential to drive the adoption of beneficial practices in Brazilian agribusiness, which is recognized as one of the world's leading producers and exporters of agricultural commodities, with an extensive area of arable land and a favorable climate for agribusiness production. Furthermore, promoting sustainability in the sector contributes to strengthening competitiveness in the global market, attracting foreign investments that see in the green agenda the opportunity to promote themselves positively in the market, while contributing to the preservation of natural resources and weakening agents causing climate change. This chapter focuses on presenting a bibliographical review of studies on the impacts of Brazilian agribusiness and the search for reducing the carbon footprint through technologies financed by the issuance of green bonds, such as Biological Fixation of Nitrogen, which aims to ensure the sequestration of GHG in the soil and the reduction of the carbon footprint. Furthermore, the objective is to demonstrate advances in public policies, such as the ABC Plan, and to raise awareness about the importance of more in-depth analysis of the topic. The relevance of this study includes the approach to topics linked to the financial market and the environment, denoting the impossibility of talking about limiting the planet's warming to 1.5 °C, not to mention investing in technologies to mitigate the effects of climate change and reduce of the carbon footprint. In addition to this introductory section, the rest of the document is organized as follows: Section 2 deals with the methodology applied in the study, in order to conduct the considerations efficiently; Section 3 contextualizes agroindustry in Brazil; Section 4 presents the sector's environmental impacts, such as deforestation and soil erosion; Section 5 correlates efforts to make agribusiness sustainable, such as the use of minimum cultivation and integrated pest management; Section 6 provides guidelines for issuing green bonds according to globally defined taxonomies; and, finally, Sect. 7 concludes the research with observations about the topic discussed and proposes guidelines for future studies on a topic as comprehensive as agribusiness and the issuance of green bonds. Authors of this chapter conclude that green bonds are an important tool for decarbonization, as it is impossible to talk about resilience and mitigating the effects of climate change without including the financial market. It is also impossible to address green issues without social concern for rural populations whose income is in decline, compromised by the consequences of climate change and the depletion of resources resulting from the expansion of borders by large industries. With this, the green seal also becomes an attractive element, increasing the companies' reputation in the eyes of consumers and the market. It is important to note that strict supervision, with defined taxonomies, and accurate disclosure of the use of funds from green finance are essential to maintain the confidence of investors and the public.

I take this opportunity to thank all the contributors for their earnest efforts to bring out this book successfully. I am sure readers of this book will find it very useful.

With best wishes,
Dr. Subramanian Senthilkannan Muthu

Part I
Carbon Footprint Assessment & Mitigation in Selected Industrial Spaces

Carbon Footprints in Agriculture: Challenges, Mitigation Strategies, and Future Prospects

M. Jagadesh, Munmun Dash, Santosh Kumar Singh, Aradhana Kumari, and Krishan Kumar Verma

1 Introduction

The intersection of agriculture and environmental sustainability has become a focal point in the ongoing global dialogue on mitigating climate change and securing a resilient future for our planet (Neufeldt et al., 2013). Amidst the imperative of feeding an ever-expanding global population, the environmental impact of agricultural activities, particularly in terms of carbon footprint, has emerged as a critical concern (Gosnell et al., 2019). The intricate dynamics of agriculture, while contributing significantly to food production, simultaneously weave a complex tapestry of emissions and sequestration, shaping the environmental equilibrium in profound ways (Macelaru & Lulić, 2023). The carbon footprint of agriculture encapsulates the totality of greenhouse gas emissions, encompassing carbon dioxide (CO_2), methane (CH_4), and nitrous oxide (N_2O), generated throughout the entire lifecycle of agricultural processes (Morris, 2023). The intricate relationships between soil health,

M. Jagadesh and Munmun Dash contributed equally with all other contributors.

M. Jagadesh · M. Dash
Tamil Nadu Agricultural University, Coimbatore, Tamil Nadu, India

S. K. Singh
Dr Rajendra Prasad Central Agricultural University, Pusa, Samastipur, Bihar, India

A. Kumari (✉)
College of Agriculture, Vidisha, M.P., India

K. K. Verma
Sugarcane Research Institute, Guangxi Academy of Agricultural Sciences/Key Laboratory of Sugarcane Biotechnology and Genetic Improvement (Guangxi), Ministry of Agriculture and Rural Affairs/Guangxi Key Laboratory of Sugarcane Genetic Improvement, Nanning, China

S. S. Muthu (ed.), *Carbon Footprint Assessments*, Environmental Footprints and Eco-design of Products and Processes,
https://doi.org/10.1007/978-3-031-70262-4_2

crop cycles, livestock management, and agricultural practices create a dynamic environment where carbon emissions and sequestration occur concurrently, influencing the overall sustainability of agriculture (Meena et al., 2020). Challenges in mitigating the carbon footprint in agriculture are manifold. Intensive farming practices, land-use changes, and the widespread use of synthetic fertilizers contribute significantly to emissions (Menegat et al., 2022). The susceptibility of agricultural systems to the impacts of climate change, such as altered weather patterns and extreme events, adds an extra layer of complexity (Altieri et al., 2015). The delicate balance between the imperative to enhance food production and the urgent need to reduce the carbon intensity of agriculture underscores the multifaceted challenges that demand innovative and holistic solutions (Neethirajan, 2023).

Mitigating the carbon footprint in agriculture is not only an environmental imperative but also an integral component of broader global efforts to achieve sustainable development goals. This chapter seeks to delve into the intricate challenges associated with the carbon footprint in agriculture, scrutinize existing mitigation strategies, and explore prospects that offer the potential for a more sustainable and resilient agricultural sector. Positioned at the confluence of environmental stewardship and food security, a nuanced exploration of the complexities surrounding the carbon footprint in agriculture becomes paramount for navigating the path toward a future characterized by sustainable and resilient food systems.

2 Sources of Carbon Footprints in Agriculture

To comprehend the intricacies of mitigating carbon footprints in agriculture, a thorough examination of the diverse sources contributing to greenhouse gas emissions is essential. Agricultural activities exert their influence on the atmosphere primarily through three main channels: enteric fermentation from livestock, soil management practices, and the extensive energy demands of various farming operations. Livestock, particularly ruminants like cattle, sheep, and goats, play a pivotal role in the agricultural emissions landscape (Marino et al., 2016). Their digestive processes involve enteric fermentation, which releases methane—a potent greenhouse gas—into the atmosphere. Agricultural activities, such as plowing, tilling, and the application of synthetic fertilizers, intricately influence soil health and carbon dynamics (Kamyab et al., 2023). Organic matter decomposition in the soil releases carbon dioxide, while nitrogen-based fertilizers contribute to nitrous oxide emissions—a potent greenhouse gas (Thomson et al., 2012). The adoption of sustainable soil management practices, encompassing reduced tillage, cover cropping, and the embrace of organic farming methods, emerges as a pivotal strategy to enhance carbon sequestration and minimize emissions (Agrawal et al., 2023). Modern agricultural practices are energy-intensive, relying heavily on machinery, irrigation systems, and transportation, often fuelled by fossil resources (Bundschuh et al., 2014). This reliance translates into substantial carbon dioxide emissions.

Transitioning to renewable energy sources, deploying precision farming technologies, and enhancing overall energy efficiency within agriculture represent critical measures to mitigate this source of emissions (Das et al., 2022). Changes in land use, particularly the conversion of natural ecosystems to agricultural land (Jagadesh et al., 2023), contribute to the release of stored carbon from trees and soil (Jagadesh et al., 2024). Sustainable land-use practices, coupled with afforestation and reforestation efforts, become imperative for mitigating emissions associated with land-use changes and fostering a harmonious coexistence between agriculture and the environment (Kamyab et al., 2023). The handling of manure from livestock presents a dual challenge—improper management can lead to the release of methane and nitrous oxide (Petersen et al., 2013). Implementing efficient manure management practices, such as anaerobic digestion or composting, not only helps capture and utilize methane but also provides valuable by-products for agricultural use, thus closing the sustainability loop (Khoshnevisan et al., 2021). As the agriculture sector transforms, addressing these multifaceted sources of carbon footprints demands a holistic and integrated approach. Technological innovations, policy frameworks, and transformative changes in farming practices are vital components of a comprehensive strategy. The pursuit of emission reduction in agriculture not only contributes to global climate goals but also plays a pivotal role in shaping resilient and sustainable food systems for the future.

Nitrogen-based fertilizers, a cornerstone of modern agricultural practices, have significantly altered the dynamics of crop production, yet their usage comes with a consequential environmental cost. Among the array of greenhouse gases released, nitrous oxide (N_2O) emerges as a particularly potent contributor to the carbon footprint in agriculture (Khoshnevisan et al., 2021). Nitrogen, a crucial element for plant growth, is commonly applied to fields through synthetic fertilizers. While enhancing crop productivity, this practice initiates a cascade of events that contribute to the carbon footprint. The breakdown of nitrogen compounds in the soil leads to the release of nitrous oxide, a greenhouse gas with a warming potential far greater than carbon dioxide. The frequency and quantity of fertilizer application, alongside the efficiency of nutrient uptake by crops, influence the extent of these emissions (Signor & Cerri, 2013). Microbial activities within the soil play a pivotal role in the transformation of nitrogen compounds. The microbial processes of nitrification and denitrification are particularly relevant, as they result in the release of nitrous oxide. Nitrification converts ammonium to nitrate, and subsequent denitrification processes can produce nitrous oxide as a byproduct (Wrage-Mönnig et al., 2018). Factors influencing these processes, including soil moisture, temperature, and organic matter content, contribute to the variability in nitrous oxide emissions (Wang et al., 2021). The timing and method of fertilizer application significantly impact nitrous oxide emissions. Practices such as split application, where fertilizer is applied in smaller doses throughout the growing season, and the incorporation of nitrification inhibitors can mitigate the release of nitrous oxide (Corrochano-Monsalve et al., 2021). Precision agriculture technologies, including sensor-based nutrient management, offer a targeted approach to optimizing fertilizer application, minimizing waste, and reducing associated emissions (Majumder et al., 2021). Exploring alternative fertilization strategies, such as the use of organic and slow-release fertilizers, represents a promising avenue for mitigating nitrous oxide emissions (Ussiri et al., 2013). These alternatives, often derived from natural sources, release nutrients more gradually, reducing the likelihood of excess nitrogen in the soil and subsequent nitrous oxide emissions (Thomson et al., 2012). Policies promoting sustainable fertilizer management, alongside educational initiatives for farmers, are crucial elements in the effort to reduce nitrous oxide emissions.

Pesticides, instrumental in modern agriculture for safeguarding crops from pests and diseases, carry a dual narrative in their environmental impact. While essential for ensuring global food security, the production, application, and persistence of pesticides contribute to the carbon footprint in agriculture, impacting ecosystems and the overall sustainability of farming practices. The synthesis and manufacturing of pesticides involve energy-intensive processes, often relying on fossil fuels. The production of active ingredients, formulation, and packaging contribute to the carbon footprint, as greenhouse gases are emitted during various stages. Minimizing the environmental impact of pesticides necessitates exploring more sustainable manufacturing practices, including the use of renewable energy sources and eco-friendly formulations (Pathak et al., 2022). The application of pesticides, whether through spraying or other methods, involves the use of energy-intensive machinery and vehicles. Precision agriculture technologies, such as sensor-based application systems, offer the potential to optimize pesticide use, reducing energy consumption and associated emissions (Sishodia et al., 2020). Pesticide residues, persisting in the environment after application, contribute to the overall carbon footprint. Residues can leach into soil and water, affecting non-target organisms and ecosystems. The degradation of certain pesticides releases carbon dioxide and other byproducts into the environment (Tudi et al., 2021). Sustainable pest management practices, integrated pest management (IPM), and the use of biopesticides aim to minimize residues and their subsequent environmental impact (Barzman et al., 2015). Pesticides, designed to control pests, can inadvertently affect soil health by disrupting microbial communities crucial for nutrient cycling (Daisley et al., 2022; Kumari et al., 2023). Altered microbial activity may lead to changes in soil carbon dynamics and emissions. Adopting agroecological approaches, which prioritize natural pest control mechanisms, helps maintain a healthy balance in soil ecosystems and reduces the reliance on chemical pesticides. Exploring alternatives to traditional pesticides, such as biological controls, crop rotation, and resistant crop varieties, offers avenues to mitigate the environmental impact (Baker et al., 2020). Integrated approaches that combine chemical, biological, and cultural methods can reduce the reliance on high-impact pesticides. Sustainable farming practices, including organic farming, emphasize minimizing synthetic inputs, thereby contributing to lower overall carbon footprints. Striking a balance between fostering innovation in pest management and enforcing measures to reduce the carbon footprint is crucial. Incentives for adopting sustainable practices, stricter regulations on high-impact pesticides, and support for research into eco-friendly alternatives contribute to a more sustainable agricultural landscape.

3 Environmental Impacts

The carbon footprints in agriculture have profound and far-reaching environmental consequences. These footprints contribute significantly to greenhouse gas emissions. The release of these gases intensifies global warming and exacerbates climate change. Additionally, the environmental impact extends to soil degradation, deforestation for agricultural expansion, and water pollution, posing challenges for sustainable and resilient ecosystems. Addressing and mitigating these agricultural carbon footprints are crucial for fostering environmentally sustainable food production systems.

3.1 Contribution of Climate Change Through Greenhouse Gas Emission

Carbon footprint vis-à-vis greenhouse gas emissions make a substantial contribution to climate change, with far-reaching consequences for the environment, ecosystems, and human societies. These emissions, primarily composed of carbon dioxide (CO_2), methane (CH_4), and nitrous oxide (N_2O), result from human activities such as fossil fuel combustion, deforestation, and industrial processes. Industries. The annual net CO_2 emissions resulting from human-induced alterations in land use averaged 0.9 [0.1 to 1.7] GtC yr.$^{-1}$ from 2002 to 2011. Their cumulative effect has led to global warming, the foundational driver of numerous climate-related outcomes. This includes more frequent and severe extreme weather events, shifts in precipitation patterns, sea level rise, ocean acidification, the melting of ice and glaciers, biodiversity loss, impacts on agriculture and food security, and health-related challenges (Malhi et al., 2021). Considering a moderate emissions scenario that mirrors current development trends, the anticipated average global temperature increase from preindustrial levels is projected to fall within the range of 2.1–3.5 °C. This surpasses the 1.5–2 °C threshold established by the 2015 Paris Agreement (Tollefson, 2021). Furthermore, there is a confirmed rise in the global mean surface temperature, indicating a warming of 0.85 [0.65 to 1.06] °C from 1880 to 2012. The extent of snow cover in the Northern Hemisphere exhibited a decrease of 1.6% per decade for March and April, and 11.7% per decade for June, spanning the period from 1967 to 2012. In certain regions, Northern Alaska experienced observed warming of up to 3 °C between the early 1980s and mid-2000s, while parts of the Russian European North saw warming of up to 2 °C from 1971 to 2010. The emission of CO_2 alone has led to a radiative forcing of 1.68 [1.33 to 2.03] W m^{-2}. When considering emissions of other carbon-containing gases that also contributed to the rise in CO_2 concentrations, the total radiative forcing of CO_2 is 1.82 [1.46 to 2.18] W m^{-2}. The increase of global temperatures leads to an increase in the atmosphere's moisture capacity, leading to a rise in both storms and heavy rainfall. Paradoxically, this also results in more pronounced dry spells, as heightened temperatures cause increased evaporation from the land and alterations in global weather patterns (World Bank, 2021).

3.2 Increased Vulnerability to Extreme Weather Events

The increasing vulnerability to extreme weather events is a pressing global concern, with a significant connection to the growing carbon footprint. As anthropogenic activities continue to release greenhouse gases into the atmosphere, they contribute to the warming of the planet, leading to alterations in weather patterns and an elevated risk of extreme weather events. These events encompass hurricanes, heatwaves, droughts, floods, and wildfires, and their frequency and intensity are on the rise (Marx et al., 2021).

The connection between the carbon footprint and extreme weather events is multifaceted. As greenhouse gas concentrations increase, so does the temperature of the earth's surface. This rise in global temperature is a key driver behind the more frequent and severe heatwaves observed in recent years (Audet et al., 2017). Marx et al. (2021) suggested that in numerous parts of the world, the threshold for human survival could be reached by the end of the twenty-first century due to the dangerous interplay of increasing temperatures and humidity levels, as measured by the "wet-bulb temperature" (WBT). Warmer temperatures also provide the energy necessary for the intensification of tropical storms and hurricanes, resulting in more destructive and devastating events (Lin et al., 2020). Vecchi and Soden (2007) also reported long-term changes in potential intensity of cyclone were closely related to the regional structure of warming. Regions that warmed more than the tropical average were characterized by increased potential intensity, and vice versa. In Sundarbans region of West Bengal, the area had already experienced several notable cyclones, such as Amphan in 2007, Aila in 2009, and Bulbul in 2019, Yaas in 2021 due to climate change (Paul et al., 2023).

Additionally, the alteration of atmospheric circulation patterns caused by climate change can lead to prolonged droughts, increased precipitation in certain regions, and shifts in storm tracks, making areas more susceptible to flooding and extreme precipitation events (Dai et al., 2018; Kourgialas & Karatzas, 2016). It is highly likely that the occurrence of intense precipitation events will rise across most regions throughout the twenty-first century, contributing to an increase in floods caused by heavy rainfall. Simultaneously, the anticipated projection indicates a growth in the proportion of land experiencing extreme drought conditions at any given time (Bates et al., 2008). From the year 2000 onward, there has been a 134% increase in flood-related disasters compared to the two preceding decades. The majority of recorded flood-related deaths and economic losses occurred in Asia. Moreover, during the same period, there was a 29% increase in both the number and duration of droughts (WMO, 2021).

Moreover, the consequences of an expanding carbon footprint extend beyond meteorological changes. Rising CO_2 in the atmosphere contribute to ocean acidification and thermal expansion (Doney et al., 2009), which, in turn, heighten the risk of coastal inundation and storm surges. This, along with the degradation of natural protective barriers such as mangroves and wetlands due to climate change, further amplifies the vulnerability of coastal communities to extreme weather events (Paul et al., 2023).

3.3 Soil Degradation, Loss of Biodiversity, and Water Pollution

The influence of carbon footprints on soil degradation is profound, with high levels of CO_2 emissions from human activities contributing to climate change, which subsequently affects soil quality. Elevated temperatures and altered precipitation patterns result in increased soil erosion, reduced microbial activity, and disruptions in nutrient cycling, leading to diminished soil fertility. Furthermore, climate change-induced droughts and flooding events exacerbate the degradation of soil by eroding topsoil and compromising its structure (Lal & Stewart, 2012). Soil degradation can also hinder its ability to sequester carbon, exacerbating the carbon footprint issue, and human activities like agriculture and construction can contribute to soil compaction, diminishing its porosity and aeration. Addressing soil degradation necessitates a reduction in carbon footprints and the implementation of sustainable land management practices to safeguard soil health (Lehman et al., 2015).

The rise in global CO_2 levels, impacting both plant species composition and the quality of organic matter entering the soil, is associated with climate change. The effects of climate change vary across different biomes, leading to diverse responses. For instance, warming may decrease soil biodiversity in arid or Mediterranean ecosystems (Orgiazzi et al., 2015). Conversely, the warming of the Arctic tundra might result in an enhancement of soil biodiversity. Thus, it is not accurate to generalize that climate change universally results in either a decrease or increase in soil biodiversity (Orgiazzi et al., 2015). Muluneh (2021) reported that as a result of elevated carbon footprint vis-à-vis climate change, species distributions have moved towards higher elevations at a median rate of 11.0 m and towards higher latitudes at 16.9 km per decade. In scenarios considering migration, extinction rates for 1103 species indicate that 21–23% would occur under conditions of unlimited migration, while 38–52% would face extinction without migration. Furthermore, Tarakanov (2022) has also reported that ocean acidification is a curse of higher C footprint. It results from the rising levels of CO_2 emissions, a greenhouse gas contributing to climate change. Approximately one-third of human-induced CO_2 is absorbed by the ocean, leading to a modification in seawater chemistry known as ocean acidification.

Mitigation efforts are crucial, focusing on reducing emissions through clean energy adoption, energy efficiency improvements, and sustainable land management practices. International collaboration and agreements are pivotal in addressing and mitigating the contribution of greenhouse gas emissions to climate change.

4 Mitigation Strategies

4.1 Sustainable Land Management

4.1.1 Promoting Agroforestry and Reforestation

Agroforestry and reforestation play critical roles in reducing the carbon footprint in agriculture. Agroforestry involves the strategic integration of trees and shrubs within agricultural landscapes, creating a multifunctional system that sequesters carbon effectively. Trees, renowned carbon sinks, absorb carbon dioxide from the atmosphere and store it in their biomass and root systems. By incorporating trees in farming systems, agroforestry increases carbon sequestration, thereby reducing the overall carbon footprint of agriculture (Chavan et al., 2021). Reforestation, on the other hand, focuses on the establishment or restoration of forests on previously deforested or degraded lands (Schäfer & Blanke, 2012). These practices not only help offset emissions from land use changes but also promote biodiversity conservation, improve soil health, enhance climate resilience, and offer diversified income streams for farmers. By stabilizing soil, reducing erosion, and improving water management, agroforestry and reforestation contribute to the long-term sustainability of agricultural landscapes (Shao et al., 2019). Additionally, reforestation projects can earn carbon credits, further incentivizing these practices in the agricultural sector (Lefebvre et al., 2021).

Kumara et al. (2023) conducted a meta-analysis to investigate the effects of four prevalent agroforestry systems, specifically agrisilviculture, silvipasture, agrihortisilviculture, and agrihorticulture, on the sequestration of soil carbon in Indian condition. They indicated that the adoption of agroforestry systems led to a substantial increase in carbon sequestration (+25.34%) compared to non-agroforestry systems. Among the different agroforestry systems, agrihorticulture exhibited the highest average soil carbon stock at 38.11 Mg C ha^{-1}, representing a remarkable increase of +31.64% when compared to conventional systems. The transition from grasslands to agroforestry systems had the most significant impact on carbon sequestration, resulting in a notable increase of +36.94%. Conversely, when shifting from forested areas to agroforestry systems, there was a reduction in soil carbon sequestration (−23.42%). This implies that agroforestry systems are more proficient in carbon sequestration outside of forested environments rather than acting as direct replacements for forest ecosystems. Moreover, agroforestry demonstrated increased carbon sequestration potential in semi-arid subtropical regions as compared to its performance in other geographical areas. On farmers' fields in Rajasthan, agroforestry practices were found to have an average estimated carbon sequestration and mitigation potential of 0.26 Mg C ha^{-1} $year^{-1}$ and 0.95 Mg CO_2 equivalent ha^{-1} $year^{-1}$. At the state level, agroforestry was determined to reduce greenhouse gas (GHG) emissions by 1.42 million tonnes annually. This reduction contributes to a decrease in the carbon footprint and aids in reaching the goals outlined in the Paris Agreement (Chavan et al., 2021). Franzluebbers et al. (2017) reported that soil carbon dioxide (CO_2) release was influenced by both temperature and moisture levels, but intriguingly, their combined effects did not significantly impact the emission. The soil beneath tree canopies exhibited a lower temperature, with a difference of 1.8 ± 1.3 °C, and lower water content of 0.043 ± 0.15 m^3 m^{-3} compared to the soil in cropped alleyways. This reduction in temperature and moisture levels beneath the trees contributed to a decrease in CO_2 emissions relative to those observed in the cropped alleyways. Jagadesh et al. (2023) also found, under different land use change in Nilgiri hills,

India, the greater presence of aboveground plant residues and the more abundant fine-root biomass were responsible for the increased carbon pool levels observed in the undisturbed evergreen and deciduous forests.

In summary, agroforestry and reforestation are integral components of climate-smart agriculture, contributing to both carbon footprint reduction and the overall environmental and economic sustainability of agricultural systems.

4.1.2 Conservation Tillage and Cover Cropping

Conservation tillage and cover cropping play a pivotal role in reducing the carbon footprint of agricultural practices (Bansal et al., 2022). Conservation tillage involves minimal disruption of the soil, preserving its organic matter and carbon content, which, in turn, minimizes CO_2 emissions associated with traditional ploughing. This practice also enhances carbon sequestration in the soil, further mitigating greenhouse gas emissions (Das et al., 2013; Mangalassery et al., 2014). Additionally, cover cropping involves planting specific crops during fallow periods or between main crops, which not only reduces soil erosion but also increases organic matter content. As these cover crops grow, they capture and store CO_2 from the atmosphere, effectively lowering carbon emissions (Lal, 2008). When integrated, conservation tillage and cover cropping serve as powerful tools to not only reduce the carbon footprint of agriculture but also promote sustainable and environmentally friendly farming practices, crucial in combating climate change and ensuring long-term food security. In Conservation Agriculture (CA), when 1 foot of crop residue was retained in the field, it resulted in an additional input of 1.6 to 2.0 t ha^{-1} of crop residue compared to conventional farming practices. This increase in residue incorporation contributed to the improvement of soil aggregation, water infiltration, organic carbon levels, and the enhancement of biological soil properties. Nevertheless, the adoption of Conservation Agriculture (CA) had the potential to sequester carbon at a rate of 300 to 600 kg C ha^{-1} $year^{-1}$, with variations influenced by factors such as soil type and local climatic conditions. In rainfed semiarid areas of Hyderabad, India Pratibha et al. (2016) studied the effect of conventional tillage (CT), reduced tillage (RT), and zero tillage (ZT), and residue treatments on net global warming potential (NGWP) in two approaches: crop-based (NGWPcrop) and soil-based (NGWPsoil) approaches. The results of this research suggested that there was potential to decrease emissions related to global warming potential (NGWP) by reducing one tillage operation, such as in the case of reduced tillage (RT) and increasing the amount of crop residue by harvesting at heights of 10 and 30 cm, while maintaining crop yields with minimal adverse effects. In another study, Abdalla et al. (2014) evaluated how combining reduced tillage with a mustard cover crop (RT-CC) could help reduce current and future greenhouse gas (GHG) emissions from a fertilized spring barley field in southeastern Ireland. The introduction of a cover crop in the RT treatment led to an increase in the predicted soil organic carbon (SOC), which more than offset the higher N_2O emissions, resulting in a lower overall GHG balance (TGGB) compared to the conventional tillage (CT) treatment. The findings suggested that the effectiveness of RT-CC in mitigating GHG emissions hinges on the extent of compensatory increases in CO_2 absorption by the cover crop, contributing to a reduction in the total GHG balance. Moreover, in a four-year experiment, Huang et al. (2023) discovered that combining no-tillage with a rye cover crop led to a significant 3.5% boost in soil organic carbon content. This increase effectively balanced out the overall CO_2 emissions, resulting in the most favorable carbon footprint (CF) outcomes. Specifically, they observed a remarkable reduction in CF, with a value of -1892.99 kg CO_2 eq ha^{-1}, and a yield-scaled CF of -1.44 kg CO_2 eq kg^{-1}. This combination of no-tillage with a rye cover crop management stood out as the most environmentally friendly and sustainable approach among all the tillage and residue management strategies investigated. Nevertheless, a comparison was made between three different tillage practices: Conventional tillage with preceding crop residue incorporation, and two other practices—permanent bed—and zero tillage flatbed with preceding crop residue retention. It was found that the conventional tillage method resulted in higher net CO_2-equivalent emissions (26.5% and 27.2%), larger carbon footprints (20.8% and 14.5%), and increased water footprints (27.3% and 18.0%) compared to the permanent bed- and zero tillage approaches (Parihar et al., 2022). Over a span of two years, from 2019 to 2021, a field experiment was conducted to evaluate the effects of different tillage methods on the grain yield, energy efficiency, carbon emissions, and economic returns within a winter wheat-summer maize crop rotation system in China's Loess Plateau. The study encompassed several treatments, namely conventional tillage (CT), no-tillage (NT), ridge cultivation in combination with no-tillage (RNT), and intermittent tillage (OT). In contrast to conventional tillage (CT), ridge cultivation with no-tillage (RNT), occasional tillage (OT), and no-tillage (NT) have demonstrated notable improvements in energy efficiency, with increases of 47.8%, 31.2%, and 35.3%, respectively. Additionally, these practices have led to substantial reductions in carbon footprint, amounting to 79.3%, 46.2%, and 73.2%, respectively, when compared to CT (Sun et al., 2022). Nevertheless, biochar, prepared from crop residue, is also gaining importance for crop residue management to enhance carbon sequestration (Purakayastha et al., 2019), reduce nutrient leaching and augment soil fertility (Dey et al., 2023).

4.1.3 Organic Farming Practices

Organic farming practices play a pivotal role in mitigating the carbon footprint associated with agriculture. They achieve this by fostering enhanced soil carbon sequestration, primarily through practices like crop rotation, reduced tillage, and the incorporation of cover crops (Holka et al., 2022). By relying on natural fertilizers such as compost and manure, organic farming reduces the energy-intensive production and application of synthetic fertilizers, which are a substantial source of greenhouse gas emissions (Diacono et al., 2019). Additionally, the avoidance of synthetic pesticides and herbicides in organic farming not only protects ecosystems but also minimizes emissions associated with their manufacturing and application (Gan et al., 2011). The emphasis on biodiversity preservation and natural pest control further contributes to carbon capture through healthier soils and diverse ecosystems (Haque, 2018). Organic farming often involves local and regional markets, decreasing the carbon emissions related to long-distance transportation (Theurl et al., 2014; Schäfer & Blanke, 2012). Moreover, the more natural and less intensive animal husbandry practices in organic livestock farming reduce methane emissions from livestock (Horrillo et al., 2020). The incorporation of agroforestry practices in some organic farming systems, with trees and crops grown together, helps sequester carbon and adds to the overall reduction of the carbon footprint. Adewale et al. (2019) utilized OFoot, a carbon footprint calculator to conduct a Life Cycle Assessment on two certified organic farms that grow various crops. These farms differ in scale and climate; one is small-scale and located in a maritime climate, while the other is large-scale and situated in a semi-arid climate. The small-scale farm's most significant carbon footprint contributors were fuel usage (47.6%), greenhouse facilities (11.3%), and net soil emissions (10.3%). On the contrary, the large-scale farm's major carbon footprint sources were electricity consumption for irrigation (47.5%), fuel usage (26.1%), and soil amendments (20.1%). In another study, Yuttitham (2019) compared carbon footprint between conventional and organic farming of Thailand. The results demonstrated that when evaluating the environmental impact of Chinese kale production, conventional farming resulted in a carbon footprint (CF) of 0.402 ± 0.47 kg of CO_2 equivalent per kg of Chinese kale. The CF for conventional farming was mainly contributed from chemical fertilizer, transportation, irrigation, tillage, organic fertilizer, and insecticide to the tune of 51%, 21%, 19%, 5%, 2%, 1%, and 1%. Respectively. In contrast, organic farming exhibited a lower CF, with a value of 0.195 ± 0.122 kg of CO_2 equivalent per kg of Chinese kale. In organic farming, the CF was primarily driven by transportation at 81%, followed by organic fertilizer usage at 12%, and the utilization of fossil fuels for irrigation at 7%. In essence, organic farming represents a holistic approach that significantly diminishes the carbon footprint in agriculture by combining these various environmentally friendly practices.

4.2 Livestock Management

4.2.1 Improved Animal Nutrition

Improved animal nutrition plays a crucial role in sustainable livestock management with the aim of reducing the carbon footprint. By optimizing the diet and feeding practices, livestock can more efficiently convert feed into muscle, milk, or other products, thus minimizing the overall emissions associated with their production (Afrouzi et al., 2023). Balanced nutrition can help mitigate enteric methane emissions, a significant contributor to the livestock sector's greenhouse gas footprint

(Caro et al., 2016). Across 183 countries grouped into 11 regions, a strategy to reduce enteric methane (CH4) emissions from dairy cattle was implemented by Caro et al. (2016). This strategy involved adjusting dairy cattle diets by increasing the lipid concentration by up to 6% and decreasing the fiber concentration while maintaining a constant total gross energy intake (GEI). As a result of these dietary modifications, the research findings showed the potential for a substantial global reduction of 15.7% in CH_4 emissions originating from dairy cattle. Similarly, Moate et al. (2015) indicated that the incorporation of dietary lipid supplements (such as cottonseed, brewers grains, cold-pressed canola, hominy meal, and grape marc) along with higher wheat levels in livestock feed had led to a significant reduction in CH_4 emissions, ultimately enhancing the environmental sustainability of milk production in Australia. This accomplishment was largely attributed to the simultaneous increase in per-cow milk yields, resulting in a reduced carbon footprint (Moate et al., 2015). Furthermore, by reducing the need for additional resources to compensate for poor nutrition, such as extra feed or veterinary care, improved animal nutrition can have a ripple effect, lowering the environmental impact of feed production, transportation, and associated supply chains (Chen et al., 2021). Nevertheless, Franzluebbers (2020) reported that highly nutritious forages can help diminish CH_4 emissions when cattle graze on them. This is due to increased feed efficiency and shorter time to reach maturity. Moreover, forages containing substantial amounts of tannins, saponins, and other biochemical elements may offer the potential to mitigate or reduce enteric CH_4 emissions. In essence, enhanced animal nutrition is a fundamental strategy in the broader effort to make livestock production more sustainable and environmentally responsible.

4.2.2 Manure Management and Methane Capture

Manure management and methane capture are essential practices in agriculture for reducing the carbon footprint. Manure, if not managed properly, can release methane and nitrous oxide, potent greenhouse gases. Strategies like anaerobic digestion, composting, and nutrient management help minimize these emissions, while methane recovery systems and biogas utilization capture and repurpose methane emissions from manure storage facilities, reducing its release into the atmosphere (Bong et al., 2017). Aguirre-Villegas and Larson (2017) revealed that extended storage of liquid manure without any treatment led to the highest GHG emissions (on a per-cow, kg of milk, and ton of manure basis) vis-à-vis C footprint. However, when authorized facilities like those employing sand separation, solid-liquid separation (SLS), and anaerobic digestion (AD) processes were introduced, emissions were notably reduced, with anaerobic digestion playing the most substantial role in emission reduction. These practices not only mitigate greenhouse gas emissions but also promote sustainable nutrient management, reducing the need for synthetic fertilizers and making agriculture more environmentally friendly and efficient.

4.2.3 Shift Towards Plant-Based Diets

The comprehensive greenhouse gas (GHG) emissions from various foods can be represented by their life-cycle carbon footprint (CF). Velarde-Guillén et al. (2022) found that the carbon footprint of milk production in Latin America ranged from 1.54 to 3.57 kg CO_2 eq per kg of fat and protein corrected milk. Xu and Lan (2016) employed a consistent methodology, using uniform life cycle definitions, data sources, and byproduct allocation methods, to calculate the CFs of 22 plant-based foods, 6 animal-based foods, and several processed foods. The findings indicated that, in general, animal-based foods had higher CFs compared to plant-based foods. Specifically, beef and mutton were identified as the highest GHG emitters among the foods analyzed, while radish production exhibited the lowest CF. Moreover, in their study, Treu et al. (2017) determined that reducing diet-related carbon footprint (CFP) can be achieved by transitioning to diets with a lower inclusion of animal-based food products, and they also noted that there was not a significant distinction in CFP between organic and conventional food production systems.

4.3 Energy Efficiency

4.3.1 Transition to Renewable Energy Sources

Renewable energy sources are environmentally friendly power options like solar, wind, hydro, and geothermal energy that harness natural processes to generate electricity without emitting greenhouse gases (Mittal et al., 2016). The transition to renewable energy sources involves shifting from fossil fuels to these sustainable alternatives. Conventional energy sources, like coal and oil, significantly contribute to carbon and greenhouse gas emissions, which exacerbate climate change. Unlike conventional fossil fuels, renewable sources produce electricity without emitting significant levels of greenhouse gases, making them an inherently low-carbon option (Siddik et al., 2021). This shift promotes a cleaner, more sustainable energy mix, reducing carbon emissions and mitigating the impact of climate change.

In a review of 79 studies analyzing the environmental impact of various renewable energy sources, it was found that offshore wind technologies had the lowest greenhouse gas (GHG) emissions, ranging from 5.3 to 13 g CO_2 eq/kWh. When comparing these results to GHG emissions from traditional fossil fuel-based electricity and heat generation, renewable sources generally had lower emissions, except for nuclear-based power generation (Amponsah et al., 2014). Peyvandi et al. (2023) revealed that renewable sources accounted for approximately 9.5% of the total energy mix and had an emission intensity of 2.2 t CO /kWh. In contrast, fossil fuel-based power plants were associated with a significantly higher estimated emission intensity of 506 t CO_2/kWh. In European Union (EU) countries, Bölük and Mert (2014) reported that the use of renewable energy results in approximately 50% fewer greenhouse gas (GHG) emissions per unit of energy consumed compared to the use of fossil energy. de Souza Grilo et al. (2018) also reported that the use of photovoltaic solar energy instead of electricity from the grid reduced CO_2 emissions by approximately 207.88 kg/year, which is a substantial environmental advantage. Nevertheless, Liu et al. (2021) reported that the carbon emissions from the wind farm, at 6.57 g/kWh, are considerably lower compared to the carbon emissions from four non-renewable energy sources: coal (approximately 148.45 times less), natural gas (approximately 72.91 times less), oil (approximately 127.85 times less), and nuclear power (approximately 3.50 times less). Thus as renewable energy technologies continue to advance and become more accessible, they offer a path towards a greener and more environmentally responsible future, helping us combat climate change and minimize the carbon footprint of our energy systems.

4.3.2 Adoption of Precision Agriculture Techniques

Precision agriculture is a farming approach that uses technology and data to optimize crop and livestock production. Its characteristics include data-driven decision-making, targeted resource management, and automation of farming processes. The adoption of precision agriculture techniques plays a pivotal role in reducing the carbon footprint of agricultural practices through several key mechanisms. By harnessing data-driven decision-making and technology, precision agriculture allows farmers to optimize the use of resources such as fertilizers, pesticides, and water, reducing overapplication and minimizing runoff into ecosystems (Singh et al., 2021). This targeted resource management not only increases the efficiency of crop and livestock production but also curtails the emissions associated with the production and application of these inputs. Furthermore, the automation of farming processes, including GPS-guided machinery and drones, enhances operational efficiency, minimizing the energy consumption and carbon emissions related to fieldwork (Soto et al., 2019). Additionally, the precise mapping and monitoring of fields enable the early detection and control of pest and disease outbreaks, reducing the need for chemical interventions (McCarthy et al., 2020). Soto et al. (2019) reported that the most promising technologies for decreasing greenhouse gas emissions vis a vis carbon footprint while also preserving or improving farm profitability are variable-rate nutrient application (VRNT), machine guidance (MG), variable-rate irrigation (VRI), and controlled traffic farming (CTF). VRNT has a greater capacity for mitigation compared to MG, constituting 1.5% of the total greenhouse gas emissions from the Agriculture sector in the EU in 2015, while MG represented only 0.3%. Nevertheless, in a two-year field study spanning 2017–2018, maize yield response to variable rate (VR) nitrogen distribution, guided by proximal vigor sensors, was evaluated in contrast to a fixed rate (FR) approach on a cereal-livestock farm located in the northern Italy. The results of the experiment suggest that employing VR technology proved to be a viable and economically beneficial technique while also aligning with the current European Common Agricultural Policy as a sustainable farming practice (Schillaci et al., 2021). The widespread integration of low-energy precision application (LEPA) technologies did not decrease the rate of water pumping; however, it led to a 19.2% reduction in energy consumption and a 15.2% decrease in greenhouse gas (GHG) emissions (McCarthy et al., 2020). Singh et al. (2021) utilized the slack-based measure (SBM) data envelopment analysis (DEA) to distinguish between efficient and inefficient farms, which are known as decision-making units (DMUs), within the rice-wheat cropping system. The findings indicated that by adopting the management practices of efficient DMUs (Decision-Making Units) while maintaining their current yield level, approximately 57–63% of carbon equivalent (CET) emissions could be reduced in both rice and wheat cultivation by inefficient DMUs. Thus precision agriculture could be a critical tool in fostering more sustainable and environmentally friendly farming systems, which in turn helps mitigate the agriculture sector's contribution to greenhouse gas emissions and climate change.

4.3.3 Energy-Efficient Machinery and Practices

Energy-efficient machinery and practices in agriculture encompass a range of technologies and approaches that collectively contribute to reducing the carbon footprint of farming operations. These practices include the adoption of modern, fuel-efficient tractors and equipment, which are designed to maximize work output while minimizing fuel consumption (Liu et al., 2016). Precision agriculture techniques, such as GPS-guided machinery and data-driven decision-making, optimize the use of resources like water, fertilizers, and pesticides, significantly reducing waste and associated emissions (Bhakta et al., 2019). The integration of renewable energy sources, such as solar panels and wind turbines, on farms further decreases reliance on fossil fuels and curtails greenhouse gas emissions (Liu et al., 2021). Additionally, implementing sustainable farming practices like no-till or reduced tillage, cover cropping, and efficient irrigation systems helps conserve energy and sequester carbon in the soil. The role of energy-efficient machinery and practices in reducing the carbon footprint in agriculture is multifaceted. By reducing energy consumption and greenhouse gas emissions, they help address the pressing issue of climate change. Moreover, the adoption of such technologies and practices often leads to cost savings for farmers, increasing economic resilience in the face of fluctuating input costs and market conditions (Liu et al., 2021). Yadav et al. (2020) conducted an experiment with two primary tillage methods: conventional tillage (CT) and no-tillage (NT) in the main plots, along with four types of organic mulches in the subplots. These mulches were rice straw (RSM), Gliricidia sepium (GLM), cowpea brown manure (BMM), and a control group with no mulch (NM). This study spanned four years. The findings demonstrated that implementing no-tillage (NT) resulted in significant reductions in energy consumption (48.50%), specific energy use (49.63%), carbon footprint (16.48%), and cultivation costs (35%) when compared to conventional tillage (CT). Moreover, it improved energy efficiency and the cost-benefit ratio. Nevertheless, Karimi et al. (2012) showed that in the past, groundwater pumping in Iran consumed 20.5 billion kWh of electricity and 2 billion liters of diesel, contributing to 3.6% of the country's total carbon emissions.

However, when improved irrigation schedules and enhanced farm application efficiencies were adopted, water productivity increased, and groundwater withdrawals for irrigation significantly decreased without any reduction in yields. This reduction in irrigation application resulted in a 40% decline in energy consumption and, consequently, a notable reduction in carbon emissions associated with groundwater usage. Nevertheless, Li et al. (2019), in their study, integrated an optimization model, utilizing Chance-Constrained Programming (CCP) within a multi-objective non-linear programming framework. This model was combined with a carbon footprint (CF) assessment, an irrigation water use efficiency (IWUE) index, a soil water balance model, and a groundwater dynamics model. Through the coordinated allocation of agricultural land and water resources in both spatial and temporal dimensions, the research observed a reduction in soil carbon emissions and an improvement in water use efficiency (Li et al., 2019). In another study, Madane et al. (2023) examined the carbon footprints of onion cultivation in the coastal Konkan region of Maharashtra, India, comparing flooded and drip irrigation methods. The irrigation method that used the most water had the highest water and carbon footprint values. These values showed an upward trend, increasing from 0.8 ETc to 1.2 ETc in both seasons (Madane et al., 2023). In contrast to conventional flood irrigation, drip irrigation resulted in significant reductions in water usage, energy consumption, and carbon emissions, showing decreases of 60%, 19.47%, and 45%, respectively. The primary factor contributing to the variations in carbon emissions among different irrigation methods was the overall impact of cultivation and water-energy consumption (Huifang et al., 2021.

4.4 Agrochemical Reduction

4.4.1 Precision Nutrient Management

Precision nutrient management plays a crucial role in the agricultural sector's efforts to reduce its carbon footprint (Jiang et al., 2019). By precisely tailoring the application of fertilizers to match the specific nutritional needs of crops (Dash et al., 2023), this approach minimizes the overuse of these inputs, thereby reducing the associated carbon emissions linked to their production, transportation, and application (Parihar et al., 2021). Furthermore, precision nutrient management helps in minimizing nitrogen losses by ensuring that the right amount of nitrogen is provided to crops at the right time. This, in turn, reduces the release of nitrous oxide, a potent greenhouse gas, which often occurs when excess nitrogen leaches into the environment (Dreyfus et al., 2023). Additionally, it fosters enhanced crop yields, increasing agricultural productivity on existing land and lessening the pressure to convert forests and other natural habitats into farmland, a process that results in significant carbon emissions (Jat et al., 2021). Improved energy efficiency in the agricultural sector, driven by precision nutrient management, further contributes to reducing the carbon footprint (Balafoutis et al., 2017). The use of resources more efficiently helps decrease emissions associated with the entire fertilizer supply chain (Sapkota et al., 2014). Moreover, this approach promotes healthier soils, which have an increased capacity to sequester carbon, thereby aiding in mitigating emissions from other sectors.

The implementation of Site-Specific Nutrient Management (SSNM) led to improvements in the overall efficiency and sustainability of the maize-wheat-mungbean cropping system in northwest India. It increased productivity, water use efficiency, and energy utilization, while concurrently reducing water and carbon footprints and net CO_2-equivalent emissions in the maize-wheat-mungbean system in northwest India (Parihar et al., 2022). Moreover, in seven districts of Haryana, India, on-farm trials spanning two consecutive years (2010–11 and 2011–12) were carried out to assess the effectiveness of three distinct strategies for Site-Specific Nutrient Management (SSNM). These strategies were based on recommendations from the Nutrient Expert (NE) decision support system and applied to both no-tillage (NT) and conventional tillage (CT) wheat production systems. Additionally, it was observed that the Global Warming Potential (GWP) vis a vis carbon footprint associated with wheat production was lower when using NE-based nutrient management compared to the fertilization practices employed by local farmers (Sapkota et al., 2014). Precise application of water is also necessary to achieve reduction in carbon footprint. In Pakistan, approximately 50 billion cubic meters (BCM) of groundwater were annually pumped for agricultural irrigation. This process consumed over 6 billion kilowatt-hours of electricity and 3.5 billion liters of diesel fuel. As a result of this energy usage, approximately 3.8 million metric tons (MMT) of CO_2 emissions were attributed to it each year. The adoption of improved irrigation schedules had the potential to enhance water productivity and reduce groundwater extraction for irrigation by 24 BCM. This reduced groundwater extraction resulted in a 62% decline in energy demand (equivalent to 1.5 billion liters of diesel, commonly used in private tubewells) and a 40% reduction in carbon emissions (Qureshi, 2014).

In summary, precision nutrient management is a multifaceted strategy that not only boosts crop yields but also plays a crucial role in making agriculture more sustainable, environmentally friendly, and aligned with the goals of reducing the carbon footprint in farming practices.

4.4.2 Organic and Low-Input Farming Systems

Organic and low-input farming systems are essential contributors to reducing the carbon footprint in agriculture. These systems prioritize practices that minimize the use of synthetic fertilizers and pesticides, thus lowering emissions associated with the production and transportation of these inputs (Tuomisto et al., 2012). Moreover, they emphasize soil health through techniques like crop rotation, cover cropping, and reduced tillage, which increase carbon storage in soils (Plaza-Bonilla et al., 2018). Organic and low-input farming also tend to use less energy than conventional agriculture, reducing carbon emissions related to energy-intensive processes (Holka et al., 2022). These systems often integrate agroecological principles, such as crop-livestock integration, leading to more efficient resource utilization and carbon sequestration. Prioritizing biodiversity through practices like hedgerows and natural habitat preservation helps in carbon sequestration as well (Holka et al., 2022). Arunrat et al. (2021) compared the carbon footprint and the value of carbon sequestration ecosystem services between organic rice farming (OF) and conventional rice farming (CF). The findings indicate that there is a significant disparity in carbon footprint between OF, which has a carbon footprint intensity of −0.13 kg CO_2 eq per kg of rice yield, and CF, which has a carbon footprint intensity of 0.82 kg CO_2eq per kg of rice yield. Moreover, they also reported that in OF, the worth of carbon sequestration ecosystem services was approximately double that of CF.

4.4.3 Integrated Pest Management

Integrated Pest Management (IPM) plays a crucial role in reducing the carbon footprint in agriculture by promoting a sustainable and environmentally friendly approach to pest and disease control. Unlike conventional farming practices that often rely on chemical pesticides, IPM focuses on a holistic and integrated strategy. By using natural predators, beneficial insects, crop rotation, and resistant crop varieties, it reduces the need for synthetic pesticides (Jørgensen et al., 2014). This results in a lower carbon footprint because the production and application of these chemicals are energy-intensive and release greenhouse gases (Lal, 2008). Additionally, IPM emphasizes the monitoring of pest populations, leading to more precise and efficient pesticide use when it is necessary, minimizing unnecessary applications. In Pimentel, 1980, Pimentel provided energy estimates for the production, formulation, packaging, and transportation of different types of pesticides, measured in Mcal/kg of the active ingredient. These estimates ranged from 63 to 100 Mcal/kg for fungicides, 61 to 87 Mcal/kg for insecticides, and 28 to 65 Mcal/kg for herbicides. On average, the energy required for pesticide production was 67 Mcal/kg of the active ingredient, with the order of energy intensity being wettable powder < granules = dust < miscible oil. Reducing the use of chemical pesticides presents a promising approach to mitigate the carbon footprint associated with agriculture, and implementing IPM is a highly effective and sustainable strategy to achieve this goal. Furthermore, by promoting healthier ecosystems, IPM helps sequester carbon in soil and vegetation, contributing to climate change mitigation (Slaney et al., 2009). Brookes and Barfoot (2018) reported that the implementation of genetically modified (GM) technology for insect resistance and herbicide tolerance has led to a significant reduction in pesticide application by 671.4 million kg (8.2%), consequently resulting in an 18.4% reduction in the environmental impact associated with the use of herbicides and insecticides on these crops, as quantified by the Environmental Impact Quotient (EIQ) indicator. Overall, IPM not only reduces the environmental and health risks associated with chemical pesticides but also helps agriculture become more sustainable and resilient in the face of climate change, making it an integral part of efforts to reduce agriculture's carbon footprint.

5 Challenges and Barriers

Implementation of environmentally friendly and sustainable practices often requires investments in new technologies, energy-efficient equipment, and renewable energy sources. Economic downturns or recessions can divert attention and resources away from sustainability efforts. During challenging economic times, businesses may prioritize short-term financial stability over long-term environmental considerations. Governments may also be less inclined to impose stringent environmental regulations during economic crises to avoid further burdening industries. The availability and cost of resources, including renewable energy sources, can impact a company's ability to transition to low-carbon technologies. Publicly traded companies may face pressure from investors and shareholders to prioritize financial returns over sustainability efforts. Companies that are perceived as environmentally responsible may attract more environmentally conscious investors, but this is not always the case. Investing in research and development for innovative, low-carbon technologies may require significant upfront costs. Companies may be hesitant to allocate resources to these endeavors without clear economic incentives or market demand.

6 Future Prospects

6.1 Technology Innovations

6.1.1 Carbon Capture and Storage (CCS)

CCS for Industrial Emissions: Carbon capture technologies applied to industrial processes help capture CO_2 emissions before they are released into the atmosphere.

Direct Air Capture (DAC): Technologies like DAC aim to remove carbon dioxide directly from the air, providing a means to reduce atmospheric carbon levels (McQueen et al., 2021).

Green Building Technologies: The development of materials with enhanced insulating properties helps reduce energy consumption in buildings (Casini, 2016).

6.1.2 Efficient Industrial Processes

The integration of IoT, artificial intelligence, and data analytics in industrial processes helps optimize energy usage, reduce waste, and enhance overall efficiency. Sustainable manufacturing practices, such as 3D printing and precision machining, can reduce material waste and energy consumption. The use of sensors, drones, and data analytics in agriculture helps optimize resource use, reduce emissions, and enhance crop yields. Plant-Based Alternatives: Innovations in plant-based proteins and alternative protein sources contribute to more sustainable and lower-carbon food production (Hassoun et al., 2022).

6.1.3 Carbon-Neutral Technologies

Green hydrogen production using renewable energy sources has the potential to replace fossil fuels in various sectors. The development of synthetic fuels produced using renewable energy can serve as a low-carbon alternative for transportation and industry (Oliveira et al., 2021). Blockchain technology is being employed to create transparent and traceable systems for tracking carbon emissions throughout supply chains (Kouhizadeh & Sarkis, 2018). Innovative approaches leverage natural processes, such as reforestation and soil carbon sequestration, to remove carbon from the atmosphere (Nave et al., 2019).

6.2 Global Collaboration

The UNFCCC is a key international treaty that aims to address climate change. It provides a framework for negotiations and agreements among countries to reduce greenhouse gas emissions, including those from the agricultural sector. The Paris Agreement, a landmark accord under the UNFCCC, sets targets for countries to limit global temperature increases. It recognizes the role of sustainable agriculture in climate mitigation and adaptation (Kinley et al., 2021). Intergovernmental Panel on Climate Change (IPCC) assesses scientific information related to climate change (Seyboth, 2013). Its reports highlight the impact of agriculture on climate change and provide recommendations for sustainable practices. Global Research Alliance on Agricultural Greenhouse Gases (GRA) is a collaborative initiative that brings together countries to address agricultural greenhouse gas emissions through research and knowledge-sharing (Sándor et al., 2016).

7 Conclusion

Carbon footprints in agriculture present a complex challenge that demands immediate attention and concerted global efforts. The agricultural sector is not only a significant contributor to greenhouse gas emissions but also vulnerable to the impacts of climate change. As the global population continues to grow, the need for sustainable agricultural practices becomes increasingly urgent. Agriculture contributes to carbon footprints through various sources, including enteric fermentation, manure management, rice cultivation, and the use of synthetic fertilizers. Deforestation and changes in land use patterns further exacerbate carbon emissions, leading to a loss of biodiversity and ecosystem services. Conventional farming practices often rely heavily on fossil fuels, synthetic inputs, and water, contributing to resource depletion and environmental degradation. Implementing agroecological practices, such as organic farming, agroforestry, and crop rotation, can enhance soil health, reduce emissions, and promote biodiversity. Utilizing technology for precision agriculture helps optimize resource use, reduce waste, and improve overall efficiency. Integrating renewable energy sources on farms, such as solar and wind power, can offset emissions associated with energy-intensive agricultural activities. Adopting practices that enhance carbon sequestration in soils, such as cover cropping and conservation tillage, contributes to mitigating emissions. As we move forward, it is imperative that the agricultural sector transitions toward a more sustainable and climate-resilient model. This requires a paradigm shift in the way we produce, consume, and perceive food. While challenges persist, the multitude of mitigation strategies and ongoing innovations provide hope for a future where agriculture not only sustains global food security but also contributes to environmental restoration. A holistic approach that considers ecological, social, and economic dimensions is vital to achieving a resilient and low-carbon agricultural system. Through collaborative efforts, technological innovation, and informed policy-making, the vision of a sustainable agricultural future, with minimized carbon footprints, can become a reality.

References

Abdalla, M., Hastings, A., Helmy, M., Prescher, A., Osborne, B., Lanigan, G., et al. (2014). Assessing the combined use of reduced tillage and cover crops for mitigating greenhouse gas emissions from arable ecosystem. *Geoderma, 223*, 9–20.

Adewale, C., Reganold, J. P., Higgins, S., Evans, R. D., & Carpenter-Boggs, L. (2019). Agricultural carbon footprint is farm specific: Case study of two organic farms. *Journal of Cleaner Production, 229*, 795–805.

Afrouzi, H. N., Ahmed, J., Siddique, B. M., Khairuddin, N., & Hassan, A. (2023). A comprehensive review on carbon footprint of regular diet and ways to improving lowered emissions. *Results in Engineering*, 101054.

Agrawal, S., Singh, D., Chauhan, A. P. S., & Kushwah, N. (2023). A comprehensive review of contemporary advances in soil science and sustainable agriculture practices. *Management, 1*, 5.

Aguirre-Villegas, H. A., & Larson, R. A. (2017). Evaluating greenhouse gas emissions from dairy manure management practices using survey data and lifecycle tools. *Journal of Cleaner Production, 143*, 169–179.

Altieri, M. A., Nicholls, C. I., Henao, A., & Lana, M. A. (2015). Agroecology and the design of climate change-resilient farming systems. *Agronomy for Sustainable Development, 35*(3), 869–890.

Amponsah, N. Y., Troldborg, M., Kington, B., Aalders, I., & Hough, R. L. (2014). Greenhouse gas emissions from renewable energy sources: A review of lifecycle considerations. *Renewable and Sustainable Energy Reviews, 39*, 461–475.

Arunrat, N., Sereenonchai, S., & Wang, C. (2021). Carbon footprint and predicting the impact of climate change on carbon sequestration ecosystem services of organic rice farming and conventional rice farming: A case study in Phichit province, Thailand. *Journal of Environmental Management, 289*, 112458.

Audet, J., Neif, É. M., Cao, Y., Hoffmann, C. C., Lauridsen, T. L., Larsen, S. E., et al. (2017). Heatwave effects on greenhouse gas emissions from shallow lake mesocosms. *Freshwater Biology, 62*(7), 1130–1142.

Baker, B. P., Green, T. A., & Loker, A. J. (2020). Biological control and integrated pest management in organic and conventional systems. *Biological Control, 140*, 104095.

Balafoutis, A., Beck, B., Fountas, S., Vangeyte, J., Van der Wal, T., Soto, I., et al. (2017). Precision agriculture technologies positively contributing to GHG emissions mitigation, farm productivity and economics. *Sustainability, 9*(8), 1339.

Bansal, S., Yin, X., Schneider, L., Sykes, V., Jagadamma, S., & Lee, J. (2022). Carbon footprint and net carbon gain of major long-term cropping systems under no-tillage. *Journal of Environmental Management, 307*, 114505.

Barzman, M., Bàrberi, P., Birch, A. N. E., Boonekamp, P., Dachbrodt-Saaydeh, S., Graf, B., et al. (2015). Eight principles of integrated pest management. *Agronomy for Sustainable Development, 35*, 1199–1215.

Bates, B. C., Kundzewicz, Z. W., Wu, S., & Palutikof, J. P. (Eds.). (2008). *Climate change and water*. Technical Paper of the Intergovernmental Panel on Climate Change, IPCC Secretariat, Geneva, 210 pp.

Bhakta, I., Phadikar, S., & Majumder, K. (2019). State-of-the-art technologies in precision agriculture: A systematic review. *Journal of the Science of Food and Agriculture, 99*(11), 4878–4888.

Bölük, G., & Mert, M. (2014). Fossil & renewable energy consumption, GHGs (greenhouse gases) and economic growth: Evidence from a panel of EU (European Union) countries. *Energy, 74*, 439–446.

Bong, C. P. C., Lim, L. Y., Ho, W. S., Lim, J. S., Klemeš, J. J., Towprayoon, S., et al. (2017). A review on the global warming potential of cleaner composting and mitigation strategies. *Journal of Cleaner Production, 146*, 149–157.

Brookes, G., & Barfoot, P. (2018). Environmental impacts of genetically modified (GM) crop use 1996-2016: Impacts on pesticide use and carbon emissions. *GM Crops & Food, 9*(3), 109–139.

Bundschuh, J., Chen, G., & Mushtaq, S. (2014). Towards a sustainable energy technologies based agriculture.

Caro, D., Kebreab, E., & Mitloehner, F. M. (2016). Mitigation of enteric methane emissions from global livestock systems through nutrition strategies. *Climatic Change, 137*, 467–480.

Casini, M. (2016). *Smart buildings: Advanced materials and nanotechnology to improve energy-efficiency and environmental performance*. Woodhead Publishing.

Chavan, S. B., Newaj, R., Rizvi, R. H., Ajit, Prasad, R., Alam, B., et al. (2021). Reduction of global warming potential vis-à-vis greenhouse gases through traditional agroforestry systems in Rajasthan, India. *Environment, Development and Sustainability, 23*, 4573–4593.

Chen, W., Jafarzadeh, S., Thakur, M., Ólafsdóttir, G., Mehta, S., Bogason, S., & Holden, N. M. (2021). Environmental impacts of animal-based food supply chains with market characteristics. *Science of the Total Environment, 783*, 147077.

Corrochano-Monsalve, M., Bozal-Leorri, A., Sánchez, C., González-Murua, C., & Estavillo, J. M. (2021). Joint application of urease and nitrification inhibitors to diminish gaseous nitrogen losses under different tillage systems. *Journal of Cleaner Production, 289*, 125701.

Dai, A., Zhao, T., & Chen, J. (2018). Climate change and drought: A precipitation and evaporation perspective. *Current Climate Change Reports, 4*, 301–312.

Daisley, B. A., Chernyshova, A. M., Thompson, G. J., & Allen-Vercoe, E. (2022). Deteriorating microbiomes in agriculture-the unintended effects of pesticides on microbial life. *Microbiome Research Reports, 1*(1).

Das, T. K., Bhattacharyya, R., Sharma, A. R., Das, S., Saad, A. A., & Pathak, H. (2013). Impacts of conservation agriculture on total soil organic carbon retention potential under an irrigated agro-ecosystem of the western Indo-Gangetic Plains. *European Journal of Agronomy, 51*, 34–42.

Das, G. P., Gould, I., Zarafshan, P., Heselden, J., Badiee, A., Wright, I., & Pearson, S. (2022). Applications of robotic and solar energy in precision agriculture and smart farming. In *Solar energy advancements in agriculture and food production systems* (pp. 351–390). Academic Press.

Dash, M., Thiyageshwari, S., Selvi, D., Anandham, R., Rajan, K., Maduraimuthu, D., et al. (2023). Unravelling the release kinetics of exchangeable magnesium in acid soil of Nilgiris. *Sustainability, 15*(12), 9848.

de Souza Grilo, M. M., Fortes, A. F. C., de Souza, R. P. G., Silva, J. A. M., & Carvalho, M. (2018). Carbon footprints for the supply of electricity to a heat pump: Solar energy vs. electric grid. *Journal of Renewable and Sustainable Energy, 10*(2).

Dey, S., Purakayastha, T. J., Sarkar, B., Rinklebe, J., Kumar, S., Chakraborty, R., et al. (2023). Enhancing cation and anion exchange capacity of rice straw biochar by chemical modification for increased plant nutrient retention. *Science of the Total Environment, 886*, 163681.

Diacono, M., Persiani, A., Testani, E., Montemurro, F., & Ciaccia, C. (2019). Recycling agricultural wastes and by-products in organic farming: Biofertilizer production, yield performance and carbon footprint analysis. *Sustainability, 11*(14), 3824.

Doney, S. C., Fabry, V. J., Feely, R. A., & Kleypas, J. A. (2009). Ocean acidification: The other CO2 problem. *Annual Review of Marine Science, 1*, 169–192.

Dreyfus, G., Frederick, C., Larkin, E., Powers, Y., Chatterjee, J., & Sharma, S. (2023). Reducing nitrous oxide emissions from smallholder farmer agriculture through site specific nutrient management.

Franzluebbers, A. J. (2020). Cattle grazing effects on the environment: Greenhouse gas emissions and carbon footprint. In *Management strategies for sustainable cattle production in southern pastures* (pp. 11–34). Academic Press.

Franzluebbers, A. J., Chappell, J. C., Shi, W., & Cubbage, F. W. (2017). Greenhouse gas emissions in an agroforestry system of the southeastern USA. *Nutrient Cycling in Agroecosystems, 108*, 85–100.

Gan, Y., Liang, C., Hamel, C., Cutforth, H., & Wang, H. (2011). Strategies for reducing the carbon footprint of field crops for semiarid areas. A review. *Agronomy for Sustainable Development, 31*, 643–656.

Gosnell, H., Gill, N., & Voyer, M. (2019). Transformational adaptation on the farm: Processes of change and persistence in transitions to 'climate-smart'regenerative agriculture. *Global Environmental Change, 59*, 101965.

Haque, E. (2018). A review on bio-pesticide for organic farming. *International Journal of Chemical Studies, 2*, 11–14.

Hassoun, A., Prieto, M. A., Carpena, M., Bouzembrak, Y., Marvin, H. J., Pallares, N., et al. (2022). Exploring the role of green and industry 4.0 technologies in achieving sustainable development goals in food sectors. *Food Research International*, 112068.

Holka, M., Kowalska, J., & Jakubowska, M. (2022). Reducing carbon footprint of agriculture—Can organic farming help to mitigate climate change? *Agriculture, 12*(9), 1383.

Horrillo, A., Gaspar, P., & Escribano, M. (2020). Organic farming as a strategy to reduce carbon footprint in dehesa agroecosystems: A case study comparing different livestock products. *Animals, 10*(1), 162.

Huang, Q., Gong, Y., Dewi, R. K., Li, P., Wang, X., Hashimi, R., & Komatsuzaki, M. (2023). Enhancing energy efficiency and reducing carbon footprint in organic soybean production through no-tillage and rye cover crop integration. *Journal of Cleaner Production, 419*, 138247.

Huifang, Z. H. A. N. G., Rongqin, Z. H. A. O., & Liangang, X. I. A. O. (2021). The effects of irrigation methods on carbon emission and water-energy consumption of crop production. *Journal of Irrigation & Drainage, 40*(12).

Jagadesh, M., Selvi, D., Thiyageshwari, S., Srinivasarao, C., Kalaiselvi, T., Lourdusamy, K., et al. (2023). Soil carbon dynamics under different ecosystems of Ooty Region in the Western Ghats biodiversity hotspot of India. *Journal of Soil Science and Plant Nutrition, 23*(1), 1374–1385.

Jagadesh, M., Selvi, D., Thiyageshwari, S., Kalaiselvi, T., Allan, V., Dash, M., et al. (2024). Exploration of microbial signature and carbon footprints of the Nilgiri Hill Region in the Western Ghats global biodiversity hotspot of India. *Applied Soil Ecology, 194*, 105176.

Jat, R. K., Bijarniya, D., Kakraliya, S. K., Sapkota, T. B., Kakraliya, M., & Jat, M. L. (2021). Precision nutrient rates and placement in conservation maize-wheat system: Effects on crop productivity, profitability, nutrient-use efficiency, and environmental footprints. *Agronomy, 11*(11), 2320.

Jiang, Z., Zhong, Y., Yang, J., Wu, Y., Li, H., & Zheng, L. (2019). Effect of nitrogen fertilizer rates on carbon footprint and ecosystem service of carbon sequestration in rice production. *Science of the Total Environment, 670*, 210–217.

Jørgensen, L. N., Hovmøller, M. S., Hansen, J. G., Lassen, P., Clark, B., Bayles, R., et al. (2014). IPM strategies and their dilemmas including an introduction to www. eurowheat. org. *Journal of Integrative Agriculture, 13*(2), 265–281.

Kamyab, H., SaberiKamarposhti, M., Hashim, H., & Yusuf, M. (2023). Carbon dynamics in agricultural greenhouse gas emissions and removals: A comprehensive review. *Carbon Letters*, 1–25.

Karimi, P., Qureshi, A. S., Bahramloo, R., & Molden, D. (2012). Reducing carbon emissions through improved irrigation and groundwater management: A case study from Iran. *Agricultural Water Management, 108*, 52–60.

Khoshnevisan, B., Duan, N., Tsapekos, P., Awasthi, M. K., Liu, Z., Mohammadi, A., et al. (2021). A critical review on livestock manure biorefinery technologies: Sustainability, challenges, and future perspectives. *Renewable and Sustainable Energy Reviews, 135*, 110033.

Kinley, R., Cutajar, M. Z., de Boer, Y., & Figueres, C. (2021). Beyond good intentions, to urgent action: Former UNFCCC leaders take stock of thirty years of international climate change negotiations. *Climate Policy, 21*(5), 593–603.

Kouhizadeh, M., & Sarkis, J. (2018). Blockchain practices, potentials, and perspectives in greening supply chains. *Sustainability, 10*(10), 3652.

Kourgialas, N. N., & Karatzas, G. P. (2016). A flood risk decision making approach for Mediterranean tree crops using GIS; climate change effects and flood-tolerant species. *Environmental Science & Policy, 63*, 132–142.

Kumara, K., Pal, S., Chand, P., & Kandpal, A. (2023). Carbon sequestration potential of agroforestry systems in Indian agricultural landscape: A meta-analysis. *Ecosystem Services, 62*, 101537.

Kumari, A., Dash, M., Singh, S. K., Jagadesh, M., Mathpal, B., Mishra, P. K., et al. (2023). Soil microbes: A natural solution for mitigating the impact of climate change. *Environmental Monitoring and Assessment, 195*(12), 1436.

Lal, R. (2008). Carbon sequestration. *Philosophical Transactions of the Royal Society B: Biological Sciences, 363*(1492), 815–830.

Lal, R., & Stewart, B. A. (2012). Soil degradation: A global threat. *Advances in Soil Science, 2.*

Lefebvre, D., Williams, A. G., Kirk, G. J., Paul, Burgess, J., Meersmans, J., et al. (2021). Assessing the carbon capture potential of a reforestation project. *Scientific Reports, 11*(1), 19907.

Lehman, R. M., Cambardella, C. A., Stott, D. E., Acosta-Martinez, V., Manter, D. K., Buyer, J. S., et al. (2015). Understanding and enhancing soil biological health: The solution for reversing soil degradation. *Sustainability, 7*(1), 988–1027.

Li, M., Li, J., Singh, V. P., Fu, Q., Liu, D., & Yang, G. (2019). Efficient allocation of agricultural land and water resources for soil environment protection using a mixed optimization-simulation approach under uncertainty. *Geoderma, 353*, 55–69.

Lin, T. C., Hogan, J. A., & Chang, C. T. (2020). Tropical cyclone ecology: A scale-link perspective. *Trends in Ecology & Evolution, 35*(7), 594–604.

Liu, C., Cutforth, H., Chai, Q., & Gan, Y. (2016). Farming tactics to reduce the carbon footprint of crop cultivation in semiarid areas. A review. *Agronomy for Sustainable Development, 36*, 1–16.

Liu, P., Liu, L., Xu, X., Zhao, Y., Niu, J., & Zhang, Q. (2021). Carbon footprint and carbon emission intensity of grassland wind farms in Inner Mongolia. *Journal of Cleaner Production, 313*, 127878.

Macelaru, M., & Lulić, M. (2023). Understanding and addressing environmental challenges. *RAIS Journal for Social Sciences, 7*(2), 104–112.

Madane, D. A., Singh, M. C., Sharma, P., & Mane, M. (2023). Water and carbon footprint assessment of onion crop cultivated under differential irrigation scenarios. *Arabian Journal of Geosciences, 16*(7), 419.

Majumder, D., Akhter, J., Mandal, A., Roy, R., Mondal, D., Bhatt, R., & Hossain, A. (2021). Precision input management for minimizing and recycling of agricultural waste. *Input Use Efficiency for Food and Environmental Security*, 567–603.

Malhi, G. S., Kaur, M., & Kaushik, P. (2021). Impact of climate change on agriculture and its mitigation strategies: A review. *Sustainability, 13*(3), 1318.

Mangalassery, S., Sjögersten, S., Sparkes, D. L., Sturrock, C. J., Craigon, J., & Mooney, S. J. (2014). To what extent can zero tillage lead to a reduction in greenhouse gas emissions from temperate soils? *Scientific Reports, 4*(1), 4586.

Marino, R., Atzori, A. S., D'Andrea, M., Iovane, G., Trabalza-Marinucci, M., & Rinaldi, L. (2016). Climate change: Production performance, health issues, greenhouse gas emissions and mitigation strategies in sheep and goat farming. *Small Ruminant Research, 135*, 50–59.

Marx, W., Haunschild, R., & Bornmann, L. (2021). Heat waves: A hot topic in climate change research. *Theoretical and Applied Climatology, 146*(1–2), 781–800.

McCarthy, B., Anex, R., Wang, Y., Kendall, A. D., Anctil, A., Haacker, E. M., & Hyndman, D. W. (2020). Trends in water use, energy consumption, and carbon emissions from irrigation: Role of shifting technologies and energy sources. *Environmental Science & Technology, 54*(23), 15329–15337.

McQueen, N., Gomes, K. V., McCormick, C., Blumanthal, K., Pisciotta, M., & Wilcox, J. (2021). A review of direct air capture (DAC): Scaling up commercial technologies and innovating for the future. *Progress in Energy, 3*(3), 032001.

Meena, R. S., Kumar, S., & Yadav, G. S. (2020). Soil carbon sequestration in crop production. In *Nutrient dynamics for sustainable crop production* (pp. 1–39).

Menegat, S., Ledo, A., & Tirado, R. (2022). Greenhouse gas emissions from global production and use of nitrogen synthetic fertilisers in agriculture. *Scientific Reports, 12*(1), 14490.

Mittal, S., Dai, H., Fujimori, S., & Masui, T. (2016). Bridging greenhouse gas emissions and renewable energy deployment target: Comparative assessment of China and India. *Applied Energy, 166*, 301–313.

Moate, P. J., Deighton, M. H., Williams, S. R. O., Pryce, J. E., Hayes, B. J., Jacobs, J. L., et al. (2015). Reducing the carbon footprint of Australian milk production by mitigation of enteric methane emissions. *Animal Production Science, 56*(7), 1017–1034.

Morris, L. (2023). *Open-path dual-frequency comb spectroscopy of methane from livestock production* (Doctoral dissertation, Kansas State University).

Muluneh, M. G. (2021). Impact of climate change on biodiversity and food security: A global perspective—A review article. *Agriculture & Food Security, 10*(1), 1–25.

Nave, L. E., Walters, B. F., Hofmeister, K. L., Perry, C. H., Mishra, U., Domke, G. M., & Swanston, C. W. (2019). The role of reforestation in carbon sequestration. *New Forests, 50*(1), 115–137.

Neethirajan, S. R. (2023). AI-driven climate neutrality in dairy farming: Benchmarking emissions for sustainable transformation.

Neufeldt, H., Jahn, M., Campbell, B. M., Beddington, J. R., DeClerck, F., De Pinto, A., et al. (2013). Beyond climate-smart agriculture: Toward safe operating spaces for global food systems. *Agriculture & Food Security, 2*(1), 1–6.

Oliveira, A. M., Beswick, R. R., & Yan, Y. (2021). A green hydrogen economy for a renewable energy society. *Current Opinion in Chemical Engineering, 33*, 100701.

Orgiazzi, A., Singh, B., Wall, D., Barrios, E., Kandeler, E., Moreira, F., De Deyn, G., Chotte, J., Six, J., Hedlund, K., Briones, M., Miko, L., Johnson, N., Ramirez, K., Fierer, N., Kaneko, N., Lavelle, P., Eggleton, P., Lemanceau, P., Bardgett, R., Jeffery, S., Fraser, T., Behan Pelletier, V., Van Der Putten, W., Montanarella, L., & Jones, A. (2015). Global Soil Biodiversity Atlas, Publications Office of the European Union, Luxembourg, ISBN 978-92-79-48168-0 (online),978-92-79-48169-7 (print),978-92-79-74163-0 (Kindle), JRC95584.

Parihar, C. M., Sahrawat, Y. S., Jat, H. S., & Jat, M. L. (2021). Nutrient management in conservation agriculture-based production systems.

Parihar, C. M., Meena, B. R., Nayak, H. S., Patra, K., Sena, D. R., Singh, R., et al. (2022). Co-implementation of precision nutrient management in long-term conservation agriculture-based systems: A step towards sustainable energy-water-food nexus. *Energy, 254*, 124243.

Pathak, V. M., Verma, V. K., Rawat, B. S., Kaur, B., Babu, N., Sharma, A., et al. (2022). Current status of pesticide effects on environment, human health and it's eco-friendly management as bioremediation: A comprehensive review. *Frontiers in Microbiology, 2833*.

Paul, S., Mishra, M., Pati, S., Acharyya, T., Santos, C. A. G., da Silva, R. M., et al. (2023). Evaluation of overwash vulnerability and shoreline dynamics in cyclone-prone Sagar Island, Sundarbans (India). *Science of the Total Environment*, 167933.

Petersen, S. O., Blanchard, M., Chadwick, D., Del Prado, A., Edouard, N., Mosquera, J., & Sommer, S. G. (2013). Manure management for greenhouse gas mitigation. *Animal, 7*(s2), 266–282.

Peyvandi, M., Hajinezhad, A., & Moosavian, S. F. (2023). Investigating the intensity of GHG emissions from electricity production in Iran using renewable sources. *Results in Engineering, 17*, 100819.

Pimentel, D. (1980). Energy inputs for the production, formulation, packaging and transport of various pesticides. In D. Pimentel (Ed.), *Handbook of energy utilization in agriculture* (pp. 45–55). CRC Press.

Plaza-Bonilla, D., Nogué-Serra, I., Raffaillac, D., Cantero-Martínez, C., & Justes, E. (2018). Carbon footprint of cropping systems with grain legumes and cover crops: A case-study in SW France. *Agricultural Systems, 167*, 92–102.

Pratibha, G., Srinivas, I., Rao, K. V., Shanker, A. K., Raju, B. M. K., Choudhary, D. K., et al. (2016). Net global warming potential and greenhouse gas intensity of conventional and conservation agriculture system in rainfed semi arid tropics of India. *Atmospheric Environment, 145*, 239–250.

Purakayastha, T. J., Bera, T., Bhaduri, D., Sarkar, B., Mandal, S., Wade, P., et al. (2019). A review on biochar modulated soil condition improvements and nutrient dynamics concerning crop yields: Pathways to climate change mitigation and global food security. *Chemosphere, 227*, 345–365.

Qureshi, A. S. (2014). Reducing carbon emissions through improved irrigation management: A case study from Pakistan. *Irrigation and Drainage, 63*(1), 132–138.

Sándor, R., Ehrhardt, F., Basso, B., Bathia, A., Bellocchi, G., Brilli, L., et al. (2016, June). Global research alliance on agricultural greenhouse gases-benchmark and ensemble crop and grassland model estimates. In *MACSUR Conference "Modelling Grassland-Livestock Systems under Climate Change"* (p. np).

Sapkota, T. B., Majumdar, K., Jat, M. L., Kumar, A., Bishnoi, D. K., McDonald, A. J., & Pampolino, M. (2014). Precision nutrient management in conservation agriculture based wheat production of Northwest India: Profitability, nutrient use efficiency and environmental footprint. *Field Crops Research, 155*, 233–244.

Schäfer, F., & Blanke, M. (2012). Farming and marketing system affects carbon and water footprint–a case study using Hokaido pumpkin. *Journal of Cleaner Production, 28*, 113–119.

Schillaci, C., Tadiello, T., Acutis, M., & Perego, A. (2021). Reducing topdressing N fertilization with variable rates does not reduce maize yield. *Sustainability, 13*(14), 8059.

Seyboth, K. (2013). Intergovernmental panel on climate change (IPCC). In *Encyclopedia of energy, natural resource, and environmental economics (Volume 1: ENERGY)*. Elsevier.

Shao, P., Liang, C., Lynch, L., Xie, H., & Bao, X. (2019). Reforestation accelerates soil organic carbon accumulation: Evidence from microbial biomarkers. *Soil Biology and Biochemistry, 131*, 182–190.

Siddik, M., Islam, M., Zaman, A. K. M. M., & Hasan, M. (2021). Current status and correlation of fossil fuels consumption and greenhouse gas emissions. *Int. J. Energy Environ. Econ, 28*, 103–119.

Signor, D., & Cerri, C. E. P. (2013). Nitrous oxide emissions in agricultural soils: A review. *Pesquisa Agropecuária Tropical, 43*, 322–338.

Singh, P., Singh, G., Sodhi, G. P. S., & Benbi, D. K. (2021). Accounting carbon footprints and applying data envelopment analysis to optimize input-induced greenhouse gas emissions under rice–wheat cropping system in North-Western India. *Journal of Soil Science and Plant Nutrition, 21*(4), 3030–3050.

Sishodia, R. P., Ray, R. L., & Singh, S. K. (2020). Applications of remote sensing in precision agriculture: A review. *Remote Sensing, 12*(19), 3136.

Slaney, G. L., Lantz, V. A., & MacLean, D. A. (2009). The economics of carbon sequestration through pest management: Application to forested landbases in New Brunswick and Saskatchewan, Canada. *Forest Policy and Economics, 11*(7), 525–534.
Soto, I., Barnes, A., Balafoutis, A., Beck, B., Sánchez, B., Vangeyte, J., et al. (2019). *The contribution of precision agriculture technologies to farm productivity and the mitigation of greenhouse gas emissions in the EU*. Publications Office of the European Union.
Sun, J., Wang, Z., Du, Y., Zhang, E., Gan, H., Sun, D., & Niu, W. (2022). Optimized tillage improves yield and energy efficiency while reducing carbon footprint in winter wheat-summer maize rotation systems. *Science of the Total Environment, 820*, 153278.
Tarakanov, V. (2022). How carbon emissions acidify our ocean. *IAEA Bulletin (Online), 63*(4), 22–23. https://www.iaea.org/bulletin/how-carbon-emissions-acidify-ourocean#:~:text=Ocean%20acidification%20is%20a%20consequence,seawater%20chemistry%20called%20ocean%20acidification
Theurl, M. C., Haberl, H., Erb, K. H., & Lindenthal, T. (2014). Contrasted greenhouse gas emissions from local versus long-range tomato production. *Agronomy for Sustainable Development, 34*, 593–602.
Thomson, A. J., Giannopoulos, G., Pretty, J., Baggs, E. M., & Richardson, D. J. (2012). Biological sources and sinks of nitrous oxide and strategies to mitigate emissions. *Philosophical Transactions of the Royal Society B: Biological Sciences, 367*(1593), 1157–1168.
Tollefson, J. (2021). Earth is warmer than it's been in 125,000 years, says landmark climate report.
Treu, H., Nordborg, M., Cederberg, C., Heuer, T., Claupein, E., Hoffmann, H., & Berndes, G. (2017). Carbon footprints and land use of conventional and organic diets in Germany. *Journal of Cleaner Production, 161*, 127–142.
Tudi, M., Daniel Ruan, H., Wang, L., Lyu, J., Sadler, R., Connell, D., et al. (2021). Agriculture development, pesticide application and its impact on the environment. *International Journal of Environmental Research and Public Health, 18*(3), 1112.
Tuomisto, H. L., Hodge, I. D., Riordan, P., & Macdonald, D. W. (2012). Does organic farming reduce environmental impacts?–A meta-analysis of European research. *Journal of Environmental Management, 112*, 309–320.
Ussiri, D., Lal, R., Ussiri, D., & Lal, R. (2013). Nitrous oxide sources and mitigation strategies. *Soil Emission of Nitrous Oxide and its Mitigation*, 243–275.
Vecchi, G. A., & Soden, B. J. (2007). Effect of remote sea surface temperature change on tropical cyclone potential intensity. *Nature, 450*(7172), 1066–1070.
Velarde-Guillén, J., Arndt, C., & Gómez, C. A. (2022). Carbon footprint in Latin American dairy systems. *Tropical Animal Health and Production, 54*(1), 15.
Wang, C., Amon, B., Schulz, K., & Mehdi, B. (2021). Factors that influence nitrous oxide emissions from agricultural soils as well as their representation in simulation models: A review. *Agronomy, 11*(4), 770.
WMO. (2021). https://public.wmo.int/en/media/press-release/wake-looming-water-crisis-report-warns
World Bank. (2021). https://www.worldbank.org/en/news/feature/2021/06/17/floods-and-droughts-an-epic-response-to-these-hazards-in-the-era-of-climate-change
Wrage-Mönnig, N., Horn, M. A., Well, R., Müller, C., Velthof, G., & Oenema, O. (2018). The role of nitrifier denitrification in the production of nitrous oxide revisited. *Soil Biology and Biochemistry, 123*, A3–A16.
Xu, X., & Lan, Y. (2016). A comparative study on carbon footprints between plant-and animal-based foods in China. *Journal of Cleaner Production, 112*, 2581–2592.
Yadav, G. S., Babu, S., Das, A., Mohapatra, K. P., Singh, R., Avasthe, R. K., & Roy, S. (2020). No-till and mulching enhance energy use efficiency and reduce carbon footprint of a direct-seeded upland rice production system. *Journal of Cleaner Production, 271*, 122700.
Yuttitham, M. (2019). Comparison of carbon footprint of organic and conventional farming of Chinese Kale. *Environment and Natural Resources Journal, 17*(1), 78–92.

An In-depth Exploration of Carbon Footprint Analysis in the Construction Sector with Emphasis on the Dam Industry

Hasan Mostafaei, Muhammad Ali Rostampour, Niyousha Fallah Chamasemani, and Chengqing Wu

1 Introduction

The emission of greenhouse gases (GHG) and their accumulation on the planet present a fundamental challenge that has manifested in significant problems in recent years (Nakicenovic, 2000; Santos et al., 2022). To gauge and comprehend these effects, researchers draw on an array of data derived from meteorology, geology, oceanography, and various other disciplines. The synthesis of these multidisciplinary datasets enables comprehensive analyses of the multifaceted impacts of global warming (Mimura et al., 1998; Mishra et al., 2021; Rehman et al., 2014).

In the present day landscape, the surge in populace coupled with heightened consumption of fossil fuels and resources has led to a drastic escalation in GHG production (Heilig, 1994; Van der Werf et al., 2013). An ongoing temperature analysis, performed with the aid of scientists at NASA's Goddard Institute for Space Studies (GISS), reveals a concerning trend. The average worldwide temperature on Earth has surged by using at the least 1.1° C considering 1880, with a discernible acceleration inside the charge of increase, in particular in recent years

H. Mostafaei (✉) · C. Wu
School of Civil and Environmental Engineering, University of Technology Sydney, Sydney, NSW, Australia
e-mail: hasan.mostafaei@student.uts.edu.au; chengqinq.wu@uts.edu.au

M. A. Rostampour
Department of Civil Engineering, Isfahan University of Technology (IUT), Isfahan, Iran
e-mail: ma.rostampour@cv.iut.ac.ir

N. F. Chamasemani
Department of Civil and Environmental Engineering, Politecnico di Milano, Milan, Italy
e-mail: niyousha.fallah@mail.polimi.it

S. S. Muthu (ed.), *Carbon Footprint Assessments*, Environmental Footprints and Eco-design of Products and Processes,
https://doi.org/10.1007/978-3-031-70262-4_3

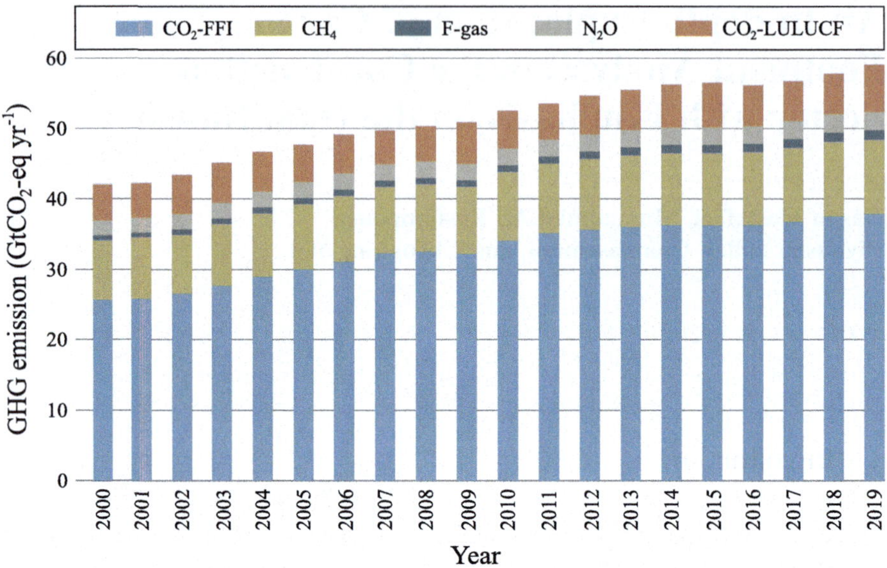

Fig. 1 Trend in global GHG emission. (Adopted from IPCC Climate Change, 2022)

(IPCC Climate Change, 2022). This data underscores the urgency of addressing the underlying reasons of GHG emissions and formulating strong mitigation techniques.

Figure 1 shows the overarching trend of GHG emissions from 2000 to 2019, showing a significant 40% increase during this period. However, when considering the concurrent population growth of 25.6%, the intensity of GHG emissions relative to the global population hovered around 12%, reaching 7.8 tCO_2-eq per person in 2019. Notably, the rate of increase in this metric has experienced a slight containment in recent years. To further contextualize the environmental impact, calculating GHG emissions in relation to the Gross Domestic Product (GDP) index unveils a consistent decline over the 20-year span, reflecting an overall reduction of 27%.

This data emphasizes the intricate interplay between population dynamics, economic activities, and GHG emissions. While the absolute increase in emissions is concerning, the per capita intensity metric provides a nuanced perspective, indicating that the emissions burden on individuals has somewhat stabilized. The observed decline in emissions relative to GDP underscores potential advancements in decoupling economic growth from environmental impact (de Souza Mendonça et al., 2020; Shi, 2001). This aligns with ongoing global efforts to transition towards more sustainable and decarbonized economies. It is imperative to delve deeper into the specific sectors contributing to this trend, exploring both drivers and mitigating factors. By understanding the intricacies of these dynamics, policymakers, researchers, and industries can formulate targeted strategies to further curb emissions, fostering a sustainable trajectory for global environmental well-being (Pisciotta et al., 2022).

Increased GHG emissions pose severe challenges to human settlements, reflecting increased air pollution, decreased air quality, public health risks, and increased climate change-related issues. In 2019, the energy sector accounted for 34% (20 $GtCO_2$-eq) of total $GtCO_2$-eq GHG emissions, with industry contributing 24% (14 $GtCO_2$-eq) and AFOLU (Agriculture, Forestry, Land and Other Users) 22% (13 $GtCO_2$-eq), transport contributes 15% (8.7 $GtCO_2$-eq), and buildings contribute 6% (3.3 $GtCO_2$-eq) (IPCC Climate Change, 2022).

The predominant perception of environmental pollution often centers on emissions from vehicles and factories, overlooking the significant environmental impact of building construction and its associated activities. Building construction and the production of related materials represent industries with substantial energy consumption and environmental pollution (Akintayo et al., 2020; Fuertes et al., 2013; Mačková et al., 2017). The production processes in these sectors contribute significantly to carbon dioxide emissions, a facet of environmental degradation that cannot be understated. The construction sector emerges as a key player responsible for a substantial proportion of global GHG emissions, both directly and indirectly. Its economic significance and extensive involvement in the construction supply chain position it as an industry with significant potential for emissions reduction. Given its role, the construction sector offers various opportunities for short-term emission reductions as compared to other major emitting industries (Chen et al., 2022; Hu, 2015).

Figure 2 illustrates the growth rates of GHG emissions for each economic sub-sector during the second decade of this century. Cement, a pivotal construction material, experienced a substantial increase of nearly 160 million tons in GHG

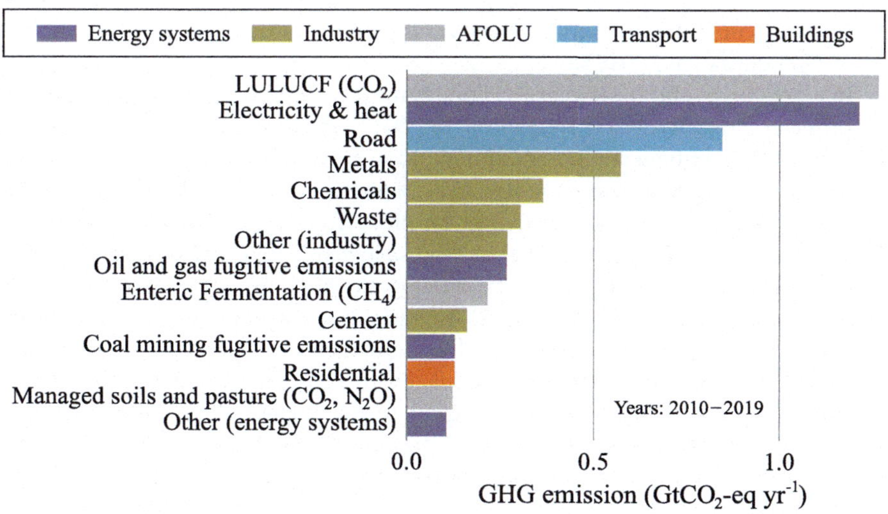

Fig. 2 Absolute GHG emissions growth by subsector. (Adopted from IPCC Climate Change, 2022)

emissions. Remarkably, the growth in emissions from the cement industry surpassed even that of the residential sub-sector. It is important to observe that in the creation industry, GHG emissions expand beyond cement, concerning different sub-sectors such as LULUCF (Land Use, Land-Use Change, and Forestry), road (transportation), metals, and waste. Addressing the environmental impact of the construction enterprise calls for a complete method, thinking about the whole existence cycle of substances and systems. Sustainable practices, progressive technology, and a shift toward low-carbon options are imperative for lowering GHG emissions. The construction area's multifaceted contribution to emissions underscores the need for strategic interventions and industry-wide projects to achieve sustainable and eco-friendly production practices. This includes promoting electricity-efficient designs, utilizing eco-friendly substances, and adopting circular economic system ideas to reduce the overall carbon footprint of production activities.

In different countries, institutions and researchers have reported varying percentages for the contribution of the construction sector to overall GHG production. Notably, the construction sector often emerges as a major carbon emitter, particularly in developing countries. While specific percentages may vary, figures approaching one-fifth are commonly cited to convey the construction sector's contribution to GHG emissions (Shi et al., 2023). Addressing the environmental effects of creation requires a comprehensive approach that considers the complete lifecycle of buildings and infrastructure. Each process from raw cloth extraction to creation, operation, maintenance, and sooner or later demolition or recycling offers opportunities for carbon emission reduction and sustainable practices. To be conscious if there are a couple of environmental factors of the construction enterprise, it is essential to develop powerful strategies to mitigate the effect on climate trade. By adopting sustainable building practices, using environmentally friendly materials, and adopting strength green technologies, the development industry plays an important position in improving environmental stewardship and reaching global weather goals (Marques et al., 2017).

In a pivotal second for our shared commitment to address weather exchange, the world came collectively on December 12, 2015, in Paris, culminating inside the historic COP21 settlement. This landmark accord, subsidized through 195 nations, set out a bold agenda to comprehensively address weather trade and worldwide warming. The agreement referred to as for proactive measures and investments in fostering a low-carbon, resilient, bendy, and sustainable financial system (Christoff, 2016; Mohammed et al., 2011). Beyond being an enormous achievement, the Paris Agreement stands as an image of global harmony within the quest to enhance the country of our planet. Its number one intention is to save the Earth's temperature from rising past pre-industrial levels, coupled with a secondary objective to reinforce our collective capability to face and adapt to the impacts of climate exchange. Specifically, the agreement outlines that the global temperature increase must be restricted to 2 stages Celsius by means of the century's stop, with a good more aspirational goal of 1.5 Celsius. Participating international locations pledged to lessen a percent of their greenhouse gasoline emissions as compared to baseline eventualities. It is essential to meet those commitments, as failure to mitigate emissions should result in a substantial three-diploma Celsius temperature upward thrust, exacerbating global demanding situations. Although the agreement lacks a specific timeframe, signatory nations are obligated to expedite efforts to reduce emissions and reach the prescribed limit. This necessitates a remarkable reduction in GHG production by the second half of this century, leveraging available technologies. Importantly, this reduction does not imply the complete elimination of GHG production but rather an attenuation of its growth, allowing natural absorption processes and technological advancements to address these emissions. The urgency of collective action underscores the need for continued international cooperation to safeguard the planet's ecological balance (Fujimori et al., 2016; Savaresi, 2016).

Over the past decades, numerous countries and governments have implemented measures to curtail GHG emissions, particularly carbon dioxide, and its associated footprint. During this period, the concept of carbon footprint has emerged as a focal point in environmental, social, and urban sciences, evolving from the notion of ecological footprint. It signifies the total GHG emissions attributed to an organization, event, product, or individual. In this chapter, following a scientific introduction to the concept of the carbon footprint, environmental considerations within the construction industry are thoroughly examined. Various segments of the construction industry are scrutinized with respect to this environmental index. Subsequently, within the same section, the principal factors contributing to carbon emissions in the construction industry are delineated individually, elucidating the magnitude of their impact. Recommendations for mitigating carbon emissions through effective control measures are then proposed to curtail environmental impact. Shifting focus to the dam construction industry, the paramount significance of its environmental role is expounded upon. Through the presentation of two comprehensive case studies on the Pine Flat and Morrow Point dams, the carbon footprint of these structures is meticulously assessed across diverse factors, measured in $kgCO_2eq$ units. The conclusion of this chapter entails a comprehensive summary of its contents, underscoring the importance of GHG emission reduction policies within the construction industry, particularly in the context of dam construction. This effort aims to chart a clear trajectory for researchers navigating this intricate field, fostering a more sustainable and environmentally conscious approach to construction practices.

This chapter first introduces the concept of the carbon footprint, emphasizing its significance in the construction industry. It then delves into the environmental concerns prevalent in construction, highlighting the need for sustainable practices. Following this, two types of dams are examined as case studies to illustrate the carbon footprint of construction projects. Subsequently, strategies for mitigating carbon emissions in the construction sector are explored. Finally, this book chapter summarizes the most important findings derived from the analysis, offering insights into the implications for sustainable construction practices.

2 Carbon Footprint Concept

The imperative of researchers and governments to quantify and discern the impact of diverse activities on global warming has given rise to the concept of the "carbon footprint." Widely recognized as a global tool, the carbon footprint analysis (CFA) serves to quantify the environmental impact of products. Given the impracticality of calculating the total carbon footprint due to the extensive data required and the influence of natural events contributing to carbon dioxide emissions, the following refined definition is proposed: "This metric represents the total GHG emissions, both direct and indirect, attributable to an individual, event, organization, or product, expressed in equivalent units of CO_2 volume. Typically assessed over a 1-year period, the carbon footprint is integral to gauging environmental sustainability."

Historically, the discourse on the carbon footprint can be contextualized as a subset derived from the broader concept of the "environmental footprint" introduced by (Wackernagel & Rees, 1998). The environmental footprint quantifies the aggregate area of fertile land and sea necessary for the sustenance of a specific human population. In this context, the carbon footprint is defined as the land area required for the complete sequestration of carbon dioxide produced throughout human life. Over time, as global warming gained significance, the carbon footprint evolved as an independent concept with a distinct definition from that of the environmental footprint.

The conceptual framework of the carbon footprint is intricately linked to financial transactions, encompassing taxes, carbon offsets, and the modulation of consumer choices. Consequently, the establishment of integrated methods for calculating carbon footprints becomes imperative. Despite variations in calculation methodologies, the unit of a carbon footprint report universally embraces the equivalent mass of carbon dioxide, denoted as CO_2-eq, based on its potential to contribute to Earth's warming over a century. Hence, the carbon footprint can be precisely defined as the equivalent mass of carbon dioxide released into the atmosphere within stipulated bounds by individuals, organizations, processes, products, or special events.

In delineating carbon, it is pertinent to note that the term extends beyond carbon dioxide; the carbon footprint encapsulates all emitted greenhouse gases. Among these, significant contributors include carbon dioxide, methane, ozone, and nitrogen oxide. Reports from international organizations delineate the emission share of each gas, with carbon dioxide claiming the lion's share, as evidenced in the (IPCC Climate Change, 2022) report, which categorizes these gases into five general classifications:

- *CO_2-FFI*: CO_2 from fossil fuel combustion and industrial processes
- *CO_2-LULUCF*: net CO_2 emissions from land use, land-use change, and forestry
- *CH_4*: methane
- *N_2O*: nitrous oxide
- *F-gases*: fluorinated gases comprising hydrofluorocarbons (HFCs), perfluorocarbons (PFCs), sulphur hexafluoride (SF_6), as well as nitrogen trifluoride (NF_3)

Based on Fig. 1 from 2019, the distribution of the GHGs including CO_2-FFI, CH_4, CO_2-LULUCF, N_2O, and F-gases stands at 64.2%, 17.9%, 11.2%, 4.4%, and 2.3%, respectively. Notably, carbon dioxide represents over 75% of the overall emissions. These GHG emissions categorized as direct emissions from facility-based activities and indirect emissions associated with electricity usage. Direct emissions involve burning fossil fuels, chemical reactions, and leaks, specially from natural fuel and petroleum systems. Indirect emissions stem from fossil gasoline combustion at strength vegetation offering power to energy commercial facilities.

The nuanced know-how of the carbon footprint serves as a foundational perspective in comprehending the wider environmental impact and facilitating knowledgeable choice-making inside the pursuit of sustainability. In the pursuit of calculating the carbon footprint of a product or activity, it will become imperative to assess the amount of GHGs emitted, removed, or trapped for the duration of its lifestyles cycle. The comprehensive life cycle entails numerous tiers, encompassing production (from raw fabric guidance to very last packaging), distribution, intake, and concluding with disposal. This holistic evaluation is often known as "cradle to grave," emphasizing the whole thing of a product's existence. Life Cycle Assessment (LCA), typically referred to as Environmental Life Cycle Assessment, offers a breathtaking angle, encompassing inputs and outputs related to air pollution, water usage, wastewater technology, electricity intake, GHG emissions, and other relevant signs.

Specifically addressing the carbon footprint, LCA facilitates the determination of GHGs emitted or trapped at each stage, colloquially termed as a "greenhouse gas audit." Once the life cycle of a product or activity is delineated, the subsequent steps involve meticulous data collection and the final computation of the carbon footprint or activity.

Among the prevailing methodologies in carbon footprint accounting, LCA and Input-Output Analysis (IOA) stand as the most commonly employed approaches. Some research studies have even amalgamated these methods in a hybrid format (Onat et al., 2014; Zhang et al., 2015). These inventory-based methodologies hinge on quantifying emissions and their corresponding emission factors. The typical progression of a CFA involves the identification of sources of GHG emissions—activities, products, or services (Wiedmann, 2009; Zhang et al., 2018). Subsequently, emissions from each source are measured or estimated, converted into carbon dioxide equivalent (CO_2-e) units using appropriate factors like emissions factors and global warming potentials. The summation of emissions from each source yields the total carbon footprint. Finally, the results are meticulously reported and communicated to relevant stakeholders after undergoing a verification process.

2.1 Standards and Protocols

Numerous international standards and guidelines govern the calculation of carbon footprints and the evaluation of a product's life cycle. However, owing to varying methodologies employed by different organizations, there is no singular universal standard for such calculations; nevertheless, several widely recognized standards facilitate the calculation and reporting of carbon footprints. These standards collectively contribute to a robust framework for evaluating and managing the carbon footprints of products and activities, fostering global efforts towards sustainable practices. Key standards include:

2.1.1 GHG Protocol

The GHG Protocol, which is developed by the World Resources Institute (WRI) and the World Business Council for Sustainable Development (WBCSD), is an extensively diagnosed framework for reporting greenhouse gas emissions. The GHG Protocol presents complete suggestions for organizations to manage, and file their GHG emissions accurately and transparently. The GHG Protocol is scientifically based on LCA and carbon accounting, providing a dependent framework to quantify GHG emissions in three scopes: Scope 1 for direct emissions, Scope 2 for indirect emissions from purchased energy, and Scope 3 for indirect emissions at some stage in the organization's value chain.

2.1.2 ISO 14064

The ISO 14064 standard, developed by the International Organization for Standardization (ISO), is a comprehensive international framework in three parts for quantifying, monitoring, and reporting GHG emissions and removals. It consists of three parts: (ISO 14064-1:2018) establishes principles for entity-level inventories, ensuring accuracy and consistency; (ISO 14064-2:2019) focuses on projects, guiding the quantification and reporting of emission reductions or removal enhancements, fostering transparency and credibility; and (ISO 14064-3:2019) addresses validation and verification processes, setting requirements for third-party entities to assess and validate GHG inventories, ensuring reliability and credibility.

2.1.3 WRI/WBCSD Corporate Standard

The WRI/WBCSD Corporate Standard, which is developed by the World Resources Institute (WRI) and the World Business Council for Sustainable Development (WBCSD), presents a complete framework for corporate-level GHG accounting. It includes two key standards: the Product Life Cycle Standard, focusing on life cycle emissions of products, and the Corporate Value Chain (Scope 3) Standard, addressing GHG emissions across an organization's entire value chain, including direct and indirect emissions. Scientifically, this standard promotes standardized and consistent GHG accounting practices, aiding organizations in accurate environmental impact assessment and informed emission reduction decisions, with methodologies aligned with scientific principles for transparent and comparable corporate sustainability reporting.

2.1.4 Sector-Specific Standards

The Sector-Specific Standards refer to guidelines and frameworks tailored to specific industries or sectors, providing methodologies for calculating and reporting GHG emissions unique to each sector. These standards acknowledge the distinct characteristics and emission sources within different industries, ensuring accurate and sector-specific assessments. By addressing the nuances of each sector, these standards contribute to a more precise understanding of GHG emissions, enabling organizations to implement targeted reduction strategies.

2.1.5 Product Life Cycle Assessment (ISO 14040/44)

The product Life Cycle Assessment under ISO 14040/44 is a systematic method for evaluating the environmental impacts of a product based on the (ISO 14040:2006) and (ISO 14044:2006) standards. Key components of the clinical framework include:

1. Defining the goals and boundaries of the evaluation
2. Considering all inputs (e.g., raw substances, energy) and outputs (e.g., emissions, waste)
3. Evaluating the capacity environmental impacts recognized in the LCI
4. Drawing conclusions and hints primarily based on the results of the LCA

The ISO 14040/forty four requirements sell a holistic know-how of a product's environmental performance, from raw material extraction and production to apply and disposal. Scientifically, this method lets in for informed decision-making, helping organizations become aware of areas for development and optimize their merchandise' sustainability for the duration of their lifestyles cycles.

2.1.6 PAS 2050

The (Publicly Available Specification, 2008), developed by the British Standards Institution (BSI), establishes a comprehensive framework for conducting LCAs with a specific focus on quantifying the GHG emissions related to goods and services. Key components include defining system boundaries, establishing a functional unit for comparison, conducting an LCI, performing impact assessments, addressing emissions allocation, and transparently reporting results. PAS 2050 aims to standardize the methodology for assessing carbon footprints, promoting comparability, encouraging emission reduction identification, and contributing to a consistent and transparent approach in support of broader sustainability initiatives.

2.1.7 IPCC Protocol/2006 Guidelines (IPCC-2006)

The Intergovernmental Panel on Climate Change (2006) or the IPCC Protocol/2006 Guidelines represent a set of guidelines established by the Intergovernmental Panel on Climate Change (IPCC) in 2006 to help nations in developing their National Greenhouse Gas Inventories. These guidelines suggest a complete and standardized approach for estimating and reporting human-induced GHG emissions and removals at the national level, overlaying various sectors. Key elements of the IPCC-2006 guidelines include sector-specific guidance, categorization of emissions into sectors, a focus on consistency and comparability for accurate global assessments, recommendations for quality assurance and control processes, and an emphasis on transparency in reporting methodologies. These guidelines play a pivotal role as a reference for countries, promoting uniform methodologies and reporting practices that enhance the global understanding of human contributions to climate change, thereby facilitating international collaboration in addressing climate-related challenges.

2.1.8 ISO 14067

(ISO 14067:2018) is an international standard developed by ISO, outlining a comprehensive framework for quantifying the carbon footprint of products. Emphasizing GHG emissions across the life cycle, it establishes key components, including defining system boundaries, establishing functional units, conducting life cycle inventories, addressing allocation in simultaneous production, conducting impact assessments, and providing guidelines for transparent communication and verification. ISO 14067 aims to standardize methodologies, enabling organizations to consistently assess and communicate the environmental impact of their products, fostering global efforts towards emission reduction and sustainable product practices.

2.2 *Software*

In quantifying the carbon footprint, various software options leverage distinct algorithms and methods to measure GHG emissions. Some noteworthy examples are included in Table 1.

For enhanced accuracy and alignment with specific requirements, users may opt for software based on different model algorithms and data. The choice of software can markedly impact the precision and reliability of carbon footprint calculations. Consideration should be given to the compatibility of chosen models with recognized standards and guidelines, ensuring the robustness of the results. Furthermore, ongoing advancements in software capabilities and updates to emission factors should be taken into account to maintain the relevance and accuracy of CFAs.

Table 1 Carbon footprint software

Name	Description
OpenLCA (Open Source)	OpenLCA employs diverse methods for LCA and carbon footprint calculation, utilizing models such as CML, Eco-indicator 99, and ILCD
Simapro	Simapro utilizes reputable LCA methods, including Eco-indicator 99, ReCiPe, and Cumulative Energy Demand (CED), to accurately calculate the carbon footprint
Carbon Trust Footprint Expert	As an online tool, Carbon Trust Footprint Expert employs valid models, including GHG Protocol and ISO 14044, to precisely calculate the carbon footprint
GaBi	GaBi utilizes various LCA methods, including Eco-indicator, CML, and ILCD, to calculate both the carbon footprint and broader environmental impacts

3 Environmental Concerns in Construction Industry

In the industrial sector, construction activities, alongside related industries, are recognized as remarkable consumers of natural resources, encompassing wood, minerals, water, and energy. As a substantial contributor to global energy consumption (36%) and GHG emissions (39%), the construction industry assumes a crucial role in addressing global warming and advancing sustainable development. Extensive research highlights China, India, the United States of America, Russia, and Japan as the foremost contributors to CO_2 emissions, underscoring the substantial impact of construction-related activities on the global carbon footprint (Zhang et al., 2019). The carbon footprint in the construction sector can be assessed across three general categories:

1. *Implementation and Construction Phase:*
   - One of the primary contributors to the carbon footprint in this phase is the production and extraction of building materials, with concrete, a globally ubiquitous material, being a major emitter of GHGs, especially through its main component, cement.
   - Material insulation in building systems markedly impacts energy consumption during the operational phase, with the use of insulating materials potentially reducing energy consumption by up to 50%.
   - Transportation, equipment, machinery, and construction waste disposal also contribute to the carbon footprint during the construction phase.
2. *Operation and Maintenance Phase*
   - The operational lifespan of civil structures influences the need for reconstruction. Regulations typically dictate the lifespan of structures, considering factors like earthquakes and wind with defined return periods.

- Prolonged operational life requires structures to withstand more intense loads, leading to increased construction costs. Balancing cost and environmental sustainability necessitate comprehensive research.
- Maintenance systems should ensure the completion of the expected service life, incorporating timely repair and retrofitting methods. Health monitoring systems can identify structural defects, enabling cost-effective and environmentally friendly repairs.

3. *Demolition and Reconstruction Phase*

   - At the end of a structure's operational life, decisions about its fate vary. Demolition, often the chosen option, can be approached sustainably through the use of recycled materials, though an LCA is essential to verify its environmental impact.
   - Recycled material usage may be considered environmentally sustainable, yet complexities arise, requiring a careful evaluation of lifetime costs. Energy consumption in material recycling, coupled with the environmental effects, needs to be compared against the production of primary materials.

These categories underscore the intricate relationship between construction practices, environmental impact, and the need for sustainable solutions. In evaluating these phases, LCAs play a pivotal role in providing a holistic understanding of the environmental implications of construction activities. In the following, the carbon footprint in various construction industries will be addressed firstly, and in the subsequent section, the key factors of carbon production in these industries will be examined.

3.1 Various Construction Industries

Civil constructions are typically categorized into two main segments: building construction and infrastructure, both of which will be explored below with regard to their carbon footprint.

3.1.1 Building Constructions

While significant efforts have been directed towards evaluating the carbon footprint in building constructions, predominant focus has been on energy-related aspects. Strategies involve the utilization of insulation materials, smart systems, and efficient energy management to mitigate carbon emissions during the building's operational phase. However, comprehensive assessment of the carbon footprint during the construction phase is crucial. To evaluate the carbon footprint in building implementation, it is essential to analyze the contribution of each component (framework, wall cladding, joinery). The total carbon produced is then calculated, and by dividing

Table 2 Recent studies on CFA within the building construction sector

Authors	Aims
Pomponi and Moncaster (2016)	Identifying 17 mitigation strategies for reducing the carbon footprint
Akbarnezhad and Xiao (2017)	Presenting five approaches for decreasing embodied carbon
Li et al. (2022)	Evaluating the carbon reduction performance in commercial buildings
Mostafaei et al. (2023a)	Investigating the carbon footprint of producing concrete and reinforced concrete buildings across different mix designs
Chamasemani et al. (2023)	Examining the impact of material selection on the carbon footprint of reinforced concrete buildings

it as the usable area of the building, the carbon footprint per unit area is determined. Research indicates that concrete structures exhibit a carbon footprint approximately 6–7 kg/m^2 higher than steel structures (Koohi Rostamkolaei & Abessi, 2018). This discrepancy is attributed to the elevated cement consumption in concrete-framed buildings, underscoring the potential environmental detriment associated with widespread construction of concrete buildings and its contribution to global warming concerns.

So far, extensive research has been conducted in the domain of carbon footprint within the construction of buildings. Mostafaei et al. (2023a) conducted a comprehensive study, employing three distinct concrete mix designs with strengths of 20, 30, and 40 MPa, wherein only the cement content was varied. They designed three concrete frame structures, maintaining identical architectural designs. The objective was to concurrently explore the dual effects of "increasing concrete strength through heightened cement consumption" and "reducing dimensions of concrete sections in the structure," aiming to identify which of these structures resulted in lower carbon emissions. The findings indicated that although the augmentation in compressive strength of concrete led to a reduction in the dimensions of concrete sections within the framework, this reduction in concrete consumption did not mitigate carbon emissions, primarily due to increased cement consumption. Table 2 highlights recent research endeavors aimed at evaluating the carbon footprint in the construction sector.

3.1.2 Infrastructure Constructions

Infrastructure, encompassing an array of structures and facilities vital for public services, useful resource supply, and societal communication, performs a pivotal function in the world. This sector of construction contains transportation, hydraulic, commercial, and energy structures. Infrastructure initiatives span a various range, inclusive of roads, bridges, tunnels, energy transmission lines, pipelines, conversation networks, water and sewage treatment centers, and airports. The environmental impact of civil infrastructures is an essential factor that warrants thorough examination. Beyond construction, the lifecycle of these projects entails ongoing operations, upkeep, and eventual decommissioning, every contributing to the overall environmental footprint. Understanding and addressing these factors is important for accomplishing sustainable and resilient infrastructure. Examining one particular place within infrastructure, the construction of dams stands out as a noteworthy instance. These mega-construction projects, crucial for hydraulics and energy generation, demand substantial material resources. The procurement of these substances, coupled with excavation and embankment activities, necessitates extensive transportation systems. The carbon footprint associated with the construction of dams underscores the environmental impact of large-scale infrastructural tasks. Roads, bridges, and tunnels, essential for transportation, involve the extraction of raw materials, manufacturing methods, and maintenance, all of which make a contribution to their carbon footprint. Water and sewage treatment facilities, critical for public health and environmental safety, require power-in depth processes, contributing to their carbon footprint. Table 3 presents some recent research studies within the domain of CFA for infrastructure construction.

Table 3 Recent studies on CFA within the infrastructure construction sector

Authors	Infrastructure	Aims
Zhang et al. (2015)	Dam	Comparing earth-rockfill dams and concrete gravity dams
Li et al. (2017)	Dam	Pre-impoundment clearance in the five largest hydro-projects
Jiang et al. (2018)	Dam	CFA of four normal size hydropower Stations in China
Martinez et al. (2018)	Dam	The carbon footprint of the end-of-life phase of a dam
Wang et al. (2019)	Dam	Evaluating the carbon footprint at a national scale
Li et al. (2019)	Dam	Comparing large- and mid-scale hydropower
Mostafaei et al. (2023c)	Dam	Investigation of the sustainability of a concrete gravity dam
Espinoza et al. (2019)	Road	Carbon footprint estimation in road construction
Keijzer et al. (2015)	Road	Comparing Carbon footprint of innovative techniques in the construction and maintenance of road infrastructure in The Netherlands
Sun et al. (2015)	Bridge	Decision support system for optimizing the maintenance of RC girder bridge superstructures in consideration of the carbon footprint
Ahn et al. (2010b)	Tunnel	Carbon footprints analysis for tunnel construction processes in the preplanning phase using collaborative simulation
Yang and Al-Qadi (2017)	Airport pavement	Development of a life-cycle assessment tool to quantify the environmental impacts of airport pavement construction

3.2 *Key Factors of Emissions in the Construction Industry*

The construction industry stands as a pivotal contributor to carbon emissions, with its influence extending across the comprehensive life cycle of construction projects. A thorough examination of the carbon footprint within this industry necessitates a holistic evaluation, encompassing various stages such as raw material extraction, transportation, construction processes, operational energy consumption, and considerations for the project's end-of-life (Bahmani et al., 2023; Mostafaei et al., 2023b). The intricacies concerned in each area underline the significance of a meticulous CFA, shedding light on the environmental repercussions of construction (Chamasemani et al., 2023). As the exploration of the key factors influencing carbon emissions in construction is delved into, it becomes evident that an in-depth know-how of these determinants is essential for the development and implementation of effective mitigation strategies. Here are some of the most important factors influencing carbon emissions in the construction industry.

3.2.1 Material Production and Extraction

The impact of materials on the carbon footprint of construction is a concern. Cement, which is an important material in construction, is produced by heating raw materials, such as limestone and clay, to high temperatures, which release carbon dioxide (CO_2) in the process, known as calcination. Similarly, lime and plaster production involve energy-intensive processes, further exacerbating the carbon footprint. Steel production, with its reliance on iron ore extraction and subsequent processing, also stands out as a substantial contributor to emissions in the construction industry. The extraction phase is a key juncture where environmental considerations come into play. Activities such as mining for minerals like stone, gravel, sand, and lime entail not only energy-intensive processes but also mechanical operations, drilling, and extensive transportation. The carbon footprint of these extraction activities, combined with subsequent processing and transportation to manufacturing facilities, remarkably influences the overall environmental impact of construction materials.

The embodied carbon content, measured in CO_2/kg, varies among common building materials as follows: concrete 0.159, bricks 0.24, marble 0.116, stainless steel 6.15, timber 0.72, MDF 0.72, plywood 1.07, glass 0.85, and ceramic tiles 0.72. Table 4 displays the estimated carbon emission percentages for various building materials. In this context, the prominence of cement stands out significantly among building materials.

Researchers and professionals are investigating alternative materials with lower carbon footprints and exploring methods to optimize the production processes of traditional materials. LCA is increasingly utilized to comprehensively evaluate the environmental impact of building materials, considering factors from raw material extraction to end-of-life disposal (Chamasemani et al., 2023). To mitigate the adverse impacts of carbon emissions in processes of the material production and

Table 4 Percentage of carbon emissions from different building materials (Kilgore, 2023)

Building material	Carbon emission percentage
Cement	30.3
Ceramic	20.3
Steel	18.7
Lime	7.9
Mortar	6.9
Gravel	2.9
Aluminum	2.3
Prefabricated concrete	2.0
Additives	1.5
Wood	1.1
PVC	1.0
Others	5.0

extraction, employing building materials characterized by high energy efficiency, adopting clean technologies, incorporating recycled materials, and optimizing production processes emerge as effective strategies. Researchers have been actively promoting environmentally friendly materials. A prominent avenue in this realm involves leveraging recycled materials in concrete compositions.

A noteworthy study conducted by Mostafaei et al. (2023b) underscores the potential carbon footprint reduction achievable by eliminating cement in the construction of high-strength concrete (HPC) and substituting it entirely with slag and lime. The results demonstrated a remarkable 61% reduction in the carbon footprint. Notably, slag, typically considered a waste material in the steel industry, finds a renewed purpose in concrete applications, diverting it from landfills and integrating it into the life cycle of a new product. Another investigation led by Bahmani et al. (Bahmani et al., 2023) delves into the evaluation of various activators for slag as a binding material in concrete. The study revealed that the incorporation of these activators in different proportions led to a substantial reduction in the carbon footprint per cubic meter of concrete production, ranging from 22% to 50%. The researchers employed the compressive strength index relative to the weight of equivalent carbon released (MPa/kg CO_2 eq) as a key criterion for selecting the optimal concrete mix. These findings underscore the transformative potential of environmentally conscious material choices and process optimizations in significantly lowering the carbon footprint within the construction industry. As sustainability becomes an integral consideration in construction practices, such research endeavors pave the way for innovative solutions, offering a more environmentally friendly path forward.

3.2.2 Construction Processes and Transportation

The transportation of materials to the site of the project and the movement of heavy machinery substantially make contribution to carbon emissions. This includes the transportation of raw materials, finished products, and workers to and from the site of construction. Additionally, strength-in depth construction approaches, which includes excavation, foundation laying, and building assembly, play a substantial role in carbon emissions. The use of machinery and equipment, powered by fossil fuels, constitutes a major parameter in the overall carbon footprint of construction. The interplay between transportation logistics and the energy-intensive processes involved in construction underscores the significance of adopting sustainable practices and exploring alternative, low-carbon technologies within the construction industry to reduce environmental impact.

3.2.3 Waste Generation and Disposal

Inadequate management of construction waste can lead to environmental degradation and make contributions to carbon emissions, posing good sized demanding situations for sustainability in the construction enterprise. One vital issue of this environmental impact stems from the decomposition of organic waste in landfills, resulting in the release of methane as a potent GHG. Scientifically, methane is approximately 25 instances greater powerful at trapping warmth inside the atmosphere than carbon dioxide over a 100-year duration, underscoring the urgency of addressing this issue.

Professional insights emphasize the need for a complete waste management strategy inside construction sites. Effective waste sorting, recycling, and reuse practices not only mitigate the environmental impact but also align with round financial system concepts, selling aid performance. Additionally, incorporating revolutionary technology, including waste-to-electricity conversion structures or advanced recycling methods, can similarly lessen the carbon footprint related to construction waste. From an LCA attitude, know-how the entire lifestyles cycle of production substances and waste is essential for making informed choices. This includes evaluating the environmental implications of material extraction, manufacturing, transportation, utilization, and give up-of-lifestyles disposal. Integrating such checks into construction undertaking planning allows for a more holistic view of the environmental effect and aids in identifying opportunities for carbon emission reduction (Barritt, 2016; Maury-Ramirez & Mesa, 2022).

3.2.4 Renovation and Demolition

Renovation and demolition activities play a superior role in carbon production, with their environmental impact heightened when involving the disposal of construction materials and the energy-intensive process of dismantling existing structures. Innovative technologies and methodologies contribute to minimizing the carbon footprint in renovation and demolition (Schützenhofer et al., 2022). Advanced deconstruction techniques, coupled with efficient waste sorting and recycling systems, enable the recovery of valuable materials and reduce the need for new resource extraction. Moreover, utilizing renewable energy sources and energy-efficient equipment during these activities can further mitigate carbon emissions. Considering the circular economy methods, in which materials are reused or recycled rather than discarded, becomes essential in solving these types of environmental impact. Incorporating salvaged materials into new construction not only reduces carbon emissions but also increases resource conservation. This matter was discussed in the section pertaining to material production and extraction (Andrić et al., 2017; Maury-Ramirez & Mesa, 2022; Zhang et al., 2021).

3.2.5 Water Use

The extraction, treatment, and distribution of water for construction play a significant role in carbon emissions, frequently owing to the energy-in depth nature of these processes (Chamasemani et al., 2023). The extraction of water resources, which includes groundwater or surface water, involves energy-consuming activities such as pumping and treatment. The transportation of treated water to sites of construction further provides to the carbon footprint, considering the energy required for transportation. Moreover, the strength-extensive processes concerned in wastewater remedy also make contributions significantly to the overall carbon footprint of construction (Ahn et al., 2010a; Farago et al., 2022). The treatment of construction-related wastewater is crucial for environmental sustainability, but the energy consumed in these treatment facilities needs to be carefully managed to minimize carbon emissions. Strategies to optimize water usage, implement more energy-efficient treatment technologies, and explore sustainable water-sourcing alternatives can play a vital role in reducing the carbon impact associated with water use in construction. Additionally, incorporating circular water management practices and recycling of water within construction can result in environmental sustainability.

3.2.6 Land Use Changes

Land-use changes, particularly those regarding deforestation for the construction, have profound implications for carbon emissions. The act of clearing land disrupts ecosystems and releases saved carbon in vegetation and soil, contributing markedly to the overall carbon footprint. Scientifically, this system is well-documented, with studies highlighting the direct correlation between land-use adjustments and GHG emissions. Deforestation not only releases CO_2 saved in trees but also triggers other GHG emissions, such as CH_4 and N_2O, which can get up from soil disturbance and microbial activity. The scientific community emphasizes the significance of quantifying these emissions accurately through advanced methodologies, like LCAs to understand the total environmental impact of land-use changes related with construction (Tan et al., 2023).

Professional views advocate for sustainable land management practices and emphasize the incorporation of green infrastructure in construction projects. Green constructing tasks sell accountable land use, integrating herbal elements into city development to mitigate environmental impacts. Sustainable creation practices, which includes reforestation efforts and the usage of reclaimed land, can help offset carbon emissions related to alterations in land use. Balancing the needs of construction with environmental conservation is vital for attaining lengthy-time period sustainability goals.

3.2.7 Inefficient Construction Practices

Inefficient construction practices, marked through factors like insufficient project management, useless remodel, and delays, play a pivotal role in escalating strength intake and carbon emissions in the creation industry. From a scientific angle, the LCA of production tasks is instrumental in quantifying the environmental effects of inefficient practices. LCA methodologies provide a comprehensive analysis of the whole production process, figuring out areas where strength performance may be progressed to mitigate carbon emissions. Industry experts endorse for the mixing of sustainable construction practices, emphasizing the importance of streamlined workflows, powerful communique, and superior technology which include Building Information Modeling (BIM). BIM permits collaborative and efficient venture management, lowering the probability of mistakes and transform. Furthermore, the adoption of modular production strategies and prefabrication is advocated for its potential to decorate efficiency, decrease waste, and consequently lessen the overall carbon footprint related to construction activities (Asad & Khalfan, 2007; Gehlot & Shrivastava, 2022).

4 Carbon Footprint in Dam Construction

The dam construction industry stands as a fundamental sector within civil and water engineering, dedicated to the construction and operation of dams for the purpose of storing and managing water resources (Can et al., 2021; Mostafaei et al., 2020, 2022; Tortajada, 2015). Globally acknowledged as a principal solution for water resource management and energy production, this industry exerts profound effects on the environment, economy, and societies (Pahl-Wostl, 2002). The dam construction process traverses various engineering stages, incorporating geological studies, hydrology, geomechanics, and the design of geometric structures (Mostafaei et al., 2017, 2018; Pourbakhshian & Ghaemian, 2015). The pivotal step in dam construction is meticulous planning and design, wherein potential impacts on rivers, local ecosystems, and communities are rigorously evaluated. The selection of dam type, dimensions, and optimal location emerges as critical in the design phase (Adams & Hughes, 1986; Mostafaei et al., 2023d; Mostafaei & Behnamfar, 2019, 2022).

As highlighted in the infrastructure section, projects like dam construction encompass diverse aspects concerning environmental effects. A central concern in dam design is its location, determining the areas of land to be submerged and the consequential impact on residential, agricultural, and industrial zones (Badenko et al., 2016; Harada & Yasuda, 2004). Socially, relocation of residents in villages surrounding the dam is often necessitated, impacting vegetation, fauna, and even altering the climate system of the region. Consequently, the ecosystem of the region undergoes changes commensurate with the dimensions of the project (Gwazani et al., 2012; Zhang et al., 2013). Integrating all these facets into dam life cycle calculations is inherently complex, with researchers frequently simplifying certain aspects in their studies. The construction phase represents a critical juncture, where environmental effects are discernible over a short period, facilitating the calculation of its carbon footprint (Xiaodong et al., 2014). In the subsequent section, carbon footprint calculations are presented for the construction of Pine Flat and Morrow Point dams, providing insights into the environmental impact of these projects (Mostafaei et al., 2023e).

5 Case Studies

5.1 Case Study 1: Pine Flat Dam

In this case study, the CFA assessment focuses on the Pine Flat dam, a concrete gravity dam constructed in 1954 on the Kings River in Fresno County, California, United States (Farrokh, 2021; Mostafaei et al., 2023c). As depicted in Fig. 3a, the dam extends over a length of 560.8 meters and is composed of 37 monoliths, each measuring 15.24 meters in width, except for one which is 12.16 meters wide. Monolith 16, identified as the tallest non-overflow monolith, is the focus of 2D dynamic analyses in this study. It has a height of approximately 122 meters, a base length of 95.80 meters, and a crest width of 9.75 meters (Naseri & Bagherzadeh Khalkhali, 2018; Salamon et al., 2020). The concrete volume of the dam was accurately determined to be 1,580,911 m^3 (Mostafaei et al., 2023c). The position of this monolith relative to others and its cross-sectional view are depicted in Fig. 3a, b, respectively.

Throughout their operational lifespan, dams encounter seismic risks, often requiring retrofitting as design standards evolve, assumptions change, and erosion occurs. Retrofitting is crucial to maintain dam safety and varies depending on the extent of damage. Concrete gravity dams offer various retrofitting techniques, including installing post-tension anchors, adding concrete to the upstream or downstream sections, increasing dam height, reducing uplift pressure via grouting and excavation, and sealing cracks with epoxy resin to prevent water leakage (Mostafaei & Behnamfar, 2019; Nahar et al., 2020; Nayak & Maity, 2013).

This study utilizes Incremental Dynamic Analysis (IDA) to evaluate damage to the Pine Flat concrete gravity dam and determine a suitable retrofitting approach. To achieve this, a 2-D finite element model is created, encompassing both the dam and reservoir. The concrete dam is modeled using a four-node brick isoparametric element (CPS4R) (Mostafaei et al., 2023b, d). The reservoir water is presumed to

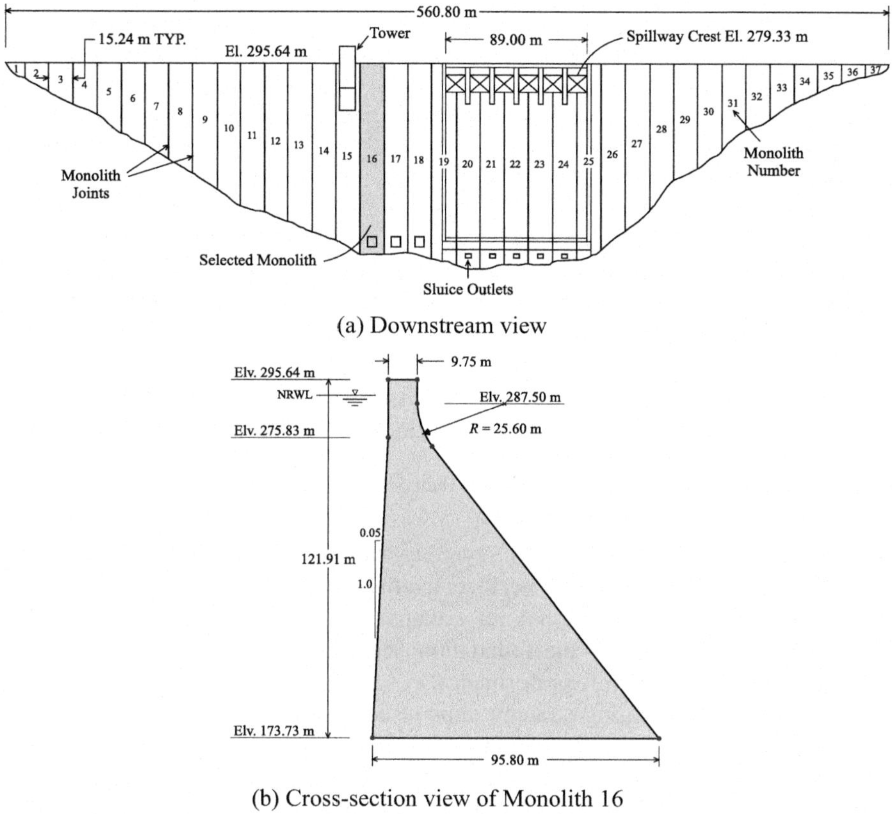

Fig. 3 2D perspective and profile of the Pine Flat dam (Mostafaei et al., 2023c). (**a**) Downstream view. (**b**) Cross-section view of Monolith 16

display linear elasticity, and acoustic elements (AC2D4) are utilized to represent it. Transmitting boundary conditions are enforced at the reservoir's far ends (Mostafaei & Behnamfar, 2022), aiding in the absorption of pressure waves as they propagate away from the domain. The present finite element model consists of 1360 elements for the dam and 5440 elements for the water.

The Finite Element Model (FEM) of the dam undergoes testing with 12 earthquake records across 20 consecutive intensity levels, offering a detailed understanding of its response from elastic threshold to collapse. Figure 4 showcases the IDA analysis outcomes for dam crest displacement across different earthquake records and spectral acceleration (Sa) levels. The illustrated curves, comprising median, mean, and mean minus/plus one sigma (16% and 84% fractiles), synthesize the findings. The median curve demonstrates a gradual reduction in slope post the elastic phase.

The failure mechanism involves the initiation of a crack from the downstream to the upstream side of the dam. Figure 5 illustrates the progression of damage in the concrete during collapse across all 12 accelerograms. Notably, collapse occurs at a

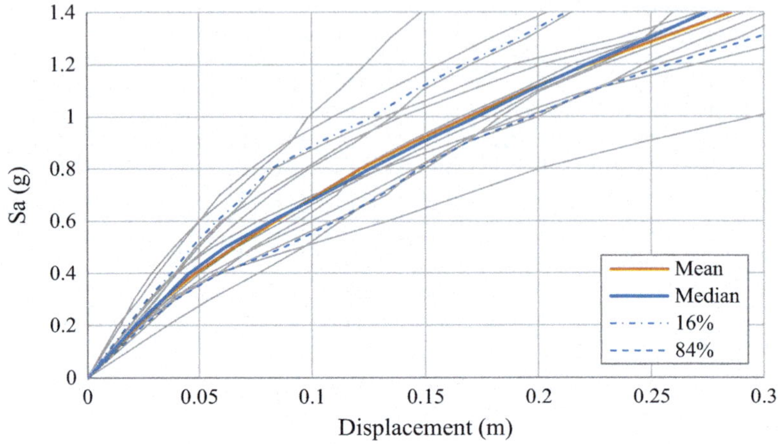

Fig. 4 IDA curves for the displacement of the dam crest. (Mostafaei et al., 2023c)

lower spectral acceleration for more severe earthquakes. At lower intensity levels, a crack area forms in the dam's neck, extending towards the upstream face with increasing spectral acceleration until collapse. Functional failure occurs in the dam's heel, where it meets the foundation. Retrofitting these areas is crucial for enhancing the dam's seismic performance.

The proposed retrofitting strategy aims to address potential dam failure modes during seismic events by incorporating a concrete buttress, post-tensioned anchors, and crack grouting with epoxy resin injection. Using CADAM software, the dam and cracks were modeled and retrofitted for a seismic event with a risk level of MCE = 0.45 g (Esmaili et al., 2012) to ensure stability. Following NRWL guidelines, the effective uplift pressure is assumed to be 40% of the total uplift value (Kazemi, 2004). Hence, a 10 m thick concrete buttress, totaling 480,000 m^3 in volume, is proposed to counter hydrostatic and hydrodynamic loads.

Figure 6 depicts the placement of two cables, each with a rupture capacity of 260 kN, spaced strategically along the dam's length, totaling 21,130 meters of anchors or around 2512 tons of post-tensioned steel strand cables (Mostafaei et al., 2023c).

To conduct a thorough assessment of the Pine Flat dam's carbon footprint, it's crucial to consider its entire life cycle, including initial construction, retrofitting, and eventual decommissioning and recycling. However, creating a comprehensive Life Cycle Inventory (LCI) database with minimal uncertainty presents a significant challenge for researchers in this field, limiting such evaluations. Due to insufficient information on the Pine Flat dam's construction, data from previous studies were utilized and adjusted to fit the project's scope (Hall, 1986; Salamon et al., 2020). Table 5 provides detailed information on materials and construction operations throughout the dam's life cycle, with a conservative 10% increase in material quantities considered. This table offers insights into material weights and transportation, both on-site and off-site, for all stages. Notably, the destruction and recycling

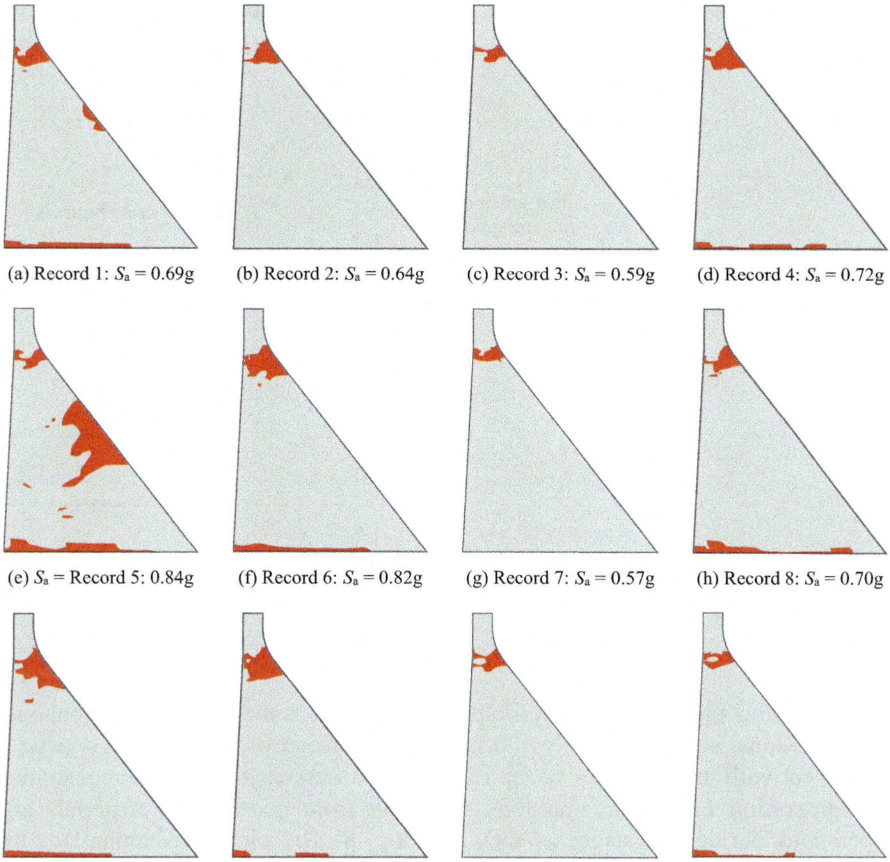

(a) Record 1: $S_a = 0.69g$ (b) Record 2: $S_a = 0.64g$ (c) Record 3: $S_a = 0.59g$ (d) Record 4: $S_a = 0.72g$

(e) S_a = Record 5: 0.84g (f) Record 6: $S_a = 0.82g$ (g) Record 7: $S_a = 0.57g$ (h) Record 8: $S_a = 0.70g$

(i) Record 9: $S_a = 0.93g$ (j) Record 10: $S_a = 0.90g$ (k) Record 11: $S_a = 0.60g$ (l) Record 12: $S_a = 0.62g$

Fig. 5 Crack propagation at the spectral acceleration of the dam collapse for all accelerograms. (Mostafaei et al., 2023c)

stage assumes a 20% concrete recycling rate, reflecting the reuse of demolished dam concrete in other projects, a scenario separately accounted for in both dam construction and retrofitting.

The analysis outcomes indicate that the overall carbon footprint during the entire construction phase is approximately 495.1 kg/m^3. Figure 7 is employed to provide a detailed breakdown of the CF contributions from various construction materials and operations. The pie chart highlights the predominant influence of concrete usage on the CF during the construction phase, accounting for 71.36%. Additionally, blasting material contributes 13.56%, while fuel consumption constitutes 12.15%. Other materials and activities collectively contribute significantly to the overall carbon footprint.

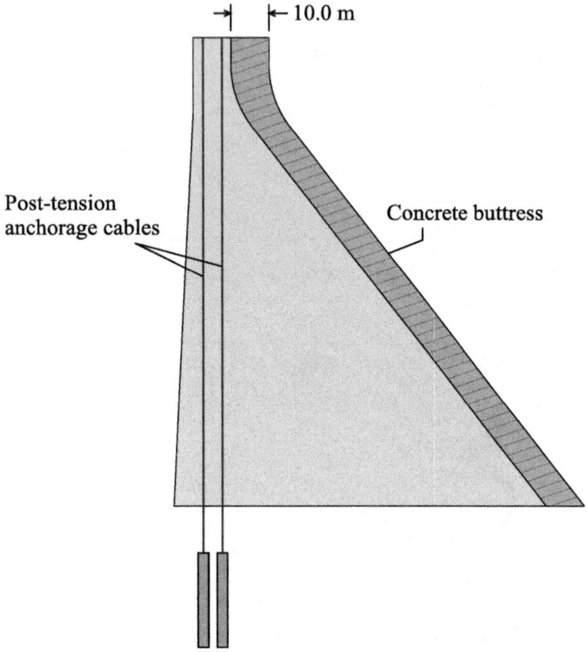

Fig. 6 The proposed retrofitting plan for Pine Flat dam. (Mostafaei et al., 2023c)

During this phase, the carbon footprint, expressed as the equivalent greenhouse gas emissions, is approximately 120 kg per cubic meter of concrete. This value is consistent with the outcomes of the Life Cycle Assessment (LCA), demonstrating the proportion of carbon emissions stemming from construction materials and operations during this stage, as shown in Fig. 8. The pie chart emphasizes the significant influence of concrete usage on the carbon footprint of the construction phase, accounting for 80.25%. Fuel consumption contributes 13.65%, while transportation accounts for 3.14%. Notably, the impact of epoxy resin usage is minimal, at 0.1%.

The decommissioning phase entails the destruction and recycling of both the non-reinforced and reinforced segments of the dam, with each aspect examined separately in this study. Carbon footprint assessments were conducted using CFA, revealing approximately 346.4 kg/m^3 and 108.1 kg/m^3 of dam concrete emissions for the decommissioned and retrofitted parts, respectively. Figure 9 depicts the contributions of transportation, electricity and diesel usage, landfilling, and concrete recycling to the overall carbon footprint. The results highlight landfilling as the most impactful factor, although implementing recycling practices led to a significant 20% reduction in greenhouse gas emissions. This underscores the potential of sustainable waste management approaches in mitigating environmental impacts, with landfilling being the most significant contributor, comprising over 50% of emissions.

In this case study, the Pine Flat dam's life cycle was analyzed through four distinct scenarios, each building upon the CFA evaluations conducted earlier. These scenarios encompass:

Table 5 LCI database

Material or construction operation	Material production (ton)	Distance (km) for out-site transportation	Distance (km) for on-site transportation
Construction stage			
Concrete	4,200,000	25	5
Steel and rebar	85	550	5
Blasting materials	30,000	620	5
Fuel oil	430,000	320	5
Wood	30,000	150	5
Earth and rock excavation	7200	0	5
Retrofitting stage			
Concrete	1,152,000	25	5
Wood	24,686	150	5
Fuel oil	117,943	320	5
Steel and rebar	2496	550	5
Epoxy resin	92	550	5
Destruction and recycling stage			
Landfill (for dam construction)	3,432,036	100	5
Recycle (for dam construction)	840,000	100	5
Landfill (for dam retrofitting)	1,179,274	100	5

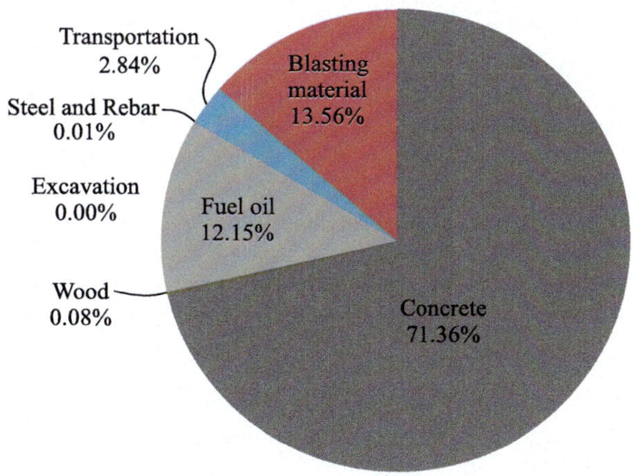

Fig. 7 Carbon footprint emissions (kg/m^3) attributable to construction materials and operations during the construction phase

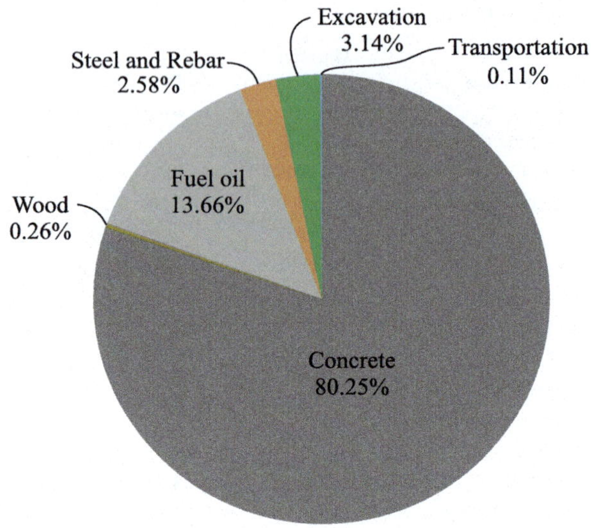

Fig. 8 Carbon footprint emissions (kg/m^3) from construction materials and operations during the retrofitting stage

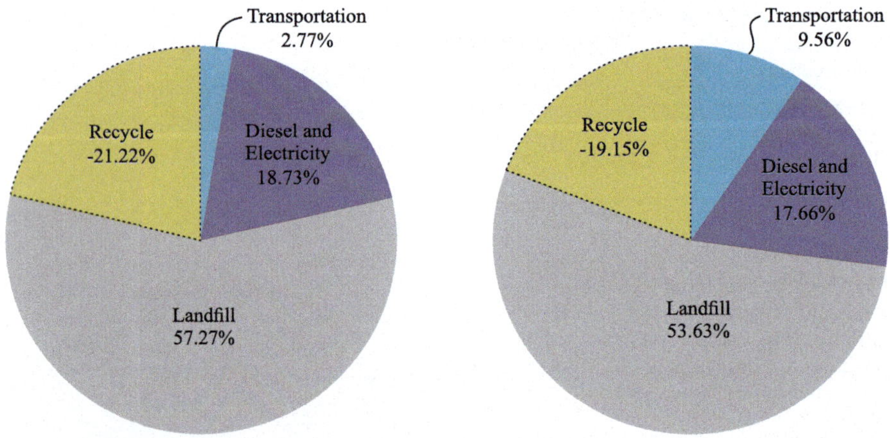

Fig. 9 Carbon footprint contribution (kg/m^3) by construction material and operations in the destruction and recycling stage. (**a**) Non-retrofitted dam (**b**) Dam's retrofitting part

1. A cradle-to-gate life cycle approach, focusing solely on the construction stage
2. A cradle-to-grave life cycle approach, covering both the construction and destruction/recycling stages
3. A cradle-to-gate life cycle approach, including both the construction and retrofitting stages
4. A cradle-to-grave life cycle approach, involving the construction, retrofitting, and destruction/recycling stages

The results show that the overall carbon footprint for the first scenario is 495 kg/m^3, while the second scenario records 841 kg/m^3. The third and fourth scenarios exhibit carbon footprints of 615 and 1070 kg/m^3, respectively. The construction stage significantly contributes to the environmental impact in terms of resource endpoint. Comparing carbon footprint values across different stages reveals that the carbon footprint during the destruction of the Pine Flat dam is approximately 70% of that generated during its construction. Retrofitting produces a 24% carbon footprint compared to the construction stage. These findings offer valuable insights for decision-makers, quantifying how retrofitting a damaged dam can mitigate the environmental impact of constructing a new one, which is crucial given the potential severe consequences of dam destruction due to earthquakes. Furthermore, despite assuming a 20% concrete recycling rate, the decommissioning stage still has a considerable detrimental impact on the environment in cradle-to-grave life cycles. In scenarios 2 and 4, the decommissioning stage yields a negative score approximately 4.9 and 6.2 times greater than the construction stage, respectively. Conversely, the environmental effects of dam retrofitting account for approximately 24% of the construction stage, significantly lower.

5.2 Case Study 2: Morrow Point Dam

In another case study, the Morrow Point arch dam undergoes a comprehensive CFA. The dam stands at a height of 143.73 m with a crest span of 221 m (Moradloo & Naiji, 2020; Mostafaei et al., 2023f). The thickness exhibits variation along its length, ranging from 3.7 m at the top to 16 m at its base. For a visual representation, refer to Fig. 10, which offers a three-dimensional view of the Morrow Point dam.

Conducting a thorough CFA focused specifically on the construction stages of the Morrow Point dam necessitates the development of a robust LCI database. Addressing the considerable challenge of mitigating uncertainties within the database presents a significant obstacle for researchers in these domains, hindering the smooth execution of such assessments. In this particular case study, the scarcity of data related to the construction of the Morrow Point dam has prompted the integration of information gleaned from prior studies, aligning with the objectives of the ongoing project (Seery, 1967).

Table 6 delineates crucial specifics concerning the materials and construction operations germane to the life cycle of the Morrow Point dam. The presented data incorporates a conservative 10% augmentation in material quantities, offering meticulous insights into material weights. Additionally, it encompasses considerations for both on-site and off-site transportation, meticulously accounting for these factors across all project phases.

The LCA reveals that the total carbon footprint attributed to the construction of the dam, utilizing the IMPACT 2002+ methodology, amounts to 89,193 tons. When

Fig. 10 3D view of Morrow Point dam

Table 6 LCI database at construction stage

Material or operation	Value (ton)	Transportation (km)
Concrete	517,200	30
Steel and rebar	85	550
Blasting	1000	620
Fuel oil	16,000	150
Wood	50,000	150
Earth and rock excavation	464,850	5

accounting for the total volume of the dam, the comprehensive carbon footprint for dam construction approximates 413 kg/m^3. Figure 11 graphically depicts the influence of six key components—namely, concrete, transportation, diesel fuel, wood, blasting materials, and steel and rebar—on the overall carbon footprint during the construction phase. The graphical representation emphasizes the substantial impact of concrete on global warming, constituting 82.42% of the total. Following closely is fuel consumption at 9.03%, transportation at 3.08%, and blasting materials at 2.66%. Wood and steel collectively contribute significantly, aggregating to 2.81%.

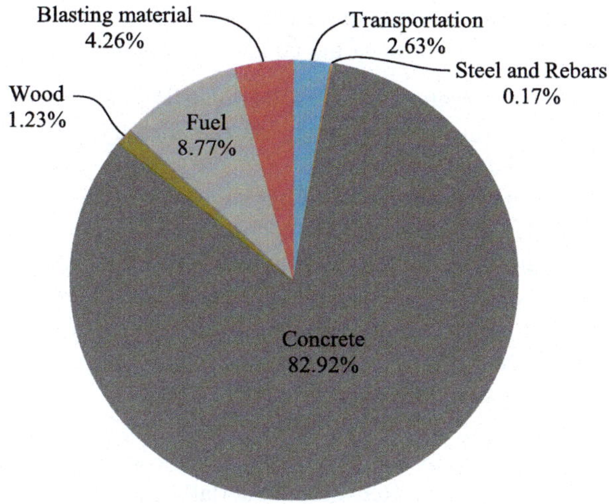

Fig. 11 Contribution of six components on the overall carbon footprint during the construction phase of the dam based on the IMPACT 2002+ method

6 Strategies for Mitigating Carbon Emissions in the Construction Sector

By delineating the pivotal factors for diminishing the carbon footprint within the construction industry, given its substantial impact on global carbon dioxide levels, we can take decisive strides towards fostering environmental sustainability. An in-depth examination of each sub-sector within the construction industry, coupled with the strategic prioritization of factors based on their influence on carbon production, controllability, and potential for substitution, holds the potential to guide governmental fund allocation to pertinent companies and effectively manage the identified factors. Governmental regulation and oversight of the construction industry and its associated economic entities are integral in curbing environmental pollutants.

Urban development extends beyond building construction and basic infrastructure projects. Urban elements like sidewalks, street tables, and urban furniture, directly linked to construction, can substantially contribute to a city's carbon footprint. Employing environmentally friendly materials in the construction of these elements emerges as an optimal solution for reducing the overall carbon impact. Complementary to this, municipalities can undertake a multifaceted approach, encompassing non-civilizational strategies like helping public transportation, fostering on foot and biking, integrating renewable strength assets, implementing powerful waste control, and expanding inexperienced spaces. Collaborative efforts among municipalities and relevant stakeholders play a crucial function in mitigating carbon emissions within the urban production region.

In a broader context, several overarching techniques may be taken into consideration to decrease the carbon footprint of the construction industry:

In a broader context, several overarching strategies can be considered to diminish the carbon footprint of the construction industry:

- Implementation of Organizational Policies and Measures: Applying organizational rules that at once or not directly impact the carbon footprint of the development enterprise
- Enhancing Structural Longevity: Prolonging the service lifestyles of systems to lessen the frequency of reconstruction
- Optimizing Structural and Architectural Designs: Utilizing optimum structural and architectural designs to limit material intake and strength utilization
- Promoting Sustainable Construction Processes: Encouraging sustainable construction practices, inclusive of the adoption of modular constructions for progressed performance
- Recycling and Waste Management: Implementing green recycling of construction materials and optimizing waste control to reduce universal waste era
- Integration of Recycled Materials: Incorporating recycled substances in each structural and non-structural additives of buildings to bringing them back into the lifestyles cycle and sell round financial system standards

These multifaceted processes collectively make contribution to the overarching goal of fostering an environmentally accountable and sustainable creation industry.

7 Conclusion

The escalating trend in construction activities, encompassing building and infrastructure construction and related endeavors, has led to an intensified utilization of natural resources within this industry. This surge in demand has set in motion a concerning depletion of these resources, signaling a trajectory towards gradual extinction. This resource consumption is meticulously assessed through the lenses of life cycle assessment and carbon footprint calculations. This chapter has explored the concept of the carbon footprint, exploring its introduction, pertinent standards and protocols, and the software instrumental in its precise calculation. Shifting the focus to environmental apprehensions within the construction industry, the discourse has extended to carbon footprint evaluations across its sub-industries, elucidating the key factors influencing carbon production. Notably, the extraction and production of materials emerge as primary contributors to GHG emissions, intertwined with factors such as land use, energy utilization, waste generation, and water consumption. This multifaceted nexus underscores the pivotal role of these elements in driving GHG emissions within the construction industry. In light of these revelations, proactive measures which were discussed at the end of the chapter are imperative, guided by the insights derived from carbon footprint evaluations in the construction sector to mitigate the environmental impact.

A substantial segment of this chapter delves into the comprehensive assessment of the overall carbon footprint of concrete dams, focusing on two specific types: the Pine Flat dam (a gravity dam) and the Morrow Point dam (a concrete arch dam). The study serves as a valuable resource for decision-makers involved in evaluating various dimensions of these dams. The first case study involves the examination of four distinct life cycle scenarios for the Pine Flat dam, covering initial construction, retrofitting, and eventual decommissioning. Environmental assessments for the demolition and recycling phases, representing two scenarios for sections with and without retrofitting, were computed, incorporating a 20% recycling rate for demolished concrete. The second case study focused solely on the construction stage. Results revealed an estimated overall carbon footprint of approximately 495.1 kg/m^3 throughout the construction phase, with concrete usage contributing significantly at 71.36%. The carbon footprint during construction equated to around 120 kg/m^3 of concrete in GHG emissions. Assessment of the carbon footprint from decommissioning and retrofitting indicated values of approximately 346.4 kg/m^3 and 108.1 kg/m^3 of dam concreting, respectively, with landfilling of destroyed materials being the primary contributor. Introduction of recycling led to a noteworthy 20% reduction in emissions, demonstrating the potential of sustainable waste management practices. Landfilling accounted for over 50% of the impact. The construction stage notably influenced environmental impact, while retrofitting generated a 24% carbon footprint compared to construction. These findings offer decision-makers quantitative insights into the environmental benefits of retrofitting damaged dams versus constructing new ones, crucial given seismic risks. The life cycle assessment revealed a total carbon footprint of approximately 426.1 kg/m^3 for the arch dam construction.

References

Adams, W. M., & Hughes, F. M. R. (1986). The environmental effects of dam construction in tropical Africa: Impacts and planning procedures. *Geoforum, 17*(3–4), 403–410.

Ahn, C., Lee, S., Peña-Mora, F., & Abourizk, S. (2010a). Toward environmentally sustainable construction processes: The US and Canada's perspective on energy consumption and GHG/CAP emissions. *Sustainability, 2*(1), 354–370.

Ahn, C., Xie, H., Lee, S., Abourizk, S., & Peña-Mora, F. (2010b). Carbon footprints analysis for tunnel construction processes in the preplanning phase using collaborative simulation. In *Construction research congress 2010: Innovation for reshaping construction practice*.

Akbarnezhad, A., & Xiao, J. (2017). Estimation and minimization of embodied carbon of buildings: A review. *Buildings, 7*(1), 5.

Akintayo, F. O., Oyebade, O. N., Songca, S. P., Adebisi, N. O., Oluwafemi, O. S., & Fadipe, O. O. (2020). Assessment of the impacts of building construction activities on the environment. *Nigerian Journal of Technology, 39*(2), 325–331.

Andrić, I., Pina, A., Ferrao, P., Lacarriere, B., & Le Corre, O. (2017). The impact of renovation measures on building environmental performance: An emergy approach. *Journal of Cleaner Production, 162*, 776–790.

Asad, S., & Khalfan, M. M. A. (2007). Integration of sustainability issues within construction processes. *Emirates Journal for Engineering Research, 12*(2), 11–21.

Badenko, V., Badenko, N., Nikonorov, A., Molodtsov, D., Terleev, V., Lednova, J., & Maslikov, V. (2016). Ecological aspect of dam design for flood regulation and sustainable urban development. *MATEC Web of Conferences, 73*, 3003.

Bahmani, H., Mostafaei, H., Ghiassi, B., Mostofinejad, D., & Wu, C. (2023). A comparative study of calcium hydroxide, calcium oxide, calcined dolomite, and metasilicate as activators for slag-based HPC. *Structure, 58*, 105653.

Barritt, J. (2016). An overview on recycling and waste in construction. *Proceedings of the Institution of Civil Engineers-Construction Materials, 169*(2), 49–53.

Can, M., Çalışır, E., Akpınar, A., & Doğan, E. (2021). The effects of the dam construction process on downstream river geomorphology. *Arabian Journal of Geosciences, 14*, 1–16.

Chamasemani, N. F., Kelishadi, M., Mostafaei, H., Najvani, M. A. D., & Mashayekhi, M. (2023). Environmental impacts of reinforced concrete buildings: Comparing common and sustainable materials: A case study. *Construction Materials, 4*(1), 1–15.

Chen, J., Shi, Q., & Zhang, W. (2022). Structural path and sensitivity analysis of the CO2 emissions in the construction industry. *Environmental Impact Assessment Review, 92*, 106679.

Christoff, P. (2016). The promissory note: COP 21 and the Paris climate agreement. *Environmental Politics, 25*(5), 765–787.

de Souza Mendonça, A. K., de Barni, G. A. C., Moro, M. F., Bornia, A. C., Kupek, E., & Fernandes, L. (2020). Hierarchical modeling of the 50 largest economies to verify the impact of GDP, population and renewable energy generation in CO2 emissions. *Sustainable Production and Consumption, 22*, 58–67.

Esmaili, R., Anvar, S. A., Talebbeydokhti, N., & Zareian, F. (2012). Seismic performance evaluation of concrete gravity dams-a case study of Pine Flat Dam. In *5th World conference on earthquake engineering (15WCEE), 2012*.

Espinoza, M., Campos, N., Yang, R., Ozer, H., Aguiar-Moya, J. P., Baldi, A., Loría-Salazar, L. G., & Al-Qadi, I. L. (2019). Carbon footprint estimation in road construction: La Abundancia–Florencia case study. *Sustainability, 11*(8), 2276.

Farago, M., Damgaard, A., Rebsdorf, M., Nielsen, P. H., & Rygaard, M. (2022). Challenges in carbon footprint evaluations of state-of-the-art municipal wastewater resource recovery facilities. *Journal of Environmental Management, 320*, 115715.

Farrokh, M. (2021). Seismic analysis and damage evaluation of pine flat concrete dam. In *Numerical analysis of dams*. https://doi.org/10.1007/978-3-030-51085-5_4

Fuertes, A., Casals, M., Gangolells, M., Forcada, N., Macarulla, M., & Roca, X. (2013). An environmental impact causal model for improving the environmental performance of construction processes. *Journal of Cleaner Production, 52*, 425–437.

Fujimori, S., Su, X., Liu, J.-Y., Hasegawa, T., Takahashi, K., Masui, T., & Takimi, M. (2016). Implication of Paris agreement in the context of long-term climate mitigation goals. *Springerplus, 5*, 1–11.

Gehlot, M., & Shrivastava, S. (2022). Sustainable construction practices: A perspective view of Indian construction industry professionals. *Materials Today: Proceedings, 61*, 315–319.

Gwazani, R., Gandiwa, E., Gandiwa, P., Mhaka, V., Hungwe, T., & Muza, M. (2012). The socio-ecological impacts of small dams: A case study of Mushandike Sanctuary, Zimbabwe. *Journal of Sustainable Development in Africa, 14*(6), 184–194.

Hall, J. F. (1986). Study of the earthquake response of Pine Flat dam. *Earthquake Engineering & Structural Dynamics, 14*(2), 281–295.

Harada, J., & Yasuda, N. (2004). Conservation and improvement of the environment in dam reservoirs. *International Journal of Water Resources Development, 20*(1), 77–96.

Heilig, G. K. (1994). The greenhouse gas methane (CH4): Sources and sinks, the impact of population growth, possible interventions. *Population and Environment, 16*(2), 109–137.

Hu, X. (2015). *Construction industry performance measurement with carbon reduction*. Deakin University.

IPCC Climate Change. (2022). *Mitigation of climate change* (Contribution of working group III to the sixth assessment report of the intergovernmental panel on climate change, Issue C). Cambridge University Press.

ISO 14040:2006. *Environmental management – Life cycle assessment: Principles and framework*: International Organization for Standardization.

ISO 14044:2006. *Environmental management – Life cycle assessment: Requirements and guidelines*: International Organization for Standardization.

ISO 14064-1:2018. *Greenhouse gases – Part 1: Specification with guidance at the organization level for quantification and reporting of greenhouse gas emissions and removals*: International Organization for Standardization.

ISO 14064-2:2019. *Greenhouse gases – Part 2: Specification with guidance at the project level for quantification, monitoring and reporting of greenhouse gas emission reductions or removal enhancements*: International Organization for Standardization.

ISO 14064-3:2019. *Greenhouse gases – Part 3: Specification with guidance for the verification and validation of greenhouse gas statements*: International Organization for Standardization.

ISO 14067:2018. *Greenhouse gases – Carbon footprint of products: Requirements and guidelines for quantification*: International Organization for Standardization.

Jiang, T., Shen, Z., Liu, Y., & Hou, Y. (2018). Carbon footprint assessment of four normal size hydropower stations in China. *Sustainability, 10*(6), 2018.

Kazemi, M. R. (2004). Reliability based analysis and design of anchor retrofitted concrete gravity dams. In *13th world conference on earthquake engineering, Vancouver, B.C., Canada.*

Keijzer, E., Leegwater, G., de Vos-Effting, S., & De Wit, M. (2015). Carbon footprint comparison of innovative techniques in the construction and maintenance of road infrastructure in The Netherlands. *Environmental Science & Policy, 54*, 218–225.

Kilgore, G. (2023). *Carbon footprint of building materials (Green Building Calculator)*. Retrieved January, 13, from https://8billiontrees.com/carbon-offsets-credits/carbon-footprint-of-building-materials/

Koohi Rostamkolaei, M., & Abessi, O. (2018). Calculating the carbon footprint in the construction industry, comparing structures with concrete and steel frames. In *International Conference on civil engineering, architecture and urban development management in Iran*. https://civilica.com/doc/845956

Li, Z., Lu, L., Lv, P., Du, H., Guo, J., He, X., & Ma, J. (2017). Carbon footprints of pre-impoundment clearance on reservoir flooded area in China's large hydro-projects: Implications for GHG emissions reduction in the hydropower industry. *Journal of Cleaner Production, 168*, 1413–1424.

Li, Z., Du, H., Xu, H., Xiao, Y., Lu, L., Guo, J., Prairie, Y., & Mercier-Blais, S. (2019). The carbon footprint of large-and mid-scale hydropower in China: Synthesis from five China's largest hydro-project. *Journal of Environmental Management, 250*, 109363.

Li, K., Ma, M., Xiang, X., Feng, W., Ma, Z., Cai, W., & Ma, X. (2022). Carbon reduction in commercial building operations: A provincial retrospection in China. *Applied Energy, 306*, 118098.

Mačková, D., Spisakova, M., Kozlovska, M., & Svajlenka, J. (2017). Environmental assessment of buildings constructed by modern methods of construction. *Applied Mechanics and Materials, 861*, 601–608.

Marques, B., Tadeu, A., De Brito, J., & Almeida, J. (2017). A perspective on the development of sustainable construction products: An eco-design approach. *International Journal of Sustainable Development and Planning, 12*(2), 304–314.

Martinez, S., del Mar Delgado, M., Marin, R. M., & Alvarez, S. (2018). The environmental footprint of the end-of-life phase of a dam through a hybrid-MRIO analysis. *Building and Environment, 146*, 143–151.

Maury-Ramirez, A. C., & Mesa, J. A. (2022). *Construction and demolition waste: Challenges and opportunities*. MDPI.

Mimura, N., Tsutsui, J., Ichinose, T., Kato, H., & Sakaki, K. (1998). *Impacts on infrastructure and socio-economic system* (pp. 165–201). The Potential Impact on Japan.

Mishra, S., Ghosh, A., Rai, K., Jaiswal, B., Yadav, D. S., Agrawal, M., & Agrawal, S. B. (2021). Dimensions of climate change and its consequences on ecosystem functioning. In *Global climate change* (pp. 109–149). Elsevier.

Mohammed, A., Mustapha, A., & Mu'azu, N. (2011). Energy efficient buildings as a tool for ensuring sustainability in the building industry. *International Journal of Energy Engineering*. https://doi.org/10.1109/ICMT.2011.6003238

Moradloo, A. J., & Naiji, A. (2020). Effects of rotational components of earthquake on seismic response of arch concrete dams. *Earthquake Engineering and Engineering Vibration, 19*, 349–362.

Mostafaei, H., & Behnamfar, F. (2019). Effect of the vertical earthquake component on nonlinear behavior of an arch dam having a foundation with discontinuities. *Journal of Numerical Methods in Civil Engineering, 4*(2), 69–78.

Mostafaei, H., & Behnamfar, F. (2022). Wedge movement effects on the nonlinear behavior of an arch dam subjected to seismic loading. *International Journal of Geomechanics, 22*(3), 04021289.

Mostafaei, H., Sohrabi Gilani, M., & Ghaemian, M. (2017). Stability analysis of arch dam abutments due to seismic loading. *Scientia Iranica, 24*(2), 467–475.

Mostafaei, H., Gilani, M. S., & Ghaemian, M. (2018). A comparative study between pseudo-static and dynamic analyses on rock wedge stability of an arch dam. *Civil Engineering Journal, 4*(1), 179–187.

Mostafaei, H., Behnamfar, F., & Alembagheri, M. (2020). Nonlinear analysis of stability of rock wedges in the abutments of an arch dam due to seismic loading. *Structural Monitoring and Maintenance, 7*(4), 295.

Mostafaei, H., Behnamfar, F., & Alembagheri, M. (2022). Reliability and sensitivity analysis of wedge stability in the abutments of an arch dam using artificial neural network. *Earthquake Engineering and Engineering Vibration, 21*(4), 1019–1033.

Mostafaei, H., Ghamami, M., & Aghabozorgi, P. (2023a). Modal identification of concrete arch dam by fully automated operational modal identification. *Structure, 32*, 228–236.

Mostafaei, H., Mousavi, H., & Barmchi, M. A. (2023b). *Finite element analysis of structures by ABAQUS: For civil engineers*. Simay-e-Danesh Publication.

Mostafaei, H., Keshavarz, Z., Rostampour, M. A., Mostofinejad, D., & Wu, C. (2023c). Sustainability evaluation of a concrete gravity dam: Life cycle assessment, carbon footprint analysis, and life cycle costing. *Structure, 53*, 279–295.

Mostafaei, H., Bahmani, H., Mostofinejad, D., & Wu, C. (2023d). A novel development of HPC without cement: Mechanical properties and sustainability evaluation. *Journal of Building Engineering, 76*, 107262.

Mostafaei, H., Badarloo, B., Chamasemani, N. F., Rostampour, M. A., & Lehner, P. (2023e). Investigating the effects of concrete mix design on the environmental impacts of reinforced concrete structures. *Buildings, 13*(5), 1313.

Mostafaei, H., Mostofinejad, D., Ghamami, M., & Wu, C. (2023f). A new approach of ensemble learning in fully automated identification of structural modal parameters of concrete gravity dams: A case study of the Koyna dam. *Structure, 50*, 255–271.

Nahar, T. T., Cao, A.-T., & Kim, D. (2020). Risk assessment of aged concrete gravity dam subjected to material deterioration under seismic excitation. *International Journal of Concrete Structures and Materials, 14*(1), 1–17.

Nakicenovic, N. (2000). Global greenhouse gas emissions scenarios: Integrated modeling approaches. *Technological Forecasting and Social Change, 63*, 105–109.

Naseri, F., & Bagherzadeh Khalkhali, A. (2018). Evaluation of seismic performance of concrete gravity dams under soil-structure-reservoir interaction exposed to vertical component of near-field earthquakes during impounding case study: Pine flat dam. *Journal of civil Engineering and Materials Application, 2*(4), 181–191.

Nayak, P., & Maity, D. (2013). Seismic damage analysis of aged concrete gravity dams. *International Journal for Computational Methods in Engineering Science and Mechanics, 14*(5), 424–439.

Onat, N. C., Kucukvar, M., & Tatari, O. (2014). Scope-based carbon footprint analysis of US residential and commercial buildings: An input–output hybrid life cycle assessment approach. *Building and Environment, 72*, 53–62.

Pahl-Wostl, C. (2002). Towards sustainability in the water sector–The importance of human actors and processes of social learning. *Aquatic Sciences, 64*, 394–411.

Pisciotta, M., Pilorge, H., Feldmann, J., Jacobson, R., Davids, J., Swett, S., Sasso, Z., & Wilcox, J. (2022). Current state of industrial heating and opportunities for decarbonization. *Progress in Energy and Combustion Science, 91*, 100982.

Pomponi, F., & Moncaster, A. (2016). Embodied carbon mitigation and reduction in the built environment–What does the evidence say? *Journal of Environmental Management, 181*, 687–700.

Pourbakhshian, S., & Ghaemian, M. (2015). Investigating stage construction in high concrete arch dams. *Indian Journal of Science and Technology, 8*(14), 1.

Publicly Available Specification. (2008). PAS 2050–Specification for the assessment of the life cycle greenhouse gas emissions of goods and services. *London: British Standards Institution, 2008*, 35–36.

Rehman, R., Hamdani, A., Naseem, A., Ashraf, M., & Kazi, A. G. (2014). Scenario of climate changes in the context of agriculture. *Physiological Mechanisms and Adaptation Strategies in Plants Under Changing Environment, 2*, 223–264.

Salamon, J. W., Wood, C., Manie, J., & Geister, A. (2020). Evaluation of seismic wave propagation in the analysis of concrete dams: Pine Flat dam benchmark study. In *Numerical analysis of Dams – Proceedings of the 15th ICOLD international benchmark workshop* (pp. 263–276). Springer.

Santos, F. D., Ferreira, P. L., & Pedersen, J. S. T. (2022). The climate change challenge: A review of the barriers and solutions to deliver a Paris solution. *Climate, 10*(5), 75.

Savaresi, A. (2016). The Paris agreement: A new beginning? *Journal of Energy & Natural Resources Law, 34*(1), 16–26.

Schützenhofer, S., Kovacic, I., Rechberger, H., & Mack, S. (2022). Improvement of environmental sustainability and circular economy through construction waste management for material reuse. *Sustainability, 14*(17), 11087.

Seery, J. D. (1967). Construction of morrow point power plant and dam. *Journal of the Construction Division, 93*(1), 47–58.

Shi, A. (2001). *Population growth and global carbon dioxide emissions, IUSSP Conference in Brazil/session-s09.*

Shi, L., Qi, X., Yang, Z., Tao, L., Li, Y., Qiu, J., & Jiang, X. (2023). Comparative study of greenhouse gas emission calculations and the environmental impact in the life cycle assessment of buildings in China, Finland, and the United States. *Journal of Building Engineering, 70*, 106396.

Sun, X.-Y., Dai, J.-G., Wang, H.-L., Dong, W.-W., & Wang, J. (2015). Decision support system for optimizing the maintenance of RC girder bridge superstructures in consideration of the carbon footprint. *Journal of Bridge Engineering, 20*(12), 04015022.

Tan, Y., Liu, Y., Chen, Y., Zhang, Z., Wu, D., Chen, H., & Han, Y. (2023). The impact of urban construction land change on carbon emissions—A case study of Wuhan City. *International Journal of Environmental Research and Public Health, 20*(2), 922.

The Intergovernmental Panel on Climate Change. (2006). *2006 IPCC guidelines for national greenhouse gas inventories*. Institute for Global Environmental Strategies (IGES).

Tortajada, C. (2015). Dams: An essential component of development. *Journal of Hydrologic Engineering, 20*(1), A4014005.

Van der Werf, G. R., Peters, W., Van Leeuwen, T. T., & Giglio, L. (2013). What could have caused pre-industrial biomass burning emissions to exceed current rates? *Climate of the Past, 9*(1), 289–306.

Wackernagel, M., & Rees, W. (1998). *Our ecological footprint: Reducing human impact on the earth* (Vol. 9). New Society Publishers.

Wang, J., Chen, X., Liu, Z., Frans, V. F., Xu, Z., Qiu, X., Xu, F., & Li, Y. (2019). Assessing the water and carbon footprint of hydropower stations at a national scale. *Science of the Total Environment, 676*, 595–612.

Wiedmann, T. (2009). *Carbon footprint and input–output analysis–an introduction*. Taylor & Francis.

Xiaodong, L., Fan, Y., Yuanxue, G., & McCarthy, J. (2014). Case study of carbon footprint of residential building construction. *Materials Research Innovations, 18*(sup4), S4–S72.

Yang, R., & Al-Qadi, I. L. (2017). Development of a life-cycle assessment tool to quantify the environmental impacts of airport pavement construction. *Transportation Research Record, 2603*(1), 89–97.

Zhang, Y., He, D., Lu, Y., Feng, Y., & Reznick, J. (2013). The influence of large dams building on resettlement in the Upper Mekong River. *Journal of Geographical Sciences, 23*, 947–957.

Zhang, S., Pang, B., & Zhang, Z. (2015). Carbon footprint analysis of two different types of hydropower schemes: Comparing earth-rockfill dams and concrete gravity dams using hybrid life cycle assessment. *Journal of Cleaner Production, 103*, 854–862.

Zhang, Q., Fang, K., Xu, M., & Liu, Q. (2018). Review of carbon footprint research based on input-output analysis. *Journal of Natural Resources, 33*(4), 696–708.

Zhang, L., Liu, B., Du, J., Liu, C., & Wang, S. (2019). CO2 emission linkage analysis in global construction sectors: Alarming trends from 1995 to 2009 and possible repercussions. *Journal of Cleaner Production, 221*, 863–877.

Zhang, C., Hu, M., Sprecher, B., Yang, X., Zhong, X., Li, C., & Tukker, A. (2021). Recycling potential in building energy renovation: A prospective study of the Dutch residential building stock up to 2050. *Journal of Cleaner Production, 301*, 126835.

A Comprehensive Review of Carbon Footprints in the Wine Supply Chain in Italy

Leonardo Agnusdei, Christian Bux, Pier Paolo Miglietta, Federica De Leo, and Giulio Paolo Agnusdei

1 Introduction

On the global scale, the agri-food sector represents one of most contributing to climate change, accounting for about one-fifth of the entire emissions (Poore & Nemecek, 2018). Nowadays, approx. 50 million km^2 of the habitable land is destined to agricultural, of which 80% to crop production and 20% to animal feed production (Ritchie & Roser, 2021). In terms of greenhouse gas (GHG) emissions, the agri-food sector is responsible for 2.8 $GtCO_2eq$ and considering the European Union (EU), it results that Italy is the third highest GHGs' emitter with about 418 $ktCO_2eq$, after Germany and France (Poore & Nemecek, 2018). Therefore, significant action plans have been implemented at the international and local level, such as the Farm to Fork Strategy envisaged in the European Green Deal, whose aim

L. Agnusdei
Department of Innovation Engineering, University of Salento, Lecce, Italy
e-mail: leonardo.agnusdei@unisalento.it

C. Bux (✉)
Department of Economics, University of Foggia, Foggia, Italy
e-mail: christian.bux@unifg.it

P. P. Miglietta
Department of Biological and Environmental Sciences and Technologies, University of Salento, Lecce, Italy
e-mail: pierpaolo.miglietta@unisalento.it

F. De Leo
Department of Economic Sciences, University of Salento, Lecce, Italy
e-mail: federica.deleo@unisalento.it

G. P. Agnusdei
Department of Wellbeing, Nutrition and Sport, Pegaso University, Napoli, Italy
e-mail: giulio.agnusdei@unipegaso.it

S. S. Muthu (ed.), *Carbon Footprint Assessments*, Environmental Footprints and Eco-design of Products and Processes,
https://doi.org/10.1007/978-3-031-70262-4_4

is to tackle climate change and chase a fair, healthy and environmentally friendly food system (European Commission, 2022). Specifically, the European Green Deal aims at making Europe climate neutral by 2050 and to realize this objective the European Parliament has introduced the European Climate law, which aims at reducing GHGs by 55% by 2020 (Official Journal of the European Union, 2021).

The wine industry generates high revenues yearly, recording 172 billion USD in 2023, and its growth is expected to reach up to 4.5% in the next years. Consumers play a significant role toward its growth, since the average per capita consumption is estimated at 2.45 L at the global scale. Specifically, Italy ranks second in terms of wine consumption worldwide, soon after Portugal, while its production exceeds 4.91 billion liters, with over 1775 enterprises at the country level. In Italy, wine is seen by consumers as an integral part of the eating (especially dining) practice, and their approach tends to experience different varieties and regions of wine, leading to an increased demand to a wider range of wines in the market (Statista Market Insight, 2023). However, although the great economic opportunities associated with the wine market, there is still an open debate on its sustainability, either from the economic, social, or environmental perspective. One of the latest trends is the growing popularity of organic and biodynamic wines, considering that consumers are becoming more aware of health and environmental issues, and they look at wines that are produced in a sustainable and environmentally friendly manner (Rugani et al., 2013; Statista Market Insight, 2023).

Historically, viticulture and winemaking in Italy dates back to the Greeks but also to pre-Roman populations, especially in Southern Italy, in Southern Piedmont, Tuscany, and Sicily. Soon after the Second World War and thanks to national and international strategies, the wine sector changed drastically, passing from vineyards intended for self-consumption to specialized winegrowing, from a fragmented to an integrated agri-food supply chain. Furthermore, the wine sector abandoned the solely domestic production and started to export wines at the global scale (Pomarici et al., 2021). From the production perspective, Italy (718 thousands of hectares, kha) ranks at third place in terms of EU vineyard area, soon after Spain (955 kha) and France (812 kha). Although non-EU countries account for a small percentage of vineyards, such as Turkey (410 kha) or Moldova (122 kha), China ranks third at the global scale, with a total area estimated at 785 kha. In the United States, the total area destined to viticulture is solely 390 kha (International Organisation of Vine and Wine, 2023).

However, one question arises spontaneously: how much is the environmental impact of the cultivation (and production) of wine globally, and especially in Italy? In literature, several tools to estimate the environmental impacts of products and processes have been developed over time, most of which adopt the lifecycle thinking (Casson et al., 2023; Finnveden & Moberg, 2005). By definition, the lifecycle thinking is "the way of thinking of the consequences in the environmental, economic and social effects of a product throughout its entire life" (Farjana et al., 2021) and consists of a theoretical approach that investigates opportunities of improvement and reductions in the environmental impacts along the supply chain, from agricultural production to transformation, distribution, use, and final destination, so-called end-of-life (Jacob-Lopes et al., 2021). Among these tools, some have been standardized at the international level, such as the life cycle assessment (LCA) by the ISO 14040: 2006 (ISO, 2006) and the carbon footprint, either for processes by the ISO14064: 2018 (ISO, 2018a) or for products by the ISO 14067:2018 (ISO, 2018b). In recent years, the application of the carbon footprint as an environmental accounting tool has been proposed in the novel technical standard UNI/TS 11820:2022 on the "Measurement of Circularity – Methods and Indicators for Measuring Circular Processes in Organizations" (UNI, 2022). Specifically, the standard underlines the importance of estimating the global warming potential (GWP) of products and processes toward sustainability, and researchers and practitioners should make some efforts to implement its calculation.

In the light of these premises, the current study presents all the research papers concerning the carbon footprint assessment of wines in Italy, discussing the environmental impacts along the entire supply chain. Based on the results of the content analysis, this work allows for the identification of the most impactful phase of the supply chain. This review contributes to the existing scientific literature by highlighting strengths and weaknesses of the current strategies on the assessment of carbon footprint and identifying further developments in this field.

The chapter is structured as follows: Sect. 2 presents the literature trends on the carbon footprint in the wine supply chain; Sect. 3 illustrates the main insights from the carbon footprints in the wine industry, distinguishing between viticulture, winemaking, bottling, distribution, and End-of-Life; last, Sect. 4 ends the chapter by discussing results and presenting future research directions.

2 Literature Trends on the Carbon Footprint in the Wine Supply Chain

2.1 Description of the Supply Chain

Typically, the wine supply chain encompasses four main stages, i.e., viticulture, winemaking, bottling, and distribution (da Silva & Esteves da Silva, 2022; Harb et al., 2021), but also the end-of-life stage, namely the disposal pathways of all by-products, co-products, and waste along the entire lifecycle must be accounted (Agnusdei et al., 2022). Figure 1 illustrates the wine supply chain, highlighting input and output.

Specifically, viticulture represents the agricultural production stage, which consists of the cultivation and harvesting of grapes and is composed of several activities, such as vine planting, soil preparation, fertilization, use of chemicals and phytosanitary products, farm management, harvesting of grapes, and transportation from the agricultural field to the industrial plant for winemaking. Subsequently, winemaking (so-called vinification) represents the core of the wine production, namely the transformation of grapes into wine through fermentation. Both in the

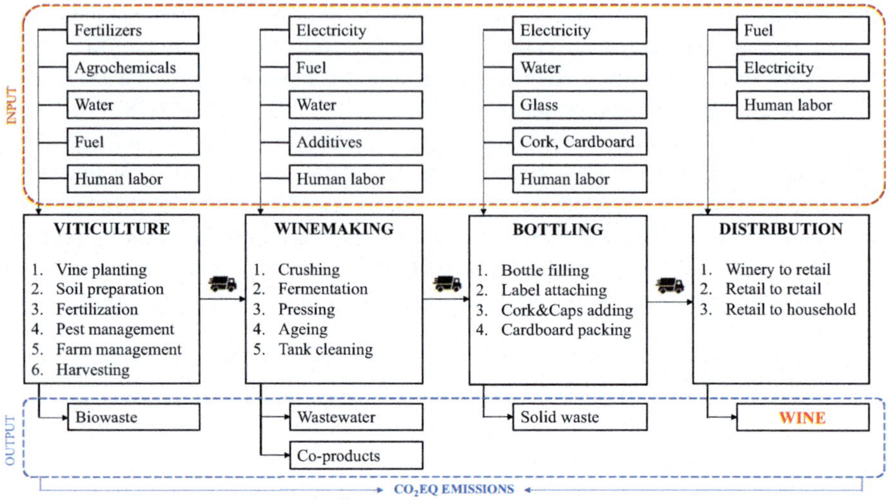

Fig. 1 Wine supply chain. (Source: Personal elaboration by the authors)

viticulture and in the winemaking several material and energy input are required, such as diesel for farm machineries and electricity for storage or refrigeration, for instance. One significant step, in terms of environmental impacts, is associated with the bottling stage, so-called packaging stage, during which wine is bottled and stored until it reaches consumers. Likewise the previous stages, also bottling requires an intensive consumption of material and energy input, respectively glass, water, and electricity (Pizzigallo et al., 2008; Vinci et al., 2022). Last, distribution regards the logistic stage, and includes different transportation means, such as boat, train, truck, or plane. However, in terms of environmental impacts associated with transportation and logistics, it should not be underestimated the consequence of the "last mile," which regards the transportation from the supermarket to consumers (Rana, 2020).

2.2 *Bibliometric Data by Using the VOSviewer Software*

The carbon footprint has been widely investigated in the academia on the global scale, and the current section reports a data synthesis of the main research trends related to the wine supply chain. Specifically, data were synthesized by using the VOSviewer software (VOSviewer, 2024), which provides insights into the structure of the research network and offers a multidimensional scaling and clustering in the research (van Eck & Waltman, 2018). The software has been largely used in the academia (Kirby, 2023; Martinez et al., 2019) and helps identify research evolution by creating suggestive maps of keywords and textual data. Figure 2 illustrates the main findings according to a co-word analysis (i.e., "network visualization").

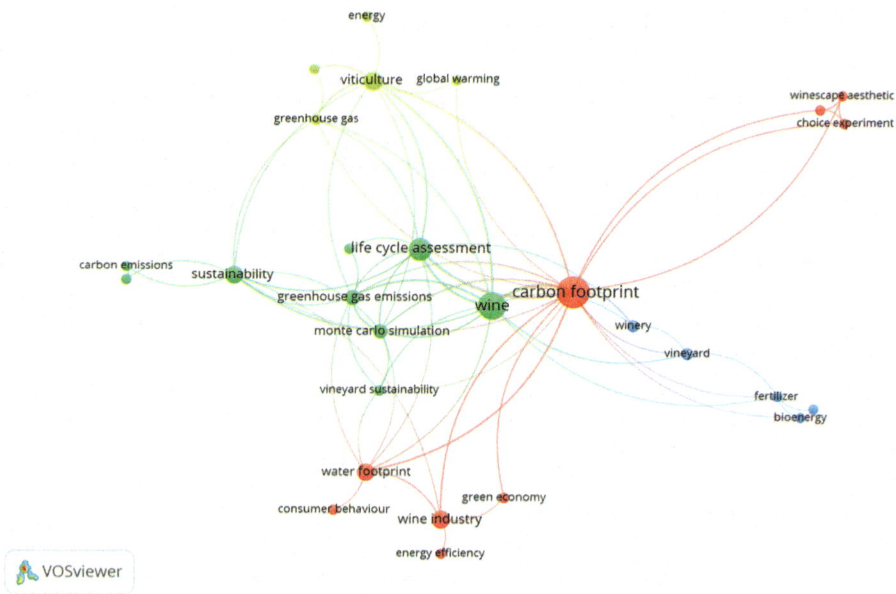

Fig. 2 VOSviewer data synthesis ("network visualization"). (Source: Personal elaboration by the authors by using VOSviewer)

In the light of the VOSviewer analysis on the environmental impacts of the wine supply chain, four clusters can be identified in research, as follows: (a) ecological footprint assessment (red cluster), which includes the analysis of the carbon and the water footprint, in the field of the so-called "green economy"; (b) bioenergy production (blue cluster), which includes the analysis of the agricultural production in the vineyard and the opportunities to generate bioenergy and biofertilizers through waste-to-energy and waste-to-bioproducts strategies; (c) life cycle assessment and sustainability (green cluster), which generally refers to the analysis of the entire lifecycle and the economic, environmental, and social sustainability of the wine industry, with specific focus on the carbon emissions and the greenhouse gas emissions; and (d) energy performance (yellow cluster), which strictly related to the green cluster in terms of carbon emissions but focuses on the energy and the electricity consumption during the agricultural stage. Specifically, it results that the viticulture and the winemaking represent the most significant topics in terms of research, whereas few contributions relate to the bottling and the distribution stage, which still have a strong contribution to climate change. However, in the red cluster, some co-occurrences relate to the consumers, in the field of the "winescape aesthetic" and the "choice experiment," highlighting the importance (in recent years) of the consumer behavior toward sustainable practices.

3 Carbon Footprints in the Wine Industry

The current research provides a comprehensive literature review on the carbon footprints in the wine industry, focusing on the four core stages of its supply chain, i.e., viticulture, winemaking, bottling, and distribution, but also including the end-of-life stage. Specifically, the authors have carried out a TITLE-ABS-KEY search in Scopus (Rana et al., 2022), which represents a reliable collector of standardized, reputable, and high-quality research in the academia (Carrera-Rivera et al., 2022). Considering the goals of the research, namely the identification of the most impactful phases of the wine supply chain, as well as the evaluation of strengths and weaknesses of the current environmental strategies in the assessment of the carbon footprint, the authors have not included a specific timeframe in the search but have limited the contributions to academic peer-reviewed research articles (not review article, nor book chapters or conference papers) published in English in Italy. By defining a comprehensive search string (i.e., "carbon footprint" AND "wine"), the authors have catalogued meta-data in Microsoft Excel sheets, highlighting (i) authors' details; (ii) title of the article; (iii) publishing year; (iv) source of the article; (v) abstract; and (vi) author keywords.

Considering the research fields proposed by Scopus, namely "agricultural and biological sciences," "environmental science," "energy," "social sciences," "business, management, and accounting," and "economics, econometrics, and finance," the authors have selected 14 contributions for in-depth review. Table 1 summarizes the selected contributions by identifying the supply chain stages analyzed, the carbon footprint method adopted in the research and the main findings, expressed in terms of CO_2eq emissions.

Specifically, out of the 14 selected contributions, 12 investigate the environmental impacts during the viticulture phase, whereas 11 during the winemaking and 10 in the bottling phase. On the other side, the impacts of distribution are estimated by solely seven studies and those related to the end-of-life by five contributions, highlighting the still high interest in upstream stages of the supply chain compared to the downstream ones, namely distribution and disposal.

Considering the object of carbon footprint evaluation, the most of comparison were done between conventional and organic wines, as well as between red and white ones. Specifically, Borsato et al. (2020) estimated the environmental impacts associated with conventional and organic viticulture, recording a decrease in CO_2eq in the organic culture (−28%), whereas other authors such as Rinaldi et al. (2016) evaluated the different carbon emissions for the production of red and white wine, highlighting the slightly higher impact of red compared to white ones (+5%).

In terms of functional unit, the vast majority of contributions adopted the 0.75-L bottle, but some authors have preferred to investigate the environmental impacts associated with 1 ha of cultivated land (Borsato et al., 2020; Morselli et al., 2023), the impacts related to the 1 kg of harvested grape (Marras et al., 2015) or the direct and indirect emissions associated with the winemaking and bottling process (Pettineo et al., 2020). Still, Rosso and Bolzonella (2009) investigated the impacts associated with the treatment of 1 m^3 of wastewater in the agricultural production.

Table 1 Synthesis of the selected 14 contributions in terms of supply chain stage, carbon footprint method, and CO_2emissions

	Supply chain stage					
Reference	VT.	WM.	BO.	D.	EOL.	CO_2eq emissions
Morselli et al. (2023)	X					CWS: 282.3 kgCO_2eq per ha per year MWS: 118.1 kgCO_2eq per ha per year TWS: 71.9 kgCO2eq per ha per year
Vinci et al. (2022)	X	X	X			Organic wine: 1.1 kgCO_2eq per 0.75 L
D'Ammaro et al. (2021a)	X	X	X	X		1.39 kgCO_2eq ± 0.25 kgCO_2eq per 0.75 L
D'Ammaro et al. (2021b)	X	X	X	X	X	Red wine: 0.97–1.97 kgCO_2eq per 0.75 L
Pettineo et al. (2020)		X	X			Direct emissions: 277 tCO2eq per year Indirect emissions: 161.5 tCO2eq per year
Borsato et al. (2020)	X					Conventional vineyard: 2534 kgCO_2eq per ha Organic vineyard: 1827 kgCO_2eq per ha
Chiriacò et al. (2019)	X	X	X			Organic wine: 0.79 ± 0. kgCO_2eq per 0.75 L
Rinaldi et al. (2016)	X	X	X	X	X	Red: 1.44 kgCO_2eq per 0.75 L White: 1.37 kgCO_2eq per 0.75 L
Bonamente et al. (2016)	X	X	X	X	X	Red wine: 1.07 ± 0.09 kgCO_2eq per 0.75 L
Marras et al. (2015)	X					Mature vineyard: 0.39 kgCO_2eq per kg of grape
Bosco et al. (2013)	X	X	X	X	X	Red wine: 0.531–0.663 kgCO_2eq per 0.75 L
Pattara et al. (2012)	X	X	X	X		Red wine: 1.29 k kgCO_2eq per 0.75 L.
Bosco et al. (2011)	X	X	X	X	X	0.6–1.3 kgCO_2eq per 0.75 L
Rosso and Bolzonella (2009)		X				9.86–17.1 kgCO_2eq per m^3 of wastewater treated

Source: Personal elaboration by the authors
Note: *VT* viticulture, *WM* winemaking, *BO* bottling, *D* distribution, *EOL* end-of-life, *CWS* chemical weeding system, *MWS* mechanical weeding system, *TWS* thermal weeding system

In the field of the carbon footprint method adopted by researchers, the vast majority adopted the ISO 14067 for the product carbon footprint assessment, but also the ISO 14064 for the assessment of the winemaking and bottling process. Moreover, previously to 2016, some authors carried out a carbon footprint analysis though the International Wine Carbon Calculator (IWCC) Protocol (Marras et al., 2015), or used the Carbon Calculator v1.3. (Pattara et al., 2012).

3.1 Viticulture

Viticulture stands as a significant and widespread agricultural system within the rural landscape of Italy, covering 636 thousand hectares (tha), with 58% located in central and southern regions (ISTAT, 2021). The agricultural phase exhibits high variability in terms of environmental impacts, influenced by both geographical and technological factors (Vázquez-Rowe et al., 2013). Consequently, assessing the carbon footprint during the viticulture phase is crucial, even though it is the third major contributor.

D'Ammaro et al. (2021a) investigated the vineyard phase, revealing carbon footprint impacts ranging from 0.03 $kgCO_2eq$ to 0.49 $kgCO_2eq$ for each bottle of 0.75-L of wine. In particular, it represents the 18% of the total emissions with an average value equal to 0.19 $kgCO_2eq$. Predominant contributors include fuel combustion during field operations and the production and utilization of fertilizers and pesticides. Basing on these results, D'Ammaro et al. (2021b) developed a multi-criteria approach to avoid potential burden shifting across different indicators.

Bosco et al. (2011) conducted an analysis in which the emissions of four different wines were considered. In this study, the agricultural phase was split into three sub-phases: vineyard planting, pre-production, and production. The results showed that the most impacting sub-phase in terms of emission is represented by "production" with 0.15 $kgCO_2eq$ per 0.75-L (67% of the total emission in the agricultural phase), while the amount of $kgCO_2eq$ per 0.7-5 L for vineyard planting and pre-production phases are respectively 0.06 (28% of the total emission in the agricultural phase) and 0.01 (5% of the total emission in the agricultural phase). Marras et al. (2015) provide direct and indirect CO_2 (N converted into CO2-eq) emissions for each agronomic activity. Among others, soil management and fossil fuels consumption represent the two main emission sources, respectively with 0.13 $kgCO_2$ per kg of grape yield (50% of the total GHG emissions) and 0.12 $kgCO_2$ per kg of grape yield (44% of the total GHG emissions). On the other hand, fertilization application contributes for 5% to the global carbon budget (0.01 $kgCO_2$ per kg of grape yield) while human work for about 2%.

Rinaldi et al. (2016) and Bonamente et al. (2016) provide results in line with the previous literature by dividing the product life cycle in a total of 11 phases, grouped into three modules: upstream, core, and downstream. The upstream phase includes the acquisition of all necessary raw materials and energy inputs for wine production; the core module involves both the production process and the packaging of the final product; the downstream module entails the transportation to a distribution platform and the management of packaging materials. In these studies, it emerges that the grapes production classified third in terms of carbon footprint (19.64%).

Borsato et al. (2020) provide an environmental comparison between organic and conventional viticulture using, among other different indicators, the carbon footprint. Conventional vineyard management exhibits higher GHG emissions, approximately 2534 $kgCO^2eq$ per hectare, compared to the 1827 $kgCO_2eq$ per hectare observed in organic vineyard management. The conventional approach contributes more significantly to air pollution, primarily due to elevated carbon footprint (CF) in fertilizers and fuel consumption, measuring 904 kgCO_2eq per hectare and 1526 kgCO_2eq per hectare, respectively. Conversely, pesticide application has a lower impact under conventional management, totalling 105 kgCO_2eq per hectare. The reduction in GHG emissions is approximately 75% for fertilizer application when there is a transition from conventional to organic vineyard management and around 66% for pesticide application. Overall, organic management results in a 15% reduction in GHG emissions in terms of CO_2eq per hectare when compared to conventional vineyard management.

Similar results were obtained through the analysis of Vinci et al. (2022), where organic grape cultivation was estimated to be the second contributor to fossil and biogenic CO_2 emissions with 0.31 CO_2eq and 0.02 CO_2eq respectively, being on average 24.4% of the entire GHG emissions. In the study of Bosco et al. (2013), taking into account changes in soil organic matter (SOM) resulted in a reduction of 20% in the greenhouse gas potential (GWP) for a 0.75 L bottle of red wine. Although the primary contributor to emissions remains the winery phase (84%), the second significant source was the subphase of vineyard planting (16%), attributed to emissions released during soil preparation.

The pre-production and production sub-phases contributed to 5% and 95% of the entire reduction, respectively. Morselli et al. (2023) provide an environmental analysis on direct and indirect CO2 emissions generated in the use of the three weeding technologies: chemical weeding system (CWS), the mechanical weeding system (MWS), and the gasification-powered thermal weeding system (TWS). It is emerged that the TWS, due to the co-produced biochar, allows for a 23.4% reduction of the emissions, recommending the opportunity to rely on a carbon neutral weeding technology.

Viticulture in Italy has significant environmental implications, particularly regarding carbon emissions. Studies highlight the vineyard phase as a one of the major sources of greenhouse gases, emphasizing the need for sustainable practices. Comparisons between organic and conventional methods show potential for lower emissions with organic practices. Innovative approaches, such as soil management adjustments and alternative weeding technologies, offer promising solutions for reducing emissions and promoting sustainability in viticulture. Overall, adopting sustainable practices is crucial for mitigating the industry's carbon footprint while maintaining economic viability.

3.2 *Winemaking*

The process of wine production results in a significant number of residues, marked by the existence of biodegradable elements and suspended solids. These residues, if not appropriately managed, can lead to adverse environmental and economic consequences such as water contamination, soil deterioration, damage to vegetation, increased energy usage, and the GHG emissions (Silva et al., 2021). For these reasons, the assessment of this wine supply chain phase is becoming fundamental in order to make it more circular.

These considerations are reflected in the analysis of D'Ammaro et al. (2021a), where cellar phases resulted as second main contributory subsystem with values ranging between 0.028 $kgCO_2eq$ per 0.75 L and 0.78 $kgCO_2eq$ per 0.75 L, accounting 27% of the overall GHGs emissions. In addition, the authors found in electricity used in the cellar stage the 14% of GHGs emissions. As for the viticulture phase described in the previous paragraph, these results were used by D'Ammaro et al. (2021b) to perform a multi-criteria approach able to better assess sustainability performances. In the studies of.

Chiriacò et al. (2019) and Rinaldi et al. (2016), the authors include in the transformation process, defined as core module by Rinaldi et al. (2016), both winemaking and bottling and it became complex to evaluate the only contribution of $kgCO_2eq$ emission due to the vinification processes. Also in the study of Bonamente et al. (2016), this phase includes both vinification and bottling, and represents the second main contributors (0.04 $kgCO_2eq$ per 0.75 L, 3.5% of the total GHGs emissions) of the carbon footprint. Bosco et al. (2013) performed a deep evaluation of the carbon footprint, providing the $kgCO_2eq$ for two different cases: one in which Soil Organic Matter (SOM) is taking into account and the other in which it is not considered. In the first case, the winemaking phase is ranked second in terms of carbon footprint with 0.088 $kgCO_2eq$ per 0.75 L; in the second case is ranked third while maintaining the same value of $kgCO_2eq$ per 0.75 L.

In addition, Bosco et al. (2011) conducted an analysis in which four different wines (two different IGT Maremma Toscana—red wine, Bianco di Pitigliano DOC—white wine and Morellino di Scansano DOCG—red wine) were considered. In this study, it can be noted that vinification represents on average the fourth most impacting phase with 0.1 $kgCO_2eq$ per 0.75 L, i.e., the 10.5% of the total GHGs emissions.

Studies highlight the importance of assessing and mitigating the carbon footprint across the production chain. Results show that the cellar and vinification phases contribute substantially to greenhouse gas emissions, emphasizing the need for sustainable practices in these areas. Overall, these findings emphasize the importance of adopting circular and sustainable approaches to minimize the environmental impact of wine production.

3.3 Bottling

In the wine production, bottling represents a complex and delicate step that requires suitable and professional skills, since a correct bottling allows wine to maintain quality during storage and shipping and avoid oxidation as long as possible (Nygaard, 2010). However, in terms of environmental impacts, the bottling stage (so-called packaging stage) has been highlighted as one of the most impactful along the wine lifecycle (Pizzigallo et al., 2008), therefore packaging decisions represent a crucial aspect for tackling climate change.

In the analysis conducted by Vinci et al. (2022), the main inputs associated with bottling are water, electricity, glass, cork, capsule (PVC), and labels, since the packaging stage includes also bottle washing. Considering 0.75-L bottle as a functional unit, the research estimated a water consumption of about 0.13 L, an electricity consumption of 0.006 kWh and an average glass bottle weight of approx. 0.55 kg. In terms of environmental impacts, the bottling stage accounts for the highest climate change emissions, with a contribution of 0.47 kgCO_2eq, also considering that 55% of the electricity required during its process and for the production of glass bottles comes from non-renewable energy sources. In general, the glass production industry represents a high-energy-demanding industrial process (Pölz & Rosner, 2020). One of the main failings is due to the fact that consumers associate the quality of a wine with the weight of the bottle, believing that the heavier the bottle the higher the quality of the wine, but recent studies (Amienyo et al., 2014; Point et al., 2012) estimated the possibility of reducing by 30% the weight of bottles without compromising the quality of the wine. Specifically, the reduction of the weight of a bottle up to 30% can reduce emissions in a range from 2 to 10%. In addition, glass bottles could also be substituted by other packaging materials, such as PET or bag-in-box (Vinci et al., 2022).

Similar results were highlighted by D'Ammaro et al. (2021a, b). In D'Ammaro et al. (2021a), the impact of packaging on the entire wine lifecycle, also including the end-of-life stage, was estimated at 29%, considering a mean weight of 0.59 kg per 0.75-L of wine (from 0.36 kg to 0.85 kg of glass), whereas Rinaldi et al. (2016) estimated its impact at 43% according to the same cradle-to-grave approach. Specifically, the CO_2eq emissions associated with bottles were estimated at about 0.51 kgCO_2eq per functional unit, with wines sold in Northern Italy having a higher glass weight (0.61 kg of glass per functional unit) compared to wines sold in Central Italy (0.55 kg of glass). Moreover, it was estimated that red wine have heavier bottles compared to white ones (+5%). In terms of possible strategies to reduce the carbon footprint of the wine lifecycle, D'Ammaro et al. (2021a) highlighted the chance of reducing by 14% the CO_2eq by adopting 32% lighter bottles, whereas D'Ammaro et al. (2021b) underlines the need to use electricity with a higher portion of renewables and to reuse hot water from bottling sterilization.

In comparison to other selected contributions, Chiriacò et al. (2019) records one of the lowest carbon footprint associated with one 0.75-L bottle of wine, namely 0.79 kgCO_2eq, of which 0.24 kgCO_2eq due to the glass bottle, 0.01 kgCO_2eq to cork and 0.05 kgCO_2eq to cardboard box containing six bottles. Not by chance the current research is based on a producer that adopts sustainable packaging solutions that include an ultralight glass bottle of 0.36 kg, that is −39% compared to D'Ammaro et al. (2021a) and −35% compared to Vinci et al. (2022) in terms of weight. Practically, the research adopted an emission factor of 0.67 kgCO_2eq per kg of glass bottle, as outlined by Hischier (2007). The research highlighted the adoption of the ultralight bottle as one of the main strategies to reduce carbon emissions, as also suggested by Martins et al. (2018) in Portugal.

Bonamente et al. (2016) evaluated the carbon footprint of a red wine along the entire lifecycle, from viticulture to end-of-life, estimating its amount at 1.07 ± 0.09 $kgCO_2eq$ per 0.75-L. In terms of packaging, the research considers a rather light bottle (0.45 kg of glass per functional unit) and highlights that the packaging accounts for 56% of the entire carbon footprint, for an amount of 0.59 $kgCO_2eq$ per functional unit. As per other studies, Bonamente et al. (2016) estimated the impact of glass weight in the carbon footprint by varying the bottle weight of ±20%, therefore considering bottles of 0.36 kg and of 0.54 kg. Specifically, it is possible to reduce emissions by 9.4% if using 0.36 kg bottles, as outlined by Chiriacò et al. (2019).

In comparison with other studies, Bosco et al. (2013) estimated the lowest carbon footprint per 0.75-L of red wine, from 0.53 to 0.66 $kgCO_2eq$. In terms of packaging, the research considers a 0.40 kg glass bottle and distinguished between two main aspects: (i) the core bottling, which accounts for 0.02 $kgCO_2eq$; and (ii) the packaging, which generates 0.32 $kgCO_2eq$, therefore contribution up to 64% of the entire carbon emissions. Similar results were reached by Bosco et al. (2011), which investigated the different impacts by adopting a green or a white glass bottle. However, independently from the typology of glass, the bottling and packaging phase still represents the highest contributor to climate change along the entire wine lifecycle, from 41% to 63%, and this variability is mainly due to the weight of the bottle, the glass type and the secondary packaging. Generally, white glass bottles have a higher carbon footprint compared to green ones (approx. + 22%).

In the field of bottling, Pattara et al. (2012) provide an interesting breakdown in Scope 3 of the wine supply chain, namely indirect emissions. Specifically, it results that the packaging stage contributes up to 93% of the indirect emissions, with glass contributing to 77%, closures to 8%, cardboard boxes to 7% and paper labels to 1%. In absolute values, bottles and containers generate 0.94 $kgCO_2eq$ per 0.75 L bottle, whereas bottle labels, closures, and fiber packaging about 0.2 $kgCO_2eq$.

3.4 Distribution and End-of-Life

Table 2 illustrates the main characteristics of the distribution in the selected carbon footprint studies, when available in the research.

D'Ammaro et al. (2021a) estimated that distribution accounts for 13%, whereas the impacts of the consumption phase are 2%, as previously discussed by solely one research conducted in Canada (Point et al., 2012). Specifically, the research highlights the hard process of collecting data related to distribution, since wine producers can use a wide range of distribution channels and logistics arrangements to provide consumers with wine. Similarly, D'Ammaro et al. (2021b) estimated the carbon emissions associated with the distribution of 15 red wines in Emilia Romagna, Toscana, Veneto, Umbria, and Marche, reporting an average impact of distribution at 12.4% of the entire carbon footprint, whereas soley 0.7% for consumption.

Table 2 Distribution characteristics in the wine supply chain

Reference	Item	km, vehicle	$kgCO_2eq$
D'Ammaro et al. (2021a)	Winery to retail	250 km, truck >32 t (E4)	0.20*
	From retail to final client	5 km, passenger car	0.01*
D'Ammaro et al. (2021b)	–	–	0.01 to 0.39*
Rinaldi et al. (2016)	–	–	– 0.05 to – 0.07*
Bonamente et al. (2016)	–	–	0.43*
Bosco et al. (2013)	Winery to retail	350 km, 0.02 kg of diesel*	0.08*
Pattara et al. (2012)	From Italy to foreign countries	Truck, 38 t lorry	–
	From Italy to Italy	Truck, 7.5 t lorry	–
	Local transportation	Truck, 3.5 t lorry	–
	Total	–	6.7*
Bosco et al. (2011)	Wine 1, from winery to retail	130 km, 0.01 kg of diesel*	0.04*
	Wine 2, from winery to retail	1600 km, 0.12 kg of diesel*	0.44*
	Wine 3, from winery to retail	60 km, 0.005 kg of diesel*	0.02*
	Wine 4, from winery to retail	350 km, 0.02 kg of diesel*	0.09*

Note: E4 euro 4; *per 0.75-L bottle

Rinaldi et al. (2016) highlighted that distribution represents the second largest contributor of carbon emissions soon after packaging and is accounts for about 30% (0.42 $kgCO_2eq$ for red wine and 0.40 gCO_2eq for white wine). In the field of the end-of-life, the authors modeled the waste management system on the basis of average national and international data, recording a net environmental impact of about −0.07 for red wines and − 0.05 for white wines. For the end-of-life emissions, the authors hypothesize the subsequent scenarios: (i) recycling at 77% for cardboard, 70% for glass and 35% for capsule; (ii) landfill at 22% for cardboard, 30% for glass and 65% for capsule; and (iii) incineration with energy recovery at 70% solely for paper labels.

Bonamente et al. (2016) evaluated the distribution from storage to consumers and estimates its impact at about 41% (0.43 $kgCO_2eq$ per 0.75-L bottle) of the entire wine supply chain, from cradle to grave. In terms of waste management, the carbon impacts have been estimated at –40.78% (−0.43 $kgCO_2eq$ per 0.75-L bottle), considering an end-of-life scenario equal to Rinaldi et al. (2016).

Bosco et al. (2013) investigated the distribution impacts along the wine supply chain, as well as the impacts associated with waste management. In terms of distribution, the carbon emissions were 0.08 $kgCO_2eq$ per 0.75-L bottle, for an amount of 12–15% of the entire supply chain, whereas the waste management emissions were evaluated at 0.02 $kgCO_2eq$ per functional unit. Considering Bosco et al. (2011), the distribution stage varies according to the different wine under research, from 0.02 to 0.44 $kgCO_2eq$ per 0.75-L bottle depending on the distance from the winery to the distribution platforms. In terms of waste management, the authors assume a 100% recycling of glass bottles, whereas the label and the stopper are assumed to be landfilled, with a total contribution to carbon emissions estimated from 0.02 to 0.03 $kgCO_2eq$ per functional unit.

Similar to previous results, Pattara et al. (2012) evaluated the carbon emissions associated with distribution at 0.08 $kgCO_2eq$ per 0.75-L bottle, but it does not consider the end-of-life pathways.

4 Conclusions

The chapter presented the main findings from a comprehensive literature review conducted on the carbon footprints of the wine industry in Italy, which ranks second in terms of wine consumption worldwide and whose production exceeds 4.91 billion liters per year. Specifically, the study focused on five main supply chain stages, namely viticulture, winemaking, bottling, distribution, and end-of-life, with the purpose of identifying the most impactful phase of the supply chain.

The chapter analyzed 14 contributions in line with the purpose of the research and it results that 12 articles consider the viticulture phase, 11 the winemaking, and 10 the bottling phase, whereas only 7 studies explore the environmental impacts associated with distribution and 5 those due to the end-of-life, with regard to reuse, recycling, and landfilling.

Starting from the first supply chain phase, the research highlights that viticulture covers about 636 tha in Italy. In the light of the selected contributions, on average 18% of the total emissions of the wine supply chain are attributable to viticulture, with fuel combustion for field operations and the use of fertilizers and pesticides as the main impacting activities. However, carbon emissions could be reduced by switching from conventional to organic vineyard, with an estimated reduction in the emissions by 15% per ha of cultivated land. Subsequently, winemaking is responsible for significant co-products and by-products, that must be managed in a suitable manner in order to avoid environmental consequences. In accordance with the selected studies, the winemaking contributes on average by 27% to the entire carbon emission of the wine supply chain, of which electricity used in the cellar stage significantly impacts. In terms of bottling, it represents a complex and delicate phase and was estimated as the most impactful stage to the environment, accounting for about 43% of the total carbon emissions from cradle-to-grave. Among several solutions, operators in the wine industry can substitute heavy bottles with lighter ones, considering that the adoption of ultralight bottles (ca. 0.36 kg of glass per bottle) can reduce emissions by 32%. Last, in the field of distribution and final consumption, the study revealed that distribution accounts for 13% of the carbon emissions, whereas the impacts of the consumption phase are approx. 2%. Considering end-of-life emissions, they vary according to recycling, incineration or landfilling.

The research revealed the carbon footprint as a suitable tool for shaping sustainable strategies and a significant decision-making guide for practitioners and public authorities. For instance, policy makers should promote the use of ultralight bottles, which can significantly reduce the environmental impacts of the entre wine supply chain, as well as alternative packaging compared to wine, like aluminum cans or bag-in-a-box. These alternatives may contribute to sustainable development without compromising the quality of wine and the customer satisfaction.

Future research directions intend to enlarge the analysis of the carbon footprints in the wine industry to other European and international realities, for instance France or Spain, which rank at second and third place in terms of wine production worldwide. Comparing Italy with other countries could help identify appropriate strategies and best practices, which can reduce the carbon emissions and enlarge the market share in terms of sustainability.

References

Agnusdei, G. P., Coluccia, B., Elia, V., & Miglietta, P. P. (2022). IoT technologies for wine supply chain traceability: potential application in the Southern Apulia Region (Italy). *Procedia Computer Science, 200*, 1125–1134.

Amienyo, D., Camilleri, C., & Azapagic, A. (2014). Environmental impacts of consumption of Australian red wine in the UK. *Journal of Cleaner Production, 72*, 110–119.

Bonamente, E., Scrucca, F., Rinaldi, S., Merico, M. C., Asdrubali, F., & Lamastra, L. (2016). Environmental impact of an Italian wine bottle: Carbon and water footprint assessment. *Science of the Total Environment, 560–561*, 274–283. https://doi.org/10.1016/j.scitotenv.2016.04.026

Borsato, E., Zucchinelli, M., D'Ammaro, D., Giubilato, E., Zabeo, A., Criscione, P., Pizzol, L., Cohen, Y., Tarolli, P., Lamastra, L., & Marinello, F. (2020). Use of multiple indicators to compare sustainability performance of organic vs conventional vineyard management. *Science of the Total Environment, 711*, 135081. https://doi.org/10.1016/j.scitotenv.2019.135081

Bosco, S., Di Bene, C., Galli, M., Remorini, D., Massai, R., & Bonari, E. (2011). Greenhouse gas emissions in the agricultural phase of wine production in the Maremma rural district in Tuscany, Italy. *Italian Journal of Agronomy, 6*(2), 15. https://doi.org/10.4081/ija.2011.e15

Bosco, S., Di Bene, C., Galli, M., Remorini, D., Massai, R., & Bonari, E. (2013). Soil organic matter accounting in the carbon footprint analysis of the wine chain. *International Journal of Life Cycle Assessment, 18*, 973–989. https://doi.org/10.1007/s11367-013-0567-3

Carrera-Rivera, A., Ochoa, W., Larrinaga, F., & Lasa, G. (2022). How-to conduct a systematic literature review: A quick guide for computer science research. *MethodsX, 9*, 101895. https://doi.org/10.1016/j.mex.2022.101895

Casson, A., Zambelli, M., Giovenzana, V., Tugnolo, A., Pampuri, A., Vignati, S., Beghi, R., & Guidetti, R. (2023). Simplified environmental impact tools for agri-food system: A systematic review on trends and future prospective. *Environmental Impact Assessment Review, 102*, 107175. https://doi.org/10.1016/j.eiar.2023.107175

Chiriacò, M. V., Belli, C., Chiti, T., Trotta, C., & Sabbatini, S. (2019). The potential carbon neutrality of sustainable viticulture showed through a comprehensive assessment of the greenhouse gas (GHG) budget of wine production. *Journal of Cleaner Production, 225*, 435–450. https://doi.org/10.1016/j.jclepro.2019.03.192

D'Ammaro, D., Capri, E., Valentino, F., Grillo, S., Fiorini, E., & Lamastra, L. (2021a). Benchmarking of carbon footprint data from the Italian wine sector: A comprehensive and extended analysis. *Science of the Total Environment, 779*, 146416. https://doi.org/10.1016/j.scitotenv.2021.146416

D'Ammaro, D., Capri, E., Valentino, F., Grillo, S., Fiorini, E., & Lamastra, L. (2021b). A multi-criteria approach to evaluate the sustainability performances of wines: the Italian red wine case study. *Science of the Total Environment, 799*, 149446. https://doi.org/10.1016/j.scitotenv.2021.149446

da Silva, L. P., & Esteves da Silva, J. C. G. (2022). Evaluation of the carbon footprint of the life cycle of wine production: A review. *Cleaner and Circular Bioeconomy, 2*, 100021. https://doi.org/10.1016/j.clcb.2022.100021

European Commission. (2022). *Farm to fork strategy for a fair, healthy and environmentally-friendly food system.* https://food.ec.europa.eu/horizontal-topics/farm-fork-strategy_en. Accessed 30 Jan 2024.

Farjana, S. H., Mahmud, M. A. P., & Huda, N. (2021). Chapter 1: Introduction to life cycle assessment. In S. H. Farjana, M. A. P. Mahmud, & N. Huda (Eds.), *Life cycle assessment for sustainable minillng* (pp. 1–13). Elsevier. https://doi.org/10.1016/B978-0-323-85451-1.00001-9

Finnveden, G., & Moberg, Å. (2005). Environmental systems analysis tools – An overview. *Journal of Cleaner Production, 13*(12), 1165–1173. https://doi.org/10.1016/j.jclepro.2004.06.004

Harb, W., Zaydan, R., & Vieira, M. (2021). Improving environmental performance in wine production by life cycle assessment: case of Lebanese wine. *International Journal of Life Cycle Assessment, 26*, 1146–1159. https://doi.org/10.1007/s11367-021-01895-0

Hischier, R. (2007). *Life cycle inventories of packaging and graphical paper.* Final report ecoinvent data v2. 0 No 11.

International Organisation of Vine and Wine. (2023). *State of the World vine and wine sector in 2022.* https://www.oiv.int/. Accessed 30 Jan 2024.

Istat (2021). Coltivazioni: Uva, vino, olive, olio. https://dati.istat.it/Index.aspx?QueryId=33706#. Accessed 30 Jan 2024.

ISO. (2006). *ISO 14040:2006. Environmental management. Life cycle assessment. Principles and framework.* https://www.iso.org/standard/37456.html. Accessed 30 Jan 2024.

ISO. (2018a). *ISO 14064-1:2018. Greenhouse gases. Part 1: Specification with guidance at the organization level for quantification and reporting of greenhouse gas emissions and removals.* https://www.iso.org/standard/66453.html. Accessed 30 Jan 2024.

ISO. (2018b). *ISO 14067:2018. Greenhouse gases. Carbon footprint of products. Requirements and guidelines for quantification.* https://www.iso.org/standard/71206.html. Accessed 30 Jan 2024.

Jacob-Lopes, E., Queiroz Zepka, L., & Costa Deprá, M. (2021). Chapter 5: Assistant's tools toward life cycle assessment. In E. Jacob-Lopes, L. Q. Zepka, & M. C. Deprá (Eds.), *Sustainability metrics and indicators of environmental impact* (pp. 77–90). https://doi.org/10.1016/B978-0-12-823411-2.00006-2

Kirby, A. (2023). Exploratory bibliometrics: using VOSviewer as a preliminary research tool. *Publications, 11*, 10. https://doi.org/10.3390/publications11010010

Marras, S., Masia, S., Duce, P., Spano, D., & Sirca, C. (2015). Carbon footprint assessment on a mature vineyard. *Agricultural and Forest Meteorology, 214–215*, 350–356. https://doi.org/10.1016/j.agrformet.2015.08.270

Martinez, S., del Mar Delgado, M., Martinez Marin, R., & Alvarez, S. (2019). Science mapping on the Environmental Footprint: A scientometric analysis-based review. *Ecological Indicators, 106*, 105543. https://doi.org/10.1016/j.ecolind.2019.105543

Martins, A. A., Araújo, A. R., Graça, A., Caetano, N. S., & Mata, T. M. (2018). Towards sustainable wine: Comparison of two Portuguese wines. *Journal of Cleaner Production, 183*, 662–676. https://doi.org/10.1016/j.jclepro.2018.02.057

Morselli, N., Boccaletti, S., Meglioraldi, S., Puglia, M., Pedrazzi, S., & Allesina, G. (2023). Biomass-powered thermal weeding in wine farms: An environmental and economic assessment. *Journal of Cleaner Production, 385*, 135684. https://doi.org/10.1016/j.jclepro.2022.135684

Nygaard, M. (2010). Oxygen management from grape to glass. *Wine Industry Journal, 25*, 24–28.

Official Journal of the European Union. (2021). Regulation (EU) 2021/1119 of the European Parliament and of the Council of 30 June 2021 establishing the framework for achieving climate neutrality and amending Regulations (EC) No 401/2009 and (EU) 2018/1999 ('European Climate Law').

Pattara, C., Raggi, A., & Cichelli, A. (2012). Life cycle assessment and carbon footprint in the wine supply-chain. *Environmental Management, 49*, 1247–1258. https://doi.org/10.1007/s00267-012-9844-3

Pettineo, G., Raciti, R., Allegro, G., Zerbo, A., & Froio, A. P. (2020). Principi Di Butera: The carbon footprint in the sicilian wine industry. In *Procedia environmental science, engineering and management* (pp. 233–242) https://tinyurl.com/29mc5p9j

Pizzigallo, A. C. I., Granai, C., & Borsa, S. (2008). The joint use of LCA and emergy evaluation for the analysis of two Italian wine farms. *Journal of Environmental Management, 86*, 396–406.

Point, E., Tyedmers, P., & Naugler, C. (2012). Life cycle environmental impacts of wine production and consumption in Nova Scotia, Canada. *Journal of Cleaner Production, 27*, 11–20.

Pölz, W., & Rosner, F. G. (2020). Sustainable strategy against climate change based on greenhouse gas emissions, energy consumption and use of material resources in Austrian wine production. *Mitteilungen Klosterneuburg, 70*, 233–246.

Pomarici, E., Corsi, A., Mazzarino, S., & Sardone, R. (2021). The Italian wine sector: Evolution, structure, competitiveness and future challenges of an enduring leader. *Italian Economic Journal, 7*, 259–295. https://doi.org/10.1007/s40797-021-00144-5

Poore, J., & Nemecek, T. (2018). Reducing food's environmental impacts through producers and consumers. *Science, 360*, 987–992. https://doi.org/10.1126/science.aaq0216

Rana, R. L. (2020). La carbon footprint nella filiera agroalimentare. In F. Contò & M. F. A. Fiore (Eds.), *Ragionando di Sviluppo Locale: Una Lettera "Nuova" di Tematiche "Antiche"* (pp. 115–139).

Rana, R. L., Bux, C., & Lombardi, M. (2022). Trends in scientific literature on the environmental sustainability of the artichoke (Cynara cardunculus L. spp) supply chain. *British Food Journal, 125*(6), 2315–2332. https://doi.org/10.1108/BFJ-07-2022-0571

Rinaldi, S., Bonamente, E., Scrucca, F., Merico, M. C., Asdrubali, F., & Cotana, F. (2016). Water and carbon footprint of wine: Methodology review and application to a case study. *Sustainability, 8*, 621. https://doi.org/10.3390/su8070621

Ritchie, H., & Roser, M. (2021). *Environmental impacts of food production*. https://ourworldindata.org/environmental-impacts-of-food#citation. Accessed 30 Jan 2024.

Rosso, D., & Bolzonella, D. (2009). Carbon footprint of aerobic biological treatment of winery wastewater. *Water Science Technology, 60*(5), 1185–1189. https://doi.org/10.2166/wst.2009.556

Rugani, B., Vázquez-Rowe, I., Benedetto, G., & Benetto, E. (2013). A comprehensive review of carbon footprint analysis as an extended environmental indicator in the wine sector. *Journal of Cleaner Production, 54*, 61–77. https://doi.org/10.1016/j.jclepro.2013.04.036

Silva, A., Silva, V., Igrejas, G., Gaivão, I., Aires, A., Klibi, N., de Enes Dapkevicius, M. L., Valentão, P., Falco, V., & Poeta, P. (2021). Valorization of winemaking by-products as a Novel source of antibacterial properties: New strategies to fight antibiotic resistance. *Molecules, 26*(8), 2331. https://doi.org/10.3390/molecules26082331

Statista Market Insight. (2023). *Wine – Italy*. https://www.statista.com/outlook/cmo/alcoholic-drinks/wine/italy. Accessed 30 Jan 2024.

UNI. (2022). *Misurazione Della Circolarità—Metodi ed Indicatori per la Misurazione dei Processi Circolari Nelle Organizzazioni*. Ministero dell'Ambiente e della Sicurezza Energetica.

van Eck, N. J., & Waltman, L. (2018). *VOSviewer manual*. https://www.vosviewer.com/documentation/Manual_VOSviewer_1.6.9.pdf. Accessed 30 Jan 2024.

Vázquez-Rowe, I., Rugani, B., & Benetto, E. (2013). Tapping carbon footprint variations in the European wine sector. *Journal of Cleaner Production, 43*, 146–155. https://doi.org/10.1016/j.jclepro.2012.12.036

Vinci, G., Prencipe, S. A., Abbafati, A., & Filippi, M. (2022). Environmental impact assessment of an organic wine production in central Italy: Case study from Lazio. *Sustainability, 14*, 15483. https://doi.org/10.3390/su142215483

VOSviewer. (2024). *Visualizing scientific landscapes*. VOSviewer. https://www.vosviewer.com/. Accessed 30 Jan 2024.

Carbon Footprint of Fashion: Assessing and Addressing Carbon Emissions in Textile Production

Philomena Mayer and Derya Tama Birkocak

1 Introduction and State-of-the-Art

Climate change caused by humans has negatively impacted agricultural productivity, forced species from their natural habitats, and is likely to hasten the extinction of various species across the planet (Peters et al., 2015; IPCC, 2014); however, approximately 10% of the worldwide carbon emission is attributed to the textile and apparel industry, which leads to the second highest polluter of the environment (Muthukumarana et al., 2018). For some time now, there have been increasing efforts to make the textile value chain more sustainable by the assistance of various countries and non-governmental organizations. In order to be able to assess the process steps ecologically, they are divided into production and disposal processes. Production processes include fiber production, textile and clothing production, and textile care. These processes consume energy and raw materials, release emissions into the atmosphere, consume large amounts of water, and damage the soil and handle waste. The disposal processes in turn deal with the various ways of disposing of textiles, such as composting, waste incineration, and landfill. Added to this is the recycling and return of textile products to the production and consumption cycle (Piegsa, 2010).

P. Mayer
Faculty of Textile and Clothing Technology, Niederrhein University of Applied Sciences, Krefeld, Germany

Textile Engineering Department, Ege University, Bornova, Izmir, Türkiye
e-mail: philomena.mayer@stud.hn.de

D. Tama Birkocak (✉)
Textile Engineering Department, Ege University, Bornova, Izmir, Türkiye
e-mail: derya.tama@ege.edu.tr

S. S. Muthu (ed.), *Carbon Footprint Assessments*, Environmental Footprints and Eco-design of Products and Processes,
https://doi.org/10.1007/978-3-031-70262-4_5

Therefore, it is becoming increasingly important to evaluate textile products based on their CO_2 emissions and thus to analyze their impact on climate change (Peters et al., 2014). One essential technique for thinking about and evaluating the relative significance of various greenhouse gases is carbon footprinting. In addition, carbon footprinting provides information about the emissions of a product at all phases of its life cycle, for instance, all emissions are considered from the time a raw material is cultivated until the textile product is disposed of (Muthu, 2020).

Carbon footprinting is based on the calculation of the product environmental footprint (PEF) and both have their base from ISO14040. The ISO 14040 standard contains the planetary regulation for life cycle assessment (LCA). Moreover, a more straightforward method of calculating environmental impact is the carbon footprint calculation, as it only includes the effects on climate change (Peters et al., 2015).

Using the life cycle assessment approach, Baydar et al. (2015) looked at the environmental impact of cotton t-shirts in relation to organic cotton t-shirts dyed organically in connection with global warming. The findings indicate that the t-shirt made of organic cotton has less of an effect on global warming. The main impact is attributed to the consumer care phase, then the cultivation and the harvesting phase and the last one is the fabric processing phase. The analysis reveals that sustainable raw materials are important in every life phase of a cotton textile product. Additionally, sustainable handling of clothing by consumers can further reduce carbon emissions. Qifeng et al. (2023) researched the impact of environmental regulations on China‘s textile industry's reduction of CO_2 emissions. In order to distinguish between command-based, market-based, and public-based environmental laws on the intensity and total carbon emissions of China‘s textile industry from 2004 to 2019, the researchers in this work used both the threshold model and the panel model. The analysis shows that on a national level command-based regulations have an effect on reducing carbon emissions and carbon emission intensity on China's textile industry. Market-base regulations have a forced emission reduction on the intensity of carbon emission whereas public-based regulations had no significant results.

Büyükkamacı et al. (2023) used the life cycle assessment to look at how a hoodie affected the environment. For the analysis, data were given from a garment factory. The current environmental impact of the hoodie was calculated with the cradle-to-gate approach, after that the limits of the gate-to-gate-approach were used to see the improvements in the garment phase. During this study, four different scenarios were developed and the environmental outcome was compared with the initial situation. The scenario last scenario achieves the greatest results. This particular scenario included biodiesel as fuel in transportation issues and the use of renewable energy in general. This reduced 74.40% of global warming compared to the initial situation.

Using a life cycle evaluation, Moazzem et al. (2021) looked into the environmental effects of two distinct clothing items that were delivered to Australia. Climate change potential is used as impact category in this analysis. The results showed that with 56.52% the biggest part of climate change potential in cotton t-shirts is consumer use. 51.63% of this is primarily attributed to energy used for clothing maintenance. The production phase of PES jackets has the highest potential influence on climate change (65.52%), which is divided in the two major polluters: fiber production (21.95%) and electricity utilized during textile processing (13.58%). According to the study's interpretation, using organic cotton, recycled PES, and recycled cotton can all help to reduce the potential effect of climate change.

In their study, Gonzalez et al. (2023) investigated the environmental impact of conventional and organic cotton, jute, silk, polyester, and flax using the life cycle assessment technique. Using the cradle-to-gate system framework, the stages of farming, spinning, weaving, and dying each type of fiber were examined. The results showed that the most significant differences occur in the agriculture phase between all fibers. The steps that affected the environment the most were preparation of yarn and silk spinning, followed by conventional cotton and jute, in that order.

Moazzem et al. (2018) conducted an analysis of the environmental impact associated with various cotton, wool, and polyester clothing items in Australia. The life cycle assessment approach was used in the study, which concentrated on the environmental effect category of climate change. The findings revealed that for cotton and polyester garments, consumer usage emerged as the primary contributor to climate change, whereas in the case of wool garments, the production process had a greater impact on climate change than the consumer use phase. Energy consumption played a crucial role in overall environmental impact. Sensitivity analyses were performed to explore various parameters, including transportation methods (from air freight to sea freight), transport distances, and washing behavior. The study demonstrated that a reduction in washing machine energy consumption by up to 40% could lead to approximately 10% savings in CO_2e emissions compared to the base case. Consequently, the use of highly efficient washing machines, especially fully automatic ones, can contribute to energy conservation and a reduction in carbon emissions.

Zhang and Chen (2019) conducted a study investigating carbon emissions in the sewing process of clothing production. Assessment models for emissions were created, taking into account factors such as materials, accessories, sewing machines, and labor. The findings indicated that labor and sewing machines are the next biggest producers of emissions, behind materials, and accessories. In order to optimize the assembly line and lower emissions, a quick and sophisticated genetic algorithm was used in the development of a multi-objective optimization model. The model and algorithm's efficacy was tested using three different workstation layouts. The results demonstrated that the assembly line with a sequenced arrangement of processes was the most efficient and had the lowest emissions. Overall, the redesigned assembly lines produced less carbon dioxide than the traditional line.

Amicarelli et al. (2022) focus on identifying hotspots in the life cycle of textile products to develop measures aimed at reducing the overall impact. The study reveals that the production and use phases bear the majority of negative impacts, with the end-of-life phase making a comparatively smaller contribution. The distribution and consumption phases have received less attention, and emerging consumption patterns, such as sharing and rental platforms, necessitate additional data for comprehensive analysis. A study was conducted by Demirdelen et al. (2023) to look at the effects of synthetic yarns on the environment. The Monte Carlo analysis method was employed to address uncertainties. The study estimates that the carbon footprints of polypropylene and polyester are 6.42 t CO_2e and 13.40 t CO_2e per ton of PP and PES, respectively. Comparing PP yarn to PES yarn, the former has a larger carbon footprint and is not as environmentally friendly. These findings can be utilized to inform greenhouse gas emission reduction policies in Turkey and provide support to firms involved in the production of these raw materials.

In their 2020 study, Jaitiang et al. focused mostly on thermal and electrical energy while examining energy consumption and the Energy Intensity Index in Thai business. The study evaluated energy-saving techniques in the textile sector by analyzing the NDC Roadmap and the Alternative Energy Development Plan 2015–2036 (AEDP2015) in accordance with Thailand's Energy Efficiency Plan 2015–2036 (EEP2015). Waste heat recovery, photovoltaics, and the adoption of energy-efficient appliances were the four scenarios that were looked at. The results indicate that by 2036, energy consumption and the Energy Intensity Index in scenarios two and four could be reduced by approximately 16.63% compared to current levels. This translates to a reduction in energy consumption of 373.41 ktoe, surpassing the EEP2015 target. It is recommended that the textile industry implements additional measures in the realm of renewable energy to meet the set targets. Regarding environmental impact, greenhouse gas emissions correlate with energy consumption. Scenarios two, three, and four could potentially reduce greenhouse gas emissions by 16.67%, 1.15%, and 17.82%, respectively, by 2036 compared to the current situation. Scenario four has the potential to exceed the NDC roadmap target by 53.75%, while scenario two can achieve savings exceeding 41.11%.

After a thorough investigation of emissions, Huang et al. (2017) found that the Chinese textile industry's main source of emissions is coal use, which is followed by electricity consumption. Production volume, energy density, energy structure, and emission factors are the main determinants of emissions. An increase in production scale leads to higher emissions, while a reduction in energy intensity and optimization of the energy structure can contribute to emission reduction. Key energy-saving measures for the Chinese textile industry are identified, particularly in the spinning, weaving, and melting processes. A scenario analysis suggests that greenhouse gas emissions could be reduced by 34.3% by 2030 through technology optimization. Policy suggestions for China‘s textile industry to reduce emissions are proposed based on the findings of this study.

Muthukumarana et al. (2018) analyzed, in their study, a framework for quantifying the environmental impact of energy consumption in garment production in Sri Lanka. To investigate the effects of energy consumption at various phases of the garment life cycle, a life cycle assessment was carried out. The study revealed that garment production has the largest environmental footprint. A case study of a production facility in Sri Lanka was carried out to identify critical processes with the highest environmental footprint. The analysis also explored how alternative energy supply scenarios could reduce the environmental impact. It was found that 90–95% of the reduction in acidification of the soil, ozone depletion, and climate change could be achieved by using photovoltaics and biomass. These findings should assist decision-makers in developing sustainable strategies for the growth of the clothing industry.

Zhang et al. (2018) conducted a study focusing on achieving more sustainable production. They developed and applied an enhanced design method based on life cycle analysis to an existing polyester-cotton production system. The results of the life cycle analysis revealed that the production of polyester-cotton has the most significant environmental impact in the areas of marine eco-toxicity, greenhouse gas emissions, and abiotic depletion. Particularly harmful were the printing and dyeing phases of the production process. Six proposals for implementing Best Available Technology were suggested, all of which were feasible and yielded both environmental and economic benefits. By adopting these technologies, significant reductions in resource consumption, environmental impact, and potential benefits to human health could be achieved. It is recommended that future studies also consider the distribution and use phases of textiles, analyzing their impact on human health, fertilizers, land use, and pesticides throughout the textile life cycle.

This chapter examines the global apparel industry and presents information on the industry's carbon impact today. Subsequently, it discusses different methodologies for measuring carbon footprints in industrial settings. The chapter concludes by exploring policy, regulation, and industry standards that aim to reduce the carbon footprint within the apparel industry, suggesting pathways for sustainable practices and compliance.

2 The Global Apparel Industry and Carbon Emissions

2.1 Global Apparel Production and Consumption Patterns

In the modern global market, changing buyer preferences, particularly with the quick-turnaround fashion industry, caused them to respond to current styles very fast. Notably, the increasing demands and varied tastes of modern consumers towards quickly changing fashion trends have led to the accumulation of discarded textile materials. On a yearly basis, the trade of garments and textile materials show an increasing trend.

Figure 1 shows the global exports of manufactured goods in 2022, denominated in U.S. dollars (billions) (WTO, 2023). It provides a clear visual representation of different sectors, with clothing (576 billion USD) and textiles (339 billion USD) forming notable segments of the chart. This indicates that the garment industry is a significant contributor to the world's manufactured goods market, though it is not the largest.

Tables 1 and 2 provide detailed insights into the top exporters and importers of textiles and clothing in 2022, with values in billions of dollars and shares in world exports/imports, along with the annual percentage change. As presented in Table 1, it is clear that, while some countries are recovering quickly in 2022 (with positive growth percentages), others are still facing negative growth, which could be due to lingering effects of the pandemic or other economic challenges.

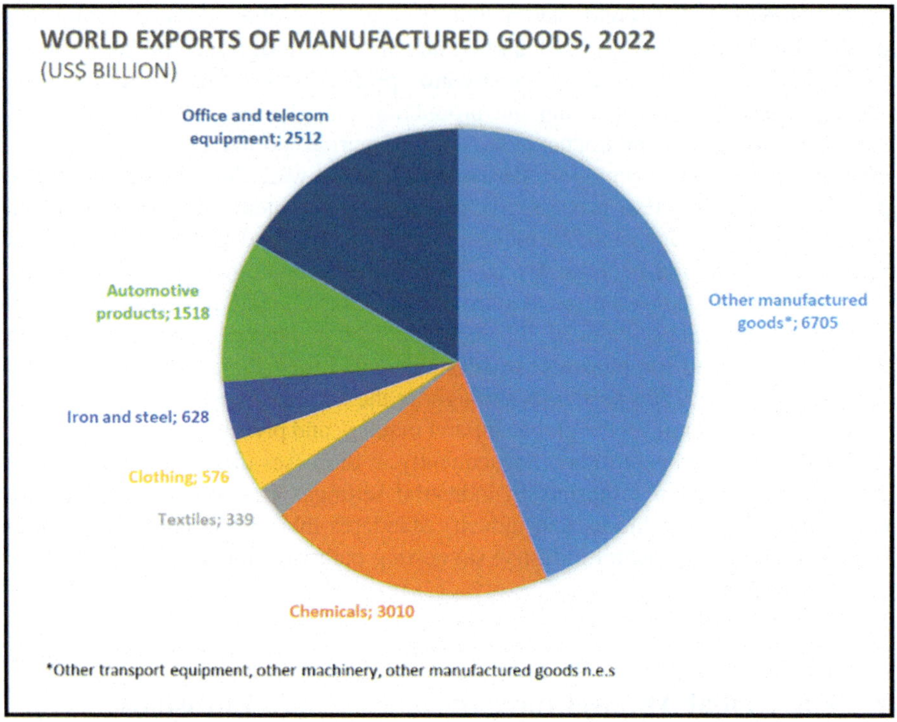

Fig. 1 World exports of manufactured goods. (*Source:* reproduced by authors from WTO, 2023)

China remains a dominant player in both textiles and clothing exports, despite the pandemic. The European Union also has a significant share in the global trade of these goods. The numbers also account for the COVID-19 pandemic's effects, which disrupted 2020. The fluctuations in annual percentage changes in subsequent years could be indicative of the market's attempt to recover. For instance, some countries show a decrease in exports in 2021, which could be due to various factors including the pandemic's disruptions in supply chains and consumer demand. Despite the pandemic, the clothing and textile markets seem to have shown resilience, with trade continuing robustly in 2022. This resilience might be attributed to the essential nature of these goods, as well as the shift towards e-commerce, which has allowed the textile and clothing industry to maintain a presence in consumer markets even during lockdowns and restrictions.

A near doubling in the global textile production happened from 2000 to 2015, yet the lifespan of clothing usage decreased by 36% within the same period. It's projected that by 2030, the global demand for clothing and footwear will rise by 63%, increasing from 62 million tonnes to 102 million tonnes. In the European Union, clothing constitutes the majority of textile consumption at 81% (European Commission, 2023a).

Table 1 Leading textile exporters and importers of 2022

Leading textile exporters and importers of 2022: rankings by billion dollars and market share									
	Value	Share in world exports/ imports				Annual percentage change			
	2022	2000	2005	2010	2022	2010–22	2020	2021	2022
Exporters									
China[a]	148	10.3	20.2	30.4	43.6	6	29	−6	2
European Union	71	33.4	32.5	25.3	21.1	1	−3	14	−3
Extra-EU exports	25	11.5	11.3	9	7.4	1	−9	16	0
India	19	3.6	4.1	5.1	5.7	4	−12	48	−13
Türkiye	15	2.4	3.5	3.5	4.3	4	−1	30	−4
United States of America	14	7	6.1	4.8	4.1	1	−15	15	5
Vietnam[b]	11	0.2	0.4	1.2	3.2	11	8	17	−6
Pakistan	9	2.9	3.5	3.1	2.6	1	−8	29	−5
Chinese Taipei	8	7.6	4.8	3.8	2.5	−1	−17	21	−2
Korea, Republic of	8	8.1	5.1	4.3	2.4	2	−15	12	−2
Japan	6	4.5	3.4	2.8	1.8	−1	−14	11	−4
Above 10	**309**	**80**	**83.5**	**84.5**	**91.1**	–	–	–	–
Importers									
European Union	78	29.7	29.2	23.8	20.7	2	30	−10	−1
Extra-EU exports	38	9.6	9.8	9.6	9.9	2	30	−10	−1
United States of America	39	9.7	10.5	8.7	10.3	4	44	−12	−1
Vietnam[b]	19	0.8	1.6	2.6	5	9	−10	21	1
Bangladesh[b]	15	0.8	1.1	1.7	3.9	10	−15	61	1
China[a]	12	7.8	7.2	6.6	3.2	−3	−10	15	−25
Japan	10	3	2.7	2.7	2.5	2	34	−21	4
Türkiye	8	1.3	2.1	2.4	2.1	2	−15	22	25
United Kingdom	8	4.4	3.5	2.7	2	1	54	−28	−3
Indonesia	8	0.8	0.4	1.6	2	5	−21	33	6
Mexico[a,c]	7	3.5	2.8	1.9	1.8	2	−14	18	8
Above 10	**203**	**61.7**	**61.1**	**54.7**	**53.6**	–	–	–	–

Source: Reproduced by authors from WTO, 2023
[a]Includes significant shipments though processing zones
[b]Secretariat estimates
[c]Imports are valued f.o.b.

Current patterns of textile consumption are unsustainable and must be reformed to facilitate a fair progression toward achieving climate neutrality. Recent studies have shown that the carbon footprint associated with fashion varies across different income groups. Moreover, synthetic fibers comprise over two-thirds (64%) of the total global fiber output. Starting in the mid-1990s, the manufacture of chemical fibers surpasses cotton production, and over the last 20 years, it has increased by more than 100%. As of 2020, synthetic fibers take part of about 62% of all fiber

Table 2 Leading clothing exporters and importers of 2022

Leading clothing exporters and importers of 2022: rankings by billion dollars and market share									
	Value	Share in world exports/ imports				Annual percentage change			
	2022	2000	2005	2010	2022	2010–22	2020	2021	2022
Exporters									
China[a]	182	18.2	26.6	31.7	31.7	3	−7	24	4
European Union	156	26.4	29.3	27.1	27.1	4	−8	20	4
Extra-EU exports	45	8.1	8.6	7.7	7.7	4	−13	14	4
Bangladesh[b]	45	2.6	2.5	7.9	7.9	10	−19	30	27
Vietnam[b]	35	0.9	1.7	6.1	6.1	11	−6	11	13
Türkiye	20	3.3	4.2	3.5	3.5	4	−6	22	6
India	18	3	3.1	3.1	3.1	4	−24	24	10
Indonesia	10	2.4	1.8	1.7	1.7	3	−12	24	8
Cambodia	9	0.5	0.8	1.6	1.6	10	−9	S	12
Pakistan	9	1.1	1.3	1.5	1.5	7	−3	37	5
United States of America	7	4.4	1.8	1.2	1.2	4	−19	27	17
Above 10	492	62.8	73.1	85.5	85.5	–	–	–	–
Importers									
European Union	215	32.7	37.4	35.5	35.5	4	−7	16	10
Extra-EU exports	111	16.4	19.4	18.4	18.4	3	−9	12	15
United States of America	116	33.1	28.7	19.2	19.2	3	−14	29	9
Japan	27	9.7	8.1	4.5	4.5	0	−12	1	2
United Kingdom	26	7.5	8.7	4.3	4.3	0	0	−12	12
Canada[c]	14	1.8	2.1	2.3	2.3	4	−7	15	15
Korea, Republic of	13	0.6	1	22	2.2	9	−12	17	15
China[a]	11	0.6	0.6	1.8	1.8	13	9	30	−12
Australia[c]	9	0.9	1.1	1.6	1.6	6	1	18	−3
Switzerland	9	1.6	1.6	1.4	1.4	4	4	11	−3
Hong Kong, China	2	–	–	–	–	−6	−31	9	−11
Retained imports[b]	2	9	–	0.3	0.3	–	6	43	44
Above 10	442	89.3	89.3	72.9	72.9	–	–	–	–

Source: Reproduced by authors from WTO, 2023
[a]Includes significant shipments though processing zones
[b]Secretariat estimates
[c]Imports are valued f.o.b.

production worldwide. Polyester was the most prevalent, claiming a 52% share of the market, while other synthetic materials such as polyamide made up 5%, and additional synthetics, including acrylic, elastane, and polypropylene, represented a combined total of 5.2%. (European Commission, 2022).

In 2022, it was globally produced approximately 116 million tonnes of fibers. Synthetic fibers form the bulk of this production, accounting for about 65%, with polyester being the dominant fiber at 54%. Plant fibers followed with 27% of the global fiber production, with cotton being the most significant at 22%. Manmade cellulosic fibers comprise 6.3%, with viscose being the most common at 5%. Animal fibers account for a smaller share at 1.7%, with wool from sheep being the most prevalent at 1% (Textile Exchange, 2022).

In the study conducted by Niinimäki et al. (2020), the environmental footprint of fiber production was analyzed, focusing on factors such as energy and freshwater consumption, as well as CO_2 emissions. The findings highlighted significant variations in these environmental impacts across different types of fibers. Natural fibers, which include cotton, non-cotton cellulosic (like viscose and lyocell), and hemp, generally use less energy for production compared to their synthetic counterparts such as polyester and polyamide. However, the cotton production requires more water, following by wool fiber. The most sustainable fiber among these fibers is hemp; however the total annual production volume of this fiber is 0.09 million tonnes (Niinimäki et al., 2020).

Considering the growth of the fiber and materials industry and its consequences, it's crucial to rethink the growth strategies (Wójcik-Karpacz et al., 2023). The textile industry accounts for about 10% of worldwide carbon emissions, ranking as the fifth largest in carbon footprint (Juanga-Labayen et al., 2022). To mitigate the environmental impact of these fibers is crucial, it has to be considered to slow down the consumption so the production. Reusing the goods has the priority; however recycling is another way to achieve sustainability in apparel industry. Various research indicates that only 20% of textile waste is either recycled or reused, while 80% ends up in landfills or is incinerated (Niinimäki et al., 2020; Pensupa et al., 2018; Wójcik-Karpacz et al., 2023).

Figure 2 presents the global fiber production data in million tonnes as well as the recycling amount of these fibers. Even though especially pre-consumer textile waste holds significant potential for recycling, only a minor portion of textile products is currently being recovered. The market shares of recycled fibers witnessed an increase, rising from 8.83% in 2020 to 9.52% in 2021, while the proportion of virgin fibers decreased from 91.17% to 90.48% during the same period. In 2021, polyester led the charge in the recycled fiber segment, accounting for about 15% of its total. This relatively higher amount of polyester was recycled from PET bottles, which is not originated as clothing product. The market share for other recycled fibers was considerably lower: recycled elastane accounted for around 3%, recycled polyamide for about 2%, and recycled polypropylene was at roughly 0.2% (Textile Exchange, 2022).

Fast fashion trends tend to generate demand for constantly changing wardrobes, leading to increase consumption, so to contribute to one of the greatest problems of textile industry, namely textile waste. However, encouraging consumers to wear the same clothes repeatedly can be crucial in decreasing the necessity for new clothing purchases, thereby mitigate the environmental impact caused by textile manufacturing processes.

Data from ThredUP presents the environmental impacts of purchasing used clothing over new (Fig. 3) (Babel et al., 2019). When an item is bought second-hand instead of new, it eliminates the CO_2 equivalent emissions that would have

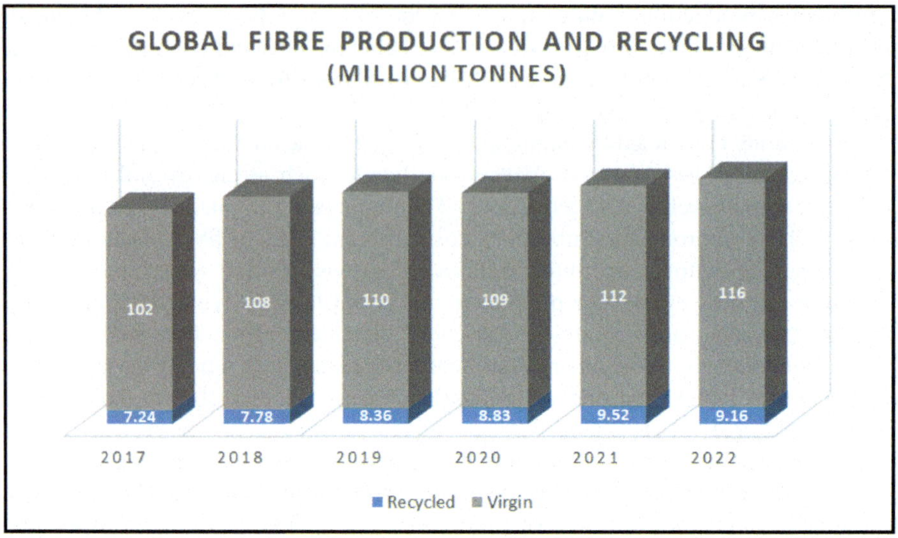

Fig. 2 Global fiber production and recycling (million tonnes). (*Source:* Reproduced by authors from Textile Exchange, 2022)

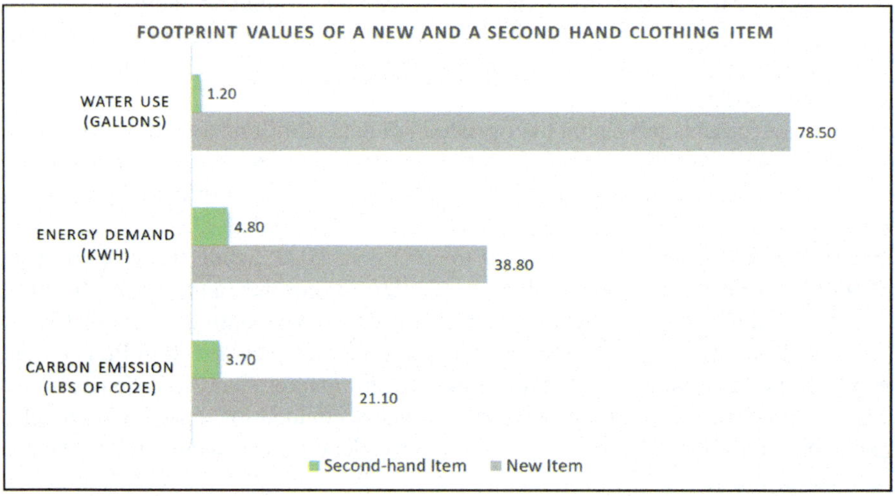

Fig. 3 The footprint values of a new garment versus a reused garment. (*Source:* reproduced by authors from ThreadUp, 2021)

been generated in the production of a new garment. Specifically, buying a used item as opposed to a new one can displace around 17.4 lbs of CO_2 emissions, effectively reducing its carbon footprint by 82%. This statistic highlights the significant potential of used clothing purchases in contributing to more sustainable fashion consumption and reducing the overall environmental impact of the apparel industry.

A typical garment is designed to withstand around 50 wash cycles before showing considerable wear. Despite this, many consumers tend to discard their clothing after using it for less than half of its intended lifespan. As stated above, the impact on global warming varies significantly between new and second-hand t-shirts worn. For a new t-shirt made of 100% cotton, the global warming potential (GWP) is 3.3783 kg CO_2e, rising to 3.7806 kg CO_2e for a polycotton blend, and 3.5546 kg CO_2e for a polyester one. In contrast, the GWP for a second-hand cotton t-shirt is only 0.0570 kg CO_2e, 0.0775 kg CO_2e for a reused polycotton t-shirt, and 0.0857 kg CO_2e for a recycled polyester t-shirt.

Regarding climate change, considering these data, the carbon footprint of a new polycotton t-shirt is roughly 47 times higher than a reused one. In the case of a new shirt, 75% of its climate change impact is attributed to the manufacturing of the t-shirt itself, while the production of polyester and cotton fibers accounts for only 24%. Similarly, the carbon footprint of a new polyester t-shirt is around 40 times greater than that of a reused one. Like with the polycotton, the major contributor to its impact on climate change is the manufacturing process of the t-shirt, with the production of the polyester fiber responsible for 27% of its total carbon footprint (Trzepacz et al., 2023).

2.2 *The Carbon Footprint of the Modern Apparel Industry*

The modern apparel industry is a significant contributor to the global carbon emissions, playing a critical role on the climate change. It is thought to contribute 10% of global carbon emissions, more than the combined emissions from international travel and maritime transportation. This statistic highlights the substantial environmental footprint of the fashion sector, encompassing the entire lifecycle from production to disposal of garments. According to the European Environment Agency, in 2020, the purchase of textiles in the EU resulted in approximately 270 kg of CO_2 emissions per person (European Commission, 2020).

Regarding the environmental impact of the modern apparel industry, with a focus on its carbon footprint, its influence on emission sources throughout the garment production cycle, from acquiring raw materials to disposing of them after their lives were presented in Fig. 4.

Figure 4 presents the data of the carbon emissions across the garment sector on a global scale. It identifies that the largest share of emissions, accounting for 36%, originates from the dyeing and finishing processes. Yarn preparation follows at 28%, fiber production contributes 15%, and fabric preparation adds another 12%. Surprisingly, the assembly stage is responsible for only 7% of emissions, while distribution and end-of-life disposal account for negligible amounts. Notably, this study excludes emissions from the use phase and transport to the customer due to the complexity in assessing consumer behavior variability.

Recent studies highlight various production stages contributing to the industry's greenhouse gas emissions. A study by Sharpe et al. (2022) examines carbon emissions across the garment sector, using life cycle assessment (LCA) and carbon

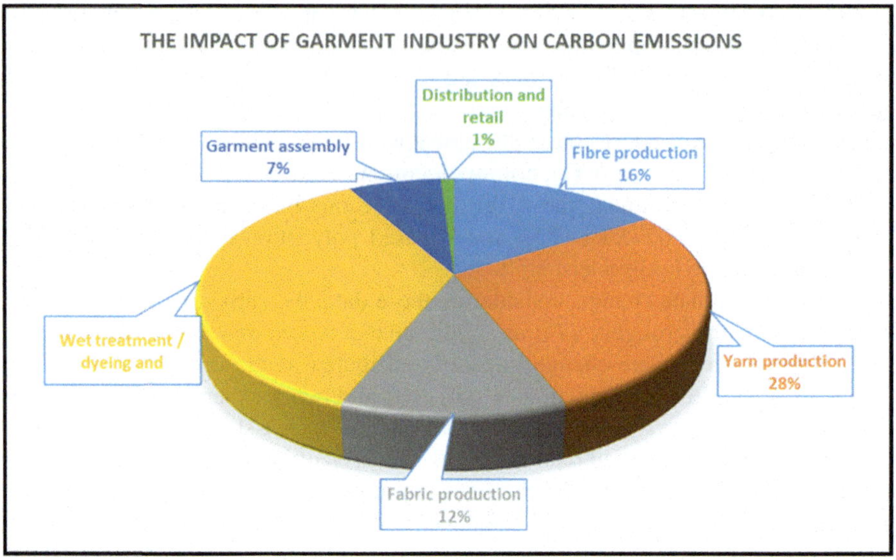

Fig. 4 The distribution of the clothing industry's effect on carbon emissions worldwide. (Source: reproduced by authors from ILO, 2021)

accounting methodologies. The paper highlights variations in carbon intensity across different stages of the supply chain and suggests prioritizing decarbonization activities accordingly (Sharpe et al., 2022). Additionally, the study by Muthukumarana et al. (2018) also emphasizes the major impact of the garment manufacturing phase to the environment, which includes processes like dyeing and finishing. This study further suggests that alternative energy sources can significantly reduce these impacts (Muthukumarana et al., 2018). After analyzing a men's shirt assembly line in China, Zhang and Chen (2019) discovered that the primary sources of carbon emissions in the sewing process of garments are accessories and fabrics. Their study emphasizes the importance of optimizing assembly line balance to reduce emissions (Zhang & Chen, 2019). Furthermore, Wang et al. (2015) investigated the carbon footprint of Chinese cotton shirts across their whole life cycle and found that the dyeing process was the main source of emissions, with significant emissions also occurring during the garment assembly stage. This study provides insights into the carbon footprint distribution across different stages of cotton shirt production (Wang et al., 2015). A study on the carbon emission reduction in the apparel material distribution process by Zhang et al. (2019) highlights the importance of optimizing material distribution paths to minimize carbon emissions in garment manufacturing (Zhang et al., 2019).

From 2019 to 2022, there has been a notable increase in greenhouse gas (GHG) emissions in the textile industry, specifically related to different materials used. Overall, the GHG emissions increased from 216 million tonnes of CO_2 equivalent (CO_2e) in 2019 to 226 million tonnes CO_2e in 2022. This increase in 2022 represents a year-on-year change of 4.6%, which contrasts with the targeted reduction of – 13.5%. Furthermore, the garment industry's ambitious target to reduce emissions by 45% has not progressed as expected (Statista, 2023). This difference highlights a substantial gap between the industry's environmental goals and the actual outcomes, underscoring the need for more effective strategies and actions to reduce its carbon footprint.

A significant portion of this increase can be attributed to the emissions from synthetic fibers. In 2019, the carbon emissions for synthetic fibers were reported at 123 million tonnes CO_2e. By 2022, it rose to 134 million tonnes CO_2e, indicating an important increase and highlighting the environmental impact of synthetic fiber production (Statista, 2023). In contrast, the emissions from plant-based, cellulosic, and animal materials remained relatively stable during this period. This stability may be caused that while these natural fibers have their environmental impacts, their contribution to the industry's overall GHG emissions has not significantly increased in the given timeframe.

3 Approaches to Measuring Carbon Footprint in Industrial Settings

3.1 Key Contributors to Carbon Emissions in Textile Production

As the textile industry is among the primary contributors to the global carbon emissions, its extensive value chain, from raw material production to the disposal of garments, has a profound environmental footprint, emphasizing the need for sustainable practices in this sector.

Throughout the entire process textile industry, the sector is one of the biggest pollutants of the environment due to the large amounts of chemical, energy, and water waste. In order to effectively mitigate this environmental impact, it is crucial to consider the entire textile process. This includes the initial phase of fiber production, progressing through to the finishing of the product, and extending to its end-of-life considerations (Aguiar et al., 2023). Thus, a comprehensive approach is essential for the industry to meaningfully reduce its ecological footprint and contribute to a more sustainable future. Figure 5 represents the major stages in garment production as well as the lifecycle of a garment. This whole process includes, material extraction and processing, yarn, fabric and garment production, consumer use, as well as the disposal. Distribution and shipping takes place in all these stages.

The cultivation and processing of natural fibers like cotton, as well as the production of synthetic fibers like polyester, are the processes require higher energy consumption. For natural fibers, the use of fertilizers, water, and land contributes to emissions, while synthetics rely heavily on fossil fuels. Dyeing, finishing, and yarn preparation are among the most emission-intensive stages as mentioned in Fig. 4.

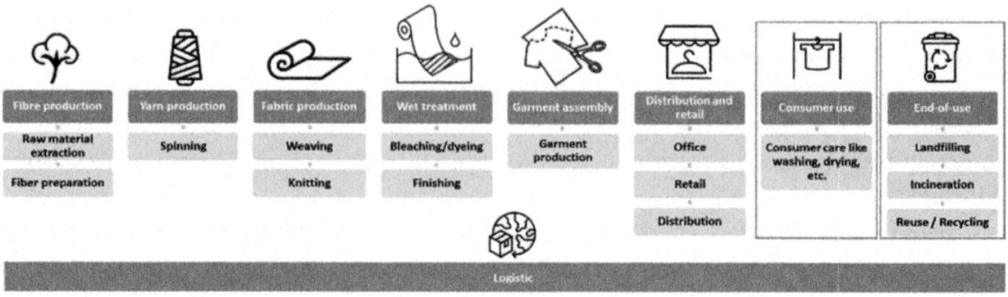

Fig. 5 The major stages of the garment industry. (*Source:* Produced by authors)

These processes often involve the consumption of large quantities of water and energy. The distribution and shipping of these materials and finished products across the world effects significantly the carbon footprint due to the reliance on fossil fuel-based transportation methods. At the end of their life, the disposal of clothing results in additional emissions, mostly due to be disposed in landfills or be incinerated. Certainly, the fast fashion business model makes this problem worse by hastening the life cycle of apparel, which increases waste.

One of the major contributors to environmental pollution is the consumer use phase, accounting for 56.52% of carbon emissions (Amicarelli et al., 2022; Moazzem et al., 2018, 2021). Specifically, within this phase, garment care energy usage constitutes 51.63% of the total carbon emissions (Moazzem et al., 2021). The subsequent significant contributor is the cultivation, harvesting, and fiber production phase, accounting for a total of 65.52% of carbon emissions. Specifically, the fiber production itself contributes 21.95% to the overall carbon emissions (Amicarelli et al., 2022; Baydar et al., 2015; Moazzem et al., 2021). This indicates that the initial stages of textile production, including the growing and processing of natural and synthetic fibers, are significant sources of emissions. Meanwhile, the carbon emissions attributed to electricity consumption during textile processing amount to 13.58% (Moazzem et al., 2021). Büyükkamacı et al. (2023) suggests that switching from conventional electricity to renewable energy sources could reduce these emissions by 74.40%, indicating a substantial opportunity for reducing the environmental impact in this area (Büyükkamacı et al., 2023). The fabric processing phase, particularly dyeing, is another critical area with a notable environmental impact. Studies by Baydar et al. (2015) and Gonzalez et al. (2023) highlight the significance of this stage, which involves extensive water and energy use, as well as chemical waste (Baydar et al., 2015; Gonzalez et al., 2023).

3.2 *Life Cycle Assessment*

Achieving sustainability necessitates adopting a life cycle perspective. The concept of a product's lifecycle is fundamental to the design of sustainable products. Life Cycle Assessment (LCA) serves as the methodology for recognizing, documenting, and controlling environmental effects throughout various stages of a product's lifecycle. This encompasses the entire process from obtaining raw materials for production to the final stages of production, distribution, and usage by consumers.

The LCA method has already been used in several studies with different materials (Baydar et al., 2015; Esteve-Turrillas & de La Guardia, 2017; Kazan et al., 2020; Manda et al., 2015), which is a comprehensive approach widely employed in environmental studies to assess the ecological footprint of various materials.

The LCA method encompasses the full life cycle of a textile product, beginning with the cultivation of raw materials, continuing through the production process, and ending with the product's disposal as waste. This method was described by LeVan (1995) and recognized by the United Nations Environment Programme and Industry (UNEP/IE) in 1996 (LeVan, 1995; UNEPIE, 1996).

As noted by the International Organization for Standardization (ISO) in 1997, the LCA method is a key tool for understanding and reducing the environmental impacts associated with products and services. Its systematic and comprehensive approach makes it an essential methodology for assessing the sustainability of textile products, guiding decision-making towards more environmentally friendly practices. This method includes several systematic procedures that consider the life cycle process and the defined research objective, the compilation of the inputs and outputs of each step of the process and quantify the associated resource consumption and emissions, and finally interpret the results in relation to the research objectives (Fig. 6) (ISO, 1997).

As presented in Fig. 7, inventory analysis is the initial step that involves creating an inventory of all energy, water, and other raw materials utilized in the manufacturing of a certain product or service as well as the environmental emissions generated during these processes.

While the impact assessment evaluates the environmental effects by taking these inputs and outputs into account to provide a more comprehensive understanding, the inventory analysis, which takes the most time, focuses on identifying the potential environmental impacts associated with the inputs (resources used) and outputs (emissions and waste) (Demirdelen et al., 2023). The final stage, interpretation, systematically evaluates and compare the results obtained from the previous three steps, and often, it necessitates revisiting some of these steps. It helps in

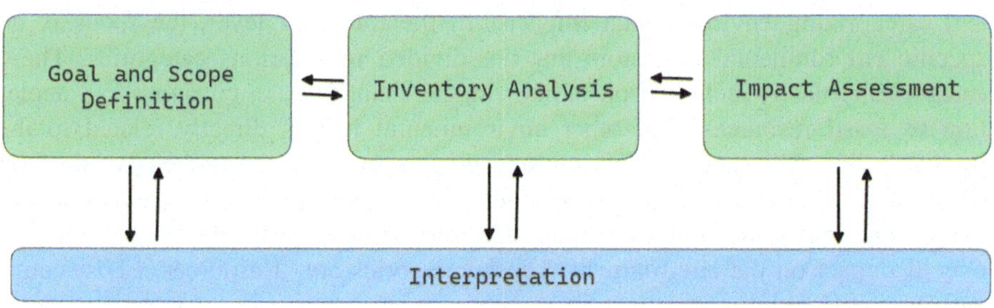

Fig. 6 The overview of life cycle assessment steps. (*Source:* Produced by authors)

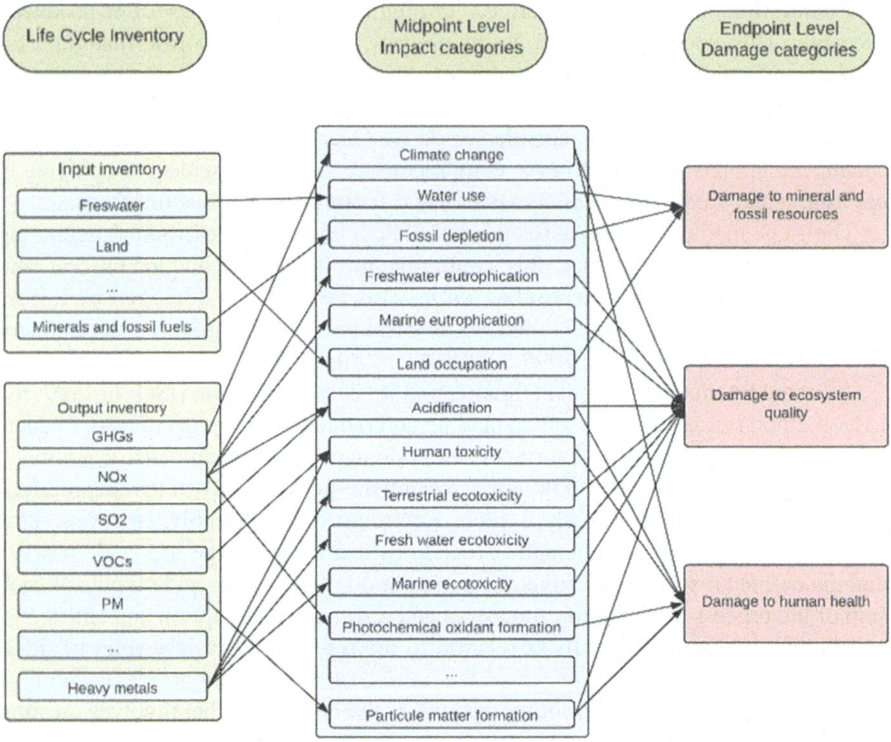

Fig. 7 LCA's environmental impact indicators. (*Source:* Produced by authors)

understanding the magnitude and significance of different environmental aspects associated with the product's life cycle (Nwodo & Anumba, 2020). In other words, the analyst is likely to learn more about the effects of choices made in previous processes at any point in the whole process. This new knowledge might highlight discrepancies that require attention or point up prospective avenues for product development (Peters et al., 2015).

In general, the LCA method is structured into two main phases, the midpoint level and the endpoint level (Fig. 7). Each phase offers a different perspective on assessing and categorizing environmental impacts. In the midpoint level, the focus is on specific environmental indicators that are divided into various categories. These categories typically include aspects like climate change, water consumption, depletion of fossil resources, and other environmental factors directly related to the product's life cycle stages. This phase's goal is to offer comprehensive data on every effect category. The endpoint level, also known as the damage assessment phase, aggregates the midpoint indicators into broader categories that reflect the overall impact on the environment. These categories are "Resources, Ecosystems, and Human Health."

The midpoint level indicators in LCA are acquired from key inventory substances and are standardized using a characterization factor. Carbon dioxide is identified as the most relevant inventory substance, and its characterization factor is the Global Warming Potential (GWP). Other greenhouse gases, such as nitrous oxide (N_2O) and methane (CH_4), can also be converted into CO_2 equivalents using GWP. This conversion enables the comparison of their impacts on an equivalent basis. Furthermore, several midpoint indicators can be further standardized to provide cumulative information, allowing for comparisons between different objects or scenarios. At the endpoint level, environmental impacts are quantified using the characterization factor specific to the endpoint (Goedkoop et al., 2009; Luo et al., 2021).

3.3 Data Collection and Analysis Techniques

The idea of a product's carbon footprint is generally understood to be the total amount of greenhouse gases released across a product's life cycle or during each of its distinct activities (Herva et al., 2012; Wang et al., 2015). Numerous carbon footprint assessment cases at the product level have emerged, particularly in the context of carbon labelling systems. However, these assessments have faced challenges due to varying system boundaries and limitations in data collection.

The assessment of carbon footprint primarily adheres to international standards and regulations. The British Standards Institute's PAS 2050, the International Standardization Organization's (ISO) ISO 14067 (Wang et al., 2015), the World Resources Institute/World Business Council for Sustainable Development (WRI/WBCSD) GHG protocol, and the European Commission's PEF are a few examples of these standards.

3.3.1 PAS 2050

The standard called PAS 2050 is used to calculate the carbon footprint of a specific service or product (Sinden, 2009; Wang et al., 2015; Zhang et al., 2020). It was developed by BSI British Standards, established with the support of the Carbon Trust and the Department for Environment, Food and Rural Affairs (Defra) (Bolwig & Gibbon, 2009a; BSI, 2008). The goal of PAS 2050 is to develop a straightforward and equivalent standard to enable comparison of the carbon footprints of different goods and services. The first established standard for a product's carbon footprint was released by the British Standards Institution on October 29, 2008. Since its introduction, PAS 2050 has gained international recognition, serving as a foundation upon which numerous other standards, derivatives, and enhancements have been developed, thereby significantly influencing the global approach to carbon footprint assessment (BSI, 2008; Sinden, 2009). Moreover, the 2011 revision of PAS 2050 incorporated insights from the development process of the Product Standard, reflecting a continuous improvement and learning approach in its methodologies for carbon footprint measurement (GHG Protocol, 2011).

An extensive lifespan analysis of a product, from raw material extraction to disposal, is part of the PAS 2050 approach for calculating carbon emissions. Initially, it's crucial to define the system boundaries. This step involves determining the scope of the analysis, including all process steps and materials to be included in the assessment. This frequently includes everything from raw material extraction to production, shipping, use, and product disposal. These emissions are then calculated. Based on the collected data, greenhouse gas emissions are calculated for each process step. This usually involves converting consumption data (e.g., energy consumption) into carbon dioxide equivalents to compare the various greenhouse gases on a consistent scale. The emissions from the various process steps are weighted depending on their relevance to the overall carbon footprint of the product. For example, emissions from the production of raw materials can make a larger contribution than emissions from transportation. The findings of the carbon footprint analysis are presented in a report that may be used to explain how the product affects the environment. This can help provide consumers and stakeholders with transparent information about the product's environmental impact. Businesses can gain a thorough understanding of the carbon emissions of their products and determine the best course of action to lessen their environmental impact by implementing the PAS 2050 methodology (Bolwig & Gibbon, 2009a; BSI, 2008; Sinden, 2009; Wang et al., 2015; Zhang et al., 2020).

Process map, limits and prioritization, data, computation, and uncertainty are the five steps of PAS 2050 for determining a product's carbon footprint (Fig. 8). As was already said, the process map creates an all-encompassing map of the product's lifespan, including all material, energy, and waste flows, from raw materials to disposal. In boundaries and prioritization step, it is determined the scope of the analysis and conducted an initial high-level footprint calculation to prioritize areas for carbon reduction efforts. Afterwards, the data process gathers detailed data on the amounts of materials, activities, and emission factors for each stage of the product's lifecycle. In the following step, calculation, the calculations are performed with the collected data to determine the product's total carbon footprint. Finally, in uncertainty step, the accuracy and precision of the footprint analysis are evaluated to understand the potential range of the results (BSI, 2008).

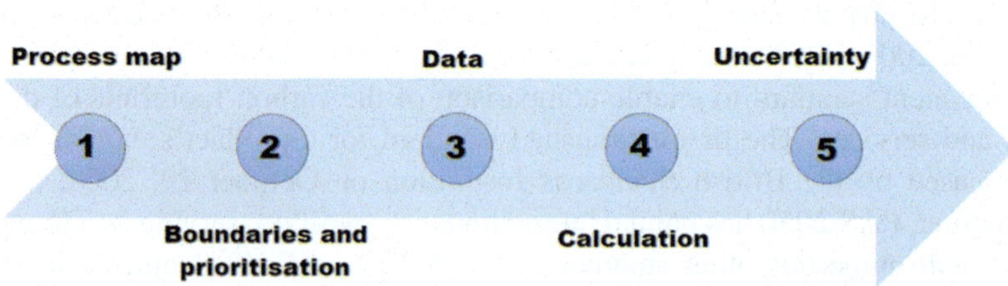

Fig. 8 The five stages of PAS 2050. (*Source:* Reproduced by authors from BSI, 2008)

3.3.2 ISO 14025

Another standard is ISO 14025 (2000) established to guide the communication of LCA results, assisting for a functional unit approach to ensure comparability rather than just mass or volume reporting. This standard emerged diverse LCA methodologies, aiming to standardize the process and enhance the credibility of LCA results as the practice became more widespread (Bolwig & Gibbon, 2009a, b).

Type III environmental declarations according to ISO 14025 (Environmental Product Declarations, EPD) provide information about the environmental properties of products (Peters et al., 2015). They explain the analyzed and documented product system since they are mostly based on life cycle assessments in accordance with ISO 14040ff. They also document results as life cycle inventory or impact indicators. EPD concept has gained traction, particularly in the construction industry. It is also a key component of the European Integrated Product Policy, serving as documentation for environmentally friendly procurement or for product developments with an environmental impact.

EPDs provide a consistent, comparable means of communicating the outcomes of a life cycle evaluation of a product in accordance with ISO 14040ff. The development and presentation of consistent and comparable data sets is outlined in ISO 14025. Three levels are described in the paper written by Schmincke and Grahl (2006), for which the standard lays out requirements: 1. Limitations for programs involving declarations and declaration verification. 2. Determining extra information needs for defined product categories in PCR documents (Product Category Rules) and detailed life cycle assessment rules in compliance with ISO 14040ff. 3. Creation and recording of EPD in compliance with PCR specifications (Schmincke & Grahl, 2006).

After several years of development-related standardization, the new ISO 14025 standard is a systematic concept for presenting life cycle assessment results in EPD. The involvement of life cycle assessment experts in ISO Type III declaration programs is essential.

3.3.3 ISO 14067

In 2008, ISO had started developing ISO/WD 14067, the "carbon footprint of products" standard. This standard, ISO 14067, aims to provide a universal framework for measuring, reporting, and verifying product-related greenhouse gas emissions. It was developed from the ISO 14000 series, particularly ISO 14040 and 14,044, and the environmental declaration methodology of ISO 14025 (Bolwig & Gibbon, 2009a). The ISO 14067 standard is divided into two segments: ISO 14067-1, which deals with the quantification aspect of a product's carbon footprint, and ISO 14067-2, which focuses on how to communicate that information effectively (Wang et al., 2015).

ISO 14067 sets out a series of principles for the assessment and communication of greenhouse gas emissions. This includes already established guidelines such as PAS 2050, as well as new principles such as coherence, preventing double counting, involvement, and equity. In order to guarantee the comparability of various resources, coherence concerns to the adoption of established assessment guidelines. Double counting should be avoided, especially when using emission-related factors for electricity. It is encouraged for interested parties to participate in communication programs. ISO 14067 only deals with climate change as an environmental impact area. One notable distinction between ISO 14067 and PAS 2050 is that the latter places more emphasis on sharing outcomes with external parties (Geraghty et al., 2018). The goal of the new guidelines is to facilitate open communication and outcome comparison. It is recommended that suppliers and subcontractors take part in the assessment study in order to receive a carbon label accreditation.

ISO 14067 is the mostly preferred as a second alternative to the LCA due to its more complex methodology (Hohenthal et al., 2019). The system boundaries are of great importance in LCA studies as they define the phases of the life cycle that should be taken into account. The sustainability assessment of, e.g., buildings should therefore include greenhouse gas emissions in all phases. The system boundaries can be set in a number of ways, including "gate-to-gate," "cradle-to-grave," "partial CFP (carbon footprint of products)," and "cradle-to-gate" (Hohenthal et al., 2019; Šerkinić et al., 2020; Wu et al., 2015). Cradle-to-grave encompasses the entire lifecycle of a product from resource extraction to disposal. Cradle-to-gate evaluations consider the effects of a product only, omitting customer transit and post-purchase use, from the point of extraction to the factory gate. Cradle-to-cradle, a form of cradle-to-grave, focuses on recycling the product at its end-of-life, ensuring materials are reused, promoting a closed-loop system (Šerkinić et al., 2020). The border of the cradle-to-grave system, which takes into account a product's whole life cycle, is advised by ISO 14067. Nevertheless, the cradle-to-gate quantification can also be applied if the end-of-life phases cannot be sufficiently represented. When information on the extraction and manufacturing of raw materials is unavailable, issues can be resolved by using gate-to-gate and partial CFP as system limits. A sensitivity analysis is another suggestion made by ISO 14067 to improve the system boundaries. This stops data manipulation when choosing the bounds of the system. To verify and bolster conclusions, sensitivity assessments have to be carried out and recorded (Wu et al., 2015). Under ISO 14067, a product or service can serve as the functional unit.

3.3.4 GHG Protocol

The World Resources Institute (WRI) and the World Business Council for Sustainable Development (WBCSD) established the GHG Protocol, also known as the Greenhouse Gas Protocol, which is a comprehensive framework for measuring and reporting greenhouse gas emissions (Bolwig & Gibbon, 2009a; GHG Protocol, 2011; Samara et al., 2022). It is a globally recognized tool to support organizations in recording, managing, and disclosing their greenhouse gas emissions. The GHG Protocol provides detailed guidance and standards for recording and reporting greenhouse gas emissions in these three scopes. It also includes specific guidance for different sectors such as industry, transport, and agriculture (GHG Protocol, 2011).

The GHG Protocol consists of three main categories, known as scopes explained below.

Scope 1: Direct emissions resulting from the burning of fuels and from operations that the corporation owns or controls, such as industrial processes or company cars.

Scope 2: Indirect emissions brought about by the production of energy that has been acquired, such as heat or electricity produced by a third party and provided to the business.

Scope 3: All additional indirect emissions related to business operations that the corporation does not directly own or control. This covers things like emissions from product manufacturing, using sold goods and services, and disposing of trash (GHG Protocol, 2011).

In other words, Scope 1 is the direct carbon, whereas the others are the indirect ones. Scope 2 and Scope 3 are related with the greenhouse gases emitted across a company's value chain, presents a complex challenge. Due to this complexity, Scope 2 and Scope 3 emissions are continuously reviewed to enhance the consistency, transparency, and accuracy of reporting these emissions (Diniz et al., 2021).

The main objectives of the GHG Protocol are to standardize and harmonize greenhouse gas reporting to improve comparability between organizations, increase transparency and support the development of effective strategies to reduce greenhouse gas emissions (Samara et al., 2022). It is used by governments, companies, non-governmental organizations, and other organizations around the world to understand their climate impact and take action to reduce emissions.

3.3.5 The Product Environmental Footprint (PEF)

First published in 2013, the Product Environmental Footprint (PEF) framework was created by the European Union (EU) to evaluate the environmental impact of products across their whole life cycle (Lehmann et al., 2016; Pedersen & Remmen, 2022). It is designed to help companies collect, assess, and communicate environmental information about their products in order to ultimately make more environmentally conscious decisions and improve the environmental performance of their products (Bach et al., 2018).

Twenty Product Environmental Footprint Category Rules (PEFCR) were created in April 2018. PEFCRs provide specific guidelines for assessing product environmental information within a category, aiming for consistency and enabling comparisons in a particular category (Pedersen & Remmen, 2022). They are designed for the PEF approach and work in a manner akin to the ISO 14025-based product category rules (PCRs), providing comprehensive guidelines for assessing a product's life cycle impact (Lehmann et al., 2016). The primary goals are to standardize environmental data calculation for products in the same category and facilitate comparative assertions when feasible. In order to further develop and apply these standards, the PEF program has entered a transition phase that is anticipated to terminate by the end of 2021. This phase began in 2018.

The PEF's foundation is the LCA approach, which permits a methodical evaluation of a product's environmental impact at every stage of its life cycle (He et al., 2019; Wu & Su, 2021). This covers the procurement of raw materials, production, shipping, usage, upkeep, and product disposal.

The PEF offers companies and consumers a transparent and comparable representation of the environmental performance of products by providing a common methodology for the assessment and communication of environmental indicators. This can help companies make greener production and design decisions, enable consumers to make informed purchasing choices, and ultimately support the transition to a more sustainable economy.

The PEF is increasingly being used by companies in the EU as the requirements for environmental labelling of products and the demand for environmentally friendly products increase. It is crucial in raising consumer knowledge of environmental issues and encouraging the application of environmental protection policies at the product level.

3.4 IPCC Emission Factor Database

Compiling emission factors from all over the world, the Intergovernmental Panel on Climate Change (IPCC) created the IPCC Emission Factors Database (EFDB). This database's objective is to offer a single, central source of emission variables for use in calculating greenhouse gas emissions from a variety of sources (IPCC, 2024).

Emission factors are numerical values that indicate the amount of greenhouse gases released from a particular activity or process. They serve as a key component in the calculation of greenhouse gas emissions from various sources such as energy production, industry, agriculture, transport, and waste management.

The EFDB serves as a key resource, offering access to well-documented emission factors and parameters. In addition, it acts as a communication platform for sharing and discussing new research and data (IPCC, 2020).

IPCC Emission Factors Database contains a variety of emission factors for different greenhouse gases, including carbon dioxide (CO_2), nitrous oxide (N_2O), methane (CH_4), and fluorinated gases. These factors are specified for different activities and sectors to enable a precise estimation of greenhouse gas emissions (IPCC, 2019).

Furthermore, the database is continuously updated and expanded to incorporate new scientific findings and data and to ensure the accuracy and reliability of the emission factors it contains. This means that it is a dynamic resource that is constantly being improved to reflect the current state of research and development.

The IPCC Emission Factors Database serves as an essential resource for governments, researchers, businesses, and other stakeholders involved in measuring, monitoring and reducing greenhouse gas emissions. It supports the development of policies to combat climate change, the conduct of inventory studies to assess greenhouse gas emissions and the implementation of emission reduction strategies at global, national, and local levels.

4 Policy, Regulation, and Industry Standards

The carbon footprint of the modern apparel industry is a multifaceted issue requiring concerted efforts from manufacturers, consumers, and policymakers. Despite the progress achieved, there are still significant challenges remained in overcoming a substantial reduction in emissions. Understanding and addressing these challenges is crucial for the industry to move towards a more sustainable future, aligning with global efforts to mitigate climate change.

4.1 Governmental and International Regulations Affecting Carbon Emissions

On June 28, 2021, the European Parliament adopted the new EU "Green Deal"climate law, setting ambitious targets to reduce emissions by 55% by 2030 under the "Fit for 55" initiative, with the ultimate goal of achieving climate neutrality by 2050. It was updated in the REPowerEU plan, a comprehensive strategy aimed at varying the energy sources and accelerating the switch to renewable energy. It is projected that the overall legislative package, impacted by these updated goals, will result in a 57% decrease in the EU's net greenhouse gas emissions by 2030 (European Commission, 2023b). The EU is leading the worldwide effort to combat climate change and strives to achieve long-term zero emissions. The targets are intended to be more effectively integrated into legislation, promising benefits like cleaner air and water, reduced energy costs, less waste, and healthier food options.

Emissions trading remains a key tool in achieving these goals, with the expectation that the cost of CO_2 in trading emissions allowances will increase. This rise will particularly impact companies in energy-intensive sectors. Due to international competition, there is a concern about carbon leakage, where companies might move their production to non-EU countries to avoid high emission costs.

In recent years, international cooperation on climate policy has been notably unstable. In 2019, the European Commission unveiled its ambitious Green Deal, a comprehensive agenda featuring over fifty measures intended to integrate climate and environmental policy with economic objectives. This agenda was published in 2020 (European Commission, 2024) and later integrated into the EU's strategies for economic recovery following the pandemic. A key aspect of the implementation plans includes a CO_2 levy on imported goods that are considered to be particularly harmful to the climate (Dröger & Schrader, 2021).

The European Emissions Trading Scheme (EU ETS) stands as an important instrument in achieving the EU's climate goals across a significant portion of its economy. Under this scheme, the total allowable annual CO_2 emissions are controlled through a progressively decrease on emission certificates. This mechanism is expected to result in an increase in the price of CO_2, thereby promoting reductions in emissions. The EU ETS represents a critical step in the EU's commitment to environmental sustainability, linking economic growth with ecological responsibility.

To mitigate carbon leakage, the EU has traditionally offered free emission allowances to companies. However, as part of the legislative implementation of the new climate goals, the European Commission proposed a novel instrument: The Carbon Border Adjustment Mechanism (CBAM). This mechanism aims to impose a levy on imports into the EU based on their CO_2 content, aligning with the EU's internal carbon pricing. The CBAM proposal, presented in draft form in July 2021 and planned for introduction in 2023, has gained significant international attention. This development is a significant step in the EU's strategy to address climate change while maintaining fair competition and preventing carbon leakage (SWP, 2021). The Carbon Border Adjustment Mechanism is designed to eliminate the potential for carbon leakage stemming from escalating CO_2 prices within the EU. Carbon leakage refers to the phenomenon where companies move their production and investments to countries outside the EU. This relocation might reduce emissions within the EU on paper, but it doesn't lead to a decrease in global emissions. To mitigate this risk, the EU Emissions Trading Directive previously allowed for the free allocation of emission allowances, particularly for companies in international competition with highly CO_2-intensive production processes, such as those in energy-intensive industries (EU-TAXUD, 2023; SWP, 2021). Essentially, the CBAM is a strategic move to ensure that the EU's ambitious climate goals contribute to a global reduction in emissions, rather than just shifting the emissions geographically.

As shown in Fig. 9, the European Union has set up a system where CBAM certificates are sold for the same price as those under the Emissions Trading System (ETS) program in order to implement the carbon border adjustment mechanism. These certificates are necessary for customs declarants to legally import specified groups of goods into the EU. Emission calculations are not only based on self-declarations; instead, they must be validated by an independent third party that is accredited by the EU's CBAM agency. This requirement for third-party verification adds a layer of credibility and objectivity to the process. If the emissions associated with imported goods are not verified, the EU will use higher levels of default regional emissions for attributing carbon costs. This approach will be encouraging for the accurate reporting and verification. The CBAM implementation includes a 3-year transition period, during which a simplified mechanism will be in place. This transitional phase allows for adjustments and preparatory measures before the full application of the CBAM (GoGlobal, 2022).

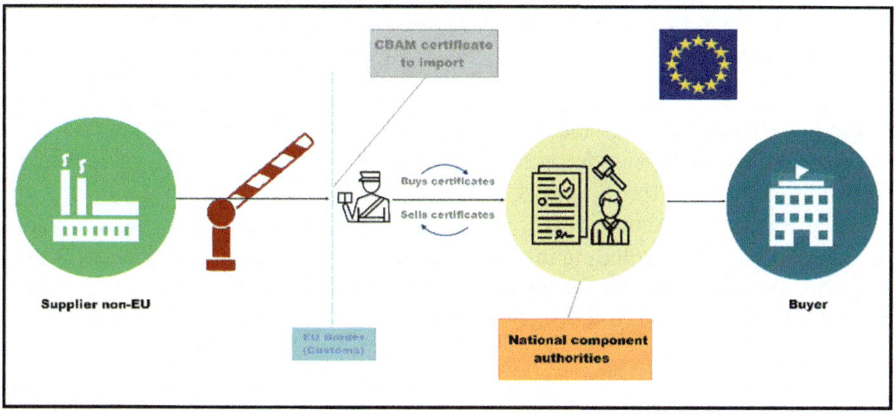

Fig. 9 European Parliament's approach to carbon border adjustment mechanism. (*Source:* Produced by authors)

The ETS would be used to determine prices for CO_2 emissions produced during the production of imported commodities, according to a specific proposal made by the European Commission. A "virtual" ETS would be used, whereby importers would report the CO_2 content of the goods to the competent authority and pay the current price incurred in the ongoing auctioning of CO_2 certificates for EU companies. However, they would not participate in emissions trading. Goods produced outside the EU—even if only from certain sectors—would thus be subject to a CO_2 price premium that corresponds to that of EU producers (EU-ETS, 2024). This proposal aims to ensure that imported goods carry a similar carbon cost burden as those produced within the EU, generating a more equitable competitive environment.

Carbon leakage triggered by climate policy measures is closely linked to the issue of economic competitiveness. In international climate policy, the different climate protection commitments set out in the Kyoto Protocol in 1997 gave rise to concerns about carbon leakage for the first time. Industrialized and transition countries (countries of the former Soviet Union) were obliged to protect the climate, while developing countries were not. It was feared that climate policy measures in the major economies, especially CO_2 prices, would cause energy prices on the global market to fall due to declining demand—and ultimately boost consumption. It was also assumed that companies would relocate their production and investments in response to rising climate policy costs to areas where there is little or no climate protection. The goal of global climate protection would be undermined by the resulting relocation of emissions.

The Paris Climate Agreement is also accompanied by various (voluntary) commitments to climate protection. The signatory states report so-called national contributions, which, however, vary greatly depending on the time frame, level of ambition and climate policy instruments (Tracker, 2021). CO_2 prices are now levied in 46 countries (WorldBank, 2024) but not at the same level. The risk of carbon leakage therefore still exists (Dröger & Schrader, 2021; European Commission, 2024).

4.2 The Role of Industry Standards in Reducing Carbon Footprint

Today's textile industry is facing with significant environmental and social challenges. The industry is under considerable pressure due to a range of legislative measures, including the Supply Chain Act, the European Green Deal, and the Circular Economy Action Plan. The European Commission has set ambitious environmental goals, including the objective for Europe to achieve climate neutrality by 2050. Beyond legislative demands, there is increasing pressure from consumers as well. The consumers interest on sustainability creates demand on sustainable products.

The expectations for CO_2 reporting are becoming more compelling. Digitalization emerges as a key solution for efficient and accurate reporting. It entails developing a digital version of a physical object that spans its whole life. This digital representation is essential for calculating a product's carbon footprint, as discussed in Sect. 3.3. Specifically, in Sect. 3.3.1 to 3.3.5, various standards and protocols such as PAS 2050, ISO 14067, the GHG Protocol, and the Product Environmental Footprint (PEF) are addressed.

Through digitalization, the textile industry can more effectively track, analyze, and provide information about the effects its product's manufacturing and distribution have on the environment, thereby meeting both legislative and consumer demands for greater sustainability and transparency. For numerous businesses, aligning with these compelling greenhouse gas reduction targets presents a huge challenge. Tools like Microsoft Excel and other spreadsheet programs are often inadequate for managing the complexity of these tasks. These conventional tools lack the necessary sophistication and integration capabilities required to accurately track and analyze the extensive data involved in comprehensive greenhouse gas emissions management. Moreover, the practice of exchanging Excel sheets globally and the tedious task of consolidating this data highlight the inefficiency of such methods. In contrast, specialized cloud solutions offer a significant improvement.

Professional carbon footprint calculators and platforms present an applicable solution to the complex task of accurately calculating carbon emissions. These tools offer comprehensive networks and emission factors, along with a calculation logic that adheres to established standards and is regularly updated. Despite the availability of these advanced tools, many companies continue to rely on traditional methods for data collection, with Excel remaining a common tool. These platforms support controlled workflow processes for collecting, validating, and calculating emissions data, utilizing certified methodologies and the latest emission factors from a single, reliable source. This modern approach streamlines the process, ensuring accuracy and efficiency in managing and reducing carbon emissions.

Large companies in the textile sector are at the forefront of adopting digital solutions to ensure their reporting practices comply with legal standards. Utilizing tools like the Emissions Calculator not only establishes a solid foundation for data but also streamlines data validation and calculation processes, significantly reducing the workload for those in charge. The move towards automated reporting, potentially on a weekly basis, further alleviates the burden on sustainability officers by making these processes more efficient.

Adopting a systematic approach is key to transitioning towards sustainable operations. The initial step involves a thorough examination of the current status, often revealing gaps in data for companies new to this level of detail. However, these gaps are not a major setback. Digital solutions available today can easily bridge the gap in internal company data. In the cases where primary data isn't accessible, tools like the Emissions Calculator can leverage secondary data from reputable sources to make accurate estimations. For instance, it can estimate the amount of fertilizer and energy a supplier in India might require for cotton cultivation-insights that many companies previously could not access. Following automatic processing, this data is organized into a report that complies with the Greenhouse Gas Protocol, assisting businesses in adhering to numerous current standards. Additionally, as a standard, the database provides great chances for comparisons within and between industries (United, 2024).

The textile industry's proactive approach in recognizing its contribution to CO_2 emissions and the increasing availability of tools and technologies to analyze the entire value chain indicates a growing commitment to identifying and seizing opportunities for improvement. This trend underscores a collective move towards enhanced sustainability and accountability across the sector.

5 Future Trend

The textile industry's engagement with carbon footprinting is increasing, emphasizing the need for carbon-related claims to be supported by accurate calculations to avoid "greenwashing." Transparency and understanding of supply chains are becoming more important, driven by several initiatives. There's a trend towards more specific standardization in carbon footprinting for textiles, highlighting the industry's move towards more comprehensive sustainability and carbon labelling schemes.

The "Fashion Industry Charter for Climate Action," published by EU in 2021, aligns with the Paris Agreement's goals, emphasizing the industry's role in achieving net-zero GHG emissions by 2050 (European Commission, 2021). It outlines commitments like setting science-based emission reduction targets, enhancing energy efficiency, sourcing renewable electricity, and advocating for climate-friendly policies. The charter also focuses on material sourcing, supplier engagement, coal phase-out, and logistics optimization towards zero emissions. Additionally, it promotes dialogue with financial institutions and aligns industry communication with sustainable pathways, supporting collective efforts to mitigate climate impact.

Addressing carbon dioxide removal (CDR) governance parallels emission reduction efforts, requiring similar policy tools like research funding and carbon pricing. Integrating CDR into climate strategies should utilize existing frameworks and address land-based mitigation governance issues, focusing on local conditions. Establishing political priorities for CDR, developing robust measurement and verification systems, enhancing synergies for co-benefits across environmental and social objectives, and accelerating innovation through demonstration projects are essential (Rogelj et al., 2023). These actions, supported by transparent reporting and international cooperation, aim to fulfil net-zero commitments while ensuring CDR's complementary role to emission reductions.

6 Conclusions

The textile industry is at a critical point, with its significant contribution to global carbon emissions underlining the urgent need for sustainable transformation. The journey towards reducing the carbon footprint involves embracing advanced digital solutions for efficient data management, stringent adherence to policy regulations, and a collective effort from all stakeholders towards eco-friendly practices. As highlighted, adopting life cycle assessments, leveraging digital tools for accurate carbon footprint calculations, and aligning with international standards like PAS 2050 and the Greenhouse Gas Protocol are crucial steps. Supported by technology and global norms, these measures aim for a sustainable future, highlighting the need for ongoing innovation, commitment to regulations, and engaging consumers in environmental objectives.

Transitioning to low emission energy not only poses challenges but also offers significant opportunities for low- and middle-income countries, including social and environmental co-benefits (Rao et al., 2023). These opportunities often align with Sustainable Development Goals, showing more synergies than trade-offs. For instance, sustainable transport can improve air quality, health, and gender equality, while clean cooking options can enhance health and reduce deforestation. Achieving these benefits requires policies that maximize co-benefits. Additionally, structural economic changes, such as demand-side efficiency and advancement in the clean energy sector, can foster sustainable growth tailored to national contexts.

References

Aguiar, G. J. A., Almeida, L. R., Fernandes, B. S., Gavazza, S., Silva, G. L., & Machado Santos, S. (2023). Use of life cycle assessment as a tool to evaluate the environmental impacts of textile effluents: A systematic review. *Environmental Science and Pollution Research, 30*, 1–16.

Amicarelli, V., Bux, C., Spinelli, M. P., & Lagioia, G. (2022). Life cycle assessment to tackle the take-make-waste paradigm in the textiles production. *Waste Management, 151*, 10–27.

Babel, N., Toma, T., & Sivanandan, A. (2019). *Comparative Life Cycle Assessment (LCA) of second-hand clothing vs new clothing*. Retrieved from https://cf-assets-tup.thredup.com/about/pwa/thredUP-Clothing-Lifecycle-Study.pdf. Accessed Jan 2024.

Bach, V., Lehmann, A., Görmer, M., & Finkbeiner, M. (2018). Product environmental footprint (PEF) pilot phase—Comparability over flexibility? *Sustainability, 10*(8), 2898.

Baydar, G., Ciliz, N., & Mammadov, A. (2015). Life cycle assessment of cotton textile products in Turkey. *Resources, Conservation and Recycling, 104*, 213–223.

Bolwig, S., & Gibbon, P. (2009a). *Emerging product carbon footprint standards and schemes and their possible trade impacts*. Danmarks Tekniske Universitet, Risø Nationallaboratoriet for Bæredygtig Energi.

Bolwig, S., & Gibbon, P. (2009b). *Global forum on trade and climate change*. OECD Conference Centre. Retrieved from https://www.oecd.org/environment/envtrade/42886201.pdf. Accessed Jan 2024.

BSI. (2008). *Guide to PAS 2050 How to assess the carbon footprint of goods and services*. Retrieved from https://www.co2-sachverstaendiger.de/pdf/BSI%20Guide%20to%20PAS2050.pdf. Accessed Jan 2024.

Büyükkamacı, N., Bahar Erdem, S., & Birol, Ş. (2023). Investigation the environmental benefits of cleaner production practices in the apparel industry. *Pamukkale University Journal of Engineering Sciences, 29*(6), 628–635.

Demirdelen, T., Aksu, İ. Ö., Yilmaz, K., Koç, D. D., Arikan, M., & Şener, A. (2023). Investigation of the carbon footprint of the textile industry: PES-and PP-based products with Monte Carlo uncertainty analysis. *Sustainability, 15*(19), 14237.

Diniz, E. H., Yamaguchi, J. A., dos Santos, T. R., de Carvalho, A. P., Alego, A. S., & Carvalho, M. (2021). Greening inventories: Blockchain to improve the GHG Protocol program in scope 2. *Journal of Cleaner Production, 291*, 125900.

Dröger, S., & Schrader, T. S. (2021). *Zurück in die Zukunft? Die internationale Klimapolitik 2021. Neue Konstellationen für die europäische Klimadiplomatie*. Stiftung Wissenschaft und Politik.

Esteve-Turrillas, F. A., & de La Guardia, M. (2017). Environmental impact of recover cotton in textile industry. *Resources, Conservation and Recycling, 116*, 107–115.

EU-ETS. (2024). *Scope of the EU emissions trading system*. Retrieved from https://climate.ec.europa.eu/eu-action/eu-emissions-trading-system-eu-ets/scope-eu-emissions-trading-system_en. Accessed Jan 2024.

European Commission. (2020). *The impact of textile production and waste on the environment (infographics)*. Retrieved from https://www.europarl.europa.eu/news/en/headlines/society/20201208STO93327/the-impact-of-textile-production-and-waste-on-the-environment-infographics. Accessed Jan 2024.

European Commission. (2021). *Fashion industry charter for climate action*. Retrieved from https://unfccc.int/sites/default/files/resource/Fashion%20Industry%20Carter%20for%20Climate%20Action_2021.pdf. Accessed Jan 2024.

European Commission. (2022). *Textiles and the environment*. Retrieved from https://www.europarl.europa.eu/RegData/etudes/BRIE/2022/729405/EPRS_BRI(2022)729405_EN.pdf. Accessed Dec 2023.

European Commission. (2023a). REPORT on an EU strategy for sustainable and circular textiles. Retrieved from https://www.europarl.europa.eu/doceo/document/A-9-2023-0176_EN.pdf. Accessed Dec 2023.

European Commission. (2023b). *Commission welcomes completion of key 'Fit for 55' legislation, putting EU on track to exceed 2030 targets*. Retrieved from https://ec.europa.eu/commission/presscorner/detail/en/ip_23_4754. Accessed Jan 2024.

European Commission. (2024). *The European Green Deal, Striving to be the first climate-neutral continent*. Retrieved from https://commission.europa.eu/strategy-and-policy/priorities-2019-2024/european-green-deal_en. Accessed Jan 2024.

EU-TAXUD. (2023). *Transitional CBAM Registry user manual for Declarants*. Retrieved from https://taxation-customs.ec.europa.eu/document/download/47322ae4-27c9-49f1-bdb7-2a1fef73d647_en?filename=CBAM-UMN-UI%20manual%20for%20Declarants%20%28incl.%20Business%20Guide%29%20-%20Release%201.2-v1.41.pdf. Accessed Jan 2024.

Geraghty, C., Snook, S. J., Alcorta, J., & Perou, N. (2018). *ISO 14067Greenhouse gases—Carbon footprint of products—Requirements and guidelines for quantification*.

GHG Protocol. (2011). *Quantifying the greenhouse gas emissions of products PAS 2050 and the GHG Protocol product standard.* GHG Protocol product standard. Retrieved from https://ghgprotocol.org/sites/default/files/2022-12/GHG%20Protocol%20PAS%202050%20Factsheet.pdf. Accessed Jan 2024.

Goedkoop, M., Heijungs, R., Huijbregts, M., De Schryver, A., Struijs, J., & Van Zelm, R. (2009). *ReCiPe 2008—A life cycle assessment method which comprises harmonised category indicators at the midpoint and the endpoint level.* The Hague.

GoGlobal. (2022). *Key content of the Border Carbon Adjustment Mechanism (CBAM).* Retrieved from https://goglobal.moit.gov.vn/en/key-content-of-the-border-carbon-adjustment-mechanism-cbam.html. Accessed Jan 2024.

Gonzalez, V., Lou, X., & Chi, T. (2023). Evaluating environmental impact of natural and synthetic fibers: A life cycle assessment approach. *Sustainability, 15*(9), 7670.

He, B., Shao, Y., Wang, S., Gu, Z., & Bai, K. (2019). Product environmental footprints assessment for product life cycle. *Journal of Cleaner Production, 233*, 446–460.

Herva, M., Álvarez, A., & Roca, E. (2012). Combined application of energy and material flow analysis and ecological footprint for the environmental evaluation of a tailoring factory. *Journal of Hazardous Materials, 237*, 231–239.

Hohenthal, C., Leon, J., Dobon, A., Kujanpää, M., Meinl, G., Ringman, J., Hortal, M., & Forsström, U. (2019). The ISO 14067 approach to open-loop recycling of paper products: Making it operational. *Journal of Cleaner Production, 224*, 264–274.

Huang, B., Zhao, J., Geng, Y., Tian, Y., & Jiang, P. (2017). Energy-related GHG emissions of the textile industry in China. *Resources, Conservation and Recycling, 119*, 69–77.

ILO. (2021). *Reducing the footprint? How to assess carbon emissions in the garment sector in Asia.* Retrieved from https://www.ilo.org/wcmsp5/groups/public/%2D%2D-asia/%2D%2D-ro-bangkok/documents/publication/wcms_781938.pdf. Accessed Dec 2023.

IPCC. (2014). *Intergovernmental panel on climate change.* Retrieved from Intergovernmental Panel on Climate Change, Retrieved from https://archive.ipcc.ch/report/ar5/syr/. Accessed Dec 2023.

IPCC. (2019). *2019 refinement to the 2006 IPCC guidelines for national greenhouse gas inventories.* Retrieved from https://www.ipcc-nggip.iges.or.jp/public/2019rf/index.html. Accessed Jan 2024.

IPCC. (2020). *Database on greenhouse gas emission factors (IPCC-EFDB), User Manual for WEB application.* Retrieved from https://www.ipcc-nggip.iges.or.jp/EFDB/documents/EFDB_User_Manual.pdf. Accessed Jan 2024.

IPCC. (2024). *Task group on data support for climate change assessments (TG-data).* IPCC.

ISO-International Organisation for Standardisation. (1997). *ISO14040: 1997 environmental management – Life cycle assessment – Principals and framework.* International Organisation for Standardisation.

Jaitiang, T., Vorayos, N., Deethayat, T., & Vorayos, N. (2020). Energy conservation tracking of Thailand's energy and GHG mitigation plan: A case of Thailand's textile industry. *Energy Reports, 6*, 467–473.

Juanga-Labayen, J. P., Labayen, I. V., & Yuan, Q. (2022). A review on textile recycling practices and challenges. *Text, 2*(1), 174–188.

Kazan, H., Akgul, D., & Kerc, A. (2020). Life cycle assessment of cotton woven shirts and alternative manufacturing techniques. *Clean Technologies and Environmental Policy, 22*, 849–864.

Lehmann, A., Bach, V., & Finkbeiner, M. (2016). EU product environmental footprint—Mid-term review of the pilot phase. *Sustainability, 8*(1), 92.

LeVan, S. L. (1995). Life cycle assessment: Measuring environmental impact. *49th annual meeting of the forest prodcust society.*

Luo, Y., Song, K., Ding, X., & Wu, X. (2021). Environmental sustainability of textiles and apparel: A review of evaluation methods. *Environmental Impact Assessment Review, 86*, 106497.

Manda, B. K., Worrell, E., & Patel, M. K. (2015). Prospective life cycle assessment of an antibacterial T-shirt and supporting business decisions to create value. *Resources, Conservation and Recycling, 103*, 47–57.

Moazzem, S., Daver, F., Crossin, E., & Wang, L. (2018). Assessing environmental impact of textile supply chain using life cycle assessment methodology. *The Journal of the Textile Institute, 109*(12), 1574–1585.

Moazzem, S., Crossin, E., Daver, F., & Wang, L. (2021). Assessing environmental impact reduction opportunities through life cycle assessment of apparel products. *Sustainable Production and Consumption, 28*, 663–674.

Muthu, S. S. (2020). Life cycle assessment (LCA) and product carbon footprint (PCF) modelling of textile products. In S. S. Muthu (Ed.), *Assessing the environmental impact of textiles and the clothing supply chain* (pp. 132–143). Woodhead Publishing.

Muthukumarana, T. T., Karunathilake, H. P., Punchihewa, H. K. G., Manthilake, M. M. I. D., & Hewage, K. N. (2018). Life cycle environmental impacts of the apparel industry in Sri Lanka: Analysis of the energy sources. *Journal of Cleaner Production, 172*, 1346–1357.

Niinimäki, K., Peters, G., Dahlbo, H., Perry, P., Rissanen, T., & Gwilt, A. (2020). The environmental price of fast fashion. *Nature Reviews Earth and Environment, 1*(4), 189–200.

Nwodo, M. N., & Anumba, C. J. (2020). Exergetic life cycle assessment: A review. *Energies, 13*(11), 2684.

Pedersen, E., & Remmen, A. (2022). Challenges with product environmental footprint: A systematic review. *The International Journal of Life Cycle Assessment, 27*(2), 342–352.

Pensupa, N., Leu, S. Y., Hu, Y., Du, C., Liu, H., Jing, H., Wang, H., & Lin, C. S. K. (2018). Recent trends in sustainable textile waste recycling methods: Current situation and future prospects. In *Chemistry and chemical technologies in waste valorization* (pp. 189–228).

Peters, G., Granberg, H., & Sweet, S. (2014). The role of science and technology in sustainable fashion. In K. Fletcher & M. Tham (Eds.), *Routlegde handbook of sustainability and fashion* (pp. 181–190). Routledge.

Peters, G., Svanström, M., Roos, S., Sandin, G., & Zamani, B. (2015). Carbon footprints in the textile industry. In S. S. Muthu (Ed.), *Handbook of life cycle assessment (LCA) of textiles and clothing* (pp. 3–30). Woodhead Publishing.

Piegsa, E. (2010). *Green Fashion: kologische Nachhaltigkeit in der Bekleidungsindustrie* (Vol. 35). Diplomica Verlag.

Qifeng, L., Jianlei, Z., Dake, Q., & Lin, H. (2023). The impact of heterogeneous environmental regulations on China's textile industry CO2 emissions. *Industria Textila, 74*(4), 419–425.

Rao, N., Mulugetta, Y., Burton, J., Saraiva, J. D., Gambhir, A., Omukuti, J., Ouedraogo, N. S., Pelz, S., Teng, F., & Tesfamichael, M. (2023). Chapter 6: Energy transitions for low-carbon development futures in low- and middle-income countries: Challenges and opportunities. In *Broken Record Temperatures hit new highs, yet world fails to cut emissions (again), EU emissions gap report*. https://doi.org/10.59117/20.500.11822/43922

Rogelj, J., Elzen, M. D., Pereira, J. P., Fransen, T., Kikstra, J., Lamboll, R., Meinshausen, M., & Tagomori, I. (2023). Chapter 4: The emissions gap in 2030 and beyond. In *Broken Record Temperatures hit new highs, yet world fails to cut emissions (again), EU emissions gap report*. https://doi.org/10.59117/20.500.11822/43922

Samara, F., Ibrahim, S., Yousuf, M. E., & Armour, R. (2022). Carbon footprint at a United Arab Emirates University: GHG Protocol. *Sustainability, 14*(5), 2522.

Schmincke, E., & Grahl, B. (2006). Umwelteigenschaften von Produkten Die Rolle der Ökobilanz in ISO Typ III Umweltdeklarationen. *Umweltwissenschaften und Schadstoff-Forschung, 18*(3), 185–192.

Šerkinić, V., Majić Renjo, M., & Ucović, V. (2020). CO2 footprint for distribution oil immersed transformers according to ISO 14067: 2018. *Journal of Energy: Energija, 69*(3), 3–9.

Sharpe, S., Dominish, E., & Martinez-Fernandez, M. C. (2022). *Taking climate action: Measuring carbon emissions in the garment sector in Asia* (No. 53). ILO Working Paper.

Sinden, G. (2009). The contribution of PAS 2050 to the evolution of international greenhouse gas emission standards. *The International Journal of Life Cycle Assessment, 14*, 195–203.

Statista. (2023). Retrieved from https://www.statista.com/topics/5091/apparel-market-worldwide/#topicOverview. Accessed Jan 2024.

SWP. (2021). *Stiftung Wissenschaft und Politik*. Retrieved from https://www.swp-berlin.org/publikation/ein-co2-grenzausgleich-fuer-den-green-deal-der-eu#hd-d53600e3640. Accessed Jan 2024.

Textile Exchange. (2022). *Preferred fiber and materials market report*. Retrieved from http://textileexchange.org/app/uploads/2022/10/Textile-Exchange_PFMR_2022.pdf. Accessed Dec 2023.

ThreadUp. (2021). *2021 Resale report*. Retrieved from https://www.thredup.com/resale/2021/#size-and-impact. Accessed Jan 2024.

Tracker, C. A. (2021). *Countries overview*. Retrieved from https://climateactiontracker.org/countries/. Accessed Jan 2024.

Trzepacz, S., Lingås, D. B., Asscherickx, L., Peeters, K., Duijn, H. V., & Akerboom, M. (2023). *LCA-based assessment of the management of European used textiles*. Norion Consult.

UNEPIE. (1996). *Life cycle assessment: what it is and how to do it?* Geneva: Publications United Nations. Retrieved from https://leap.unep.org/en/countries/ie/national-legislation/access-information-environment-regulations-1996. Accessed Dec 2023.

United, F. (2024, January 25). Retrieved from FashionUnited: https://fashionunited.de/nachrichten/business/tief-durchatmen-bitte-wie-textilunternehmen-ihre-co2-emissionen-und-die-ihrer-lieferanten-senken-koennen/2021082742508. Accessed Jan 2024.

Wang, C., Wang, L., Liu, X., Du, C., Ding, D., Jia, J., Yan, Y., & Wu, G. (2015). Carbon footprint of textile throughout its life cycle: A case study of Chinese cotton shirts. *Journal of Cleaner Production, 108*, 464–475.

Wójcik-Karpacz, A., Karpacz, J., Brzeziński, P., Pietruszka-Ortyl, A., & Ziębicki, B. (2023). Barriers and drivers for changes in circular business models in a textile recycling sector: Results of qualitative empirical research. *Energies, 16*(1), 490.

WorldBank. (2024). Retrieved from carbon pricing dashboard. Up-to-date Overview of Carbon Pricing Initiatives: https://carbonpricingdashboard.worldbank.org/. Accessed Jan 2024.

WTO. (2023). *World trade statistical review 2023*. Retrieved from https://www.wto.org/english/res_e/booksp_e/wtsr_2023_e.pdf. Accessed Dec 2023.

Wu, Y., & Su, D. (2021). LCA of an industrial luminaire using product environmental footprint method. *Journal of Cleaner Production, 305*, 127159.

Wu, P., Xia, B., & Wang, X. (2015). The contribution of ISO 14067 to the evolution of global greenhouse gas standards—A review. *Renewable and Sustainable Energy Reviews, 47*, 142–150.

Zhang, X., & Chen, Y. (2019). Carbon emission evaluation based on multi-objective balance of sewing assembly line in apparel industry. *Energies, 12*(14), 2783.

Zhang, Y., Kang, H., Hou, H., Shao, S., Sun, X., Qin, C., & Zhang, S. (2018). Improved design for textile production process based on life cycle assessment. *Clean Technologies and Environmental Policy, 20*, 1355–1365.

Zhang, X., Wang, L., & Chen, Y. (2019). Carbon emission reduction of apparel material distribution based on multi-objective genetic algorithm (NSGA-II). *Sustainability, 11*(9), 2571.

Zhang, J., Qian, X., & Feng, J. (2020). Review of carbon footprint assessment in textile industry. *Ecofeminism and Climate Change, 1*(1), 51–56.

Strategies to Reduce the Carbon Footprint of Protective Face Masks

Pasquale Giungato, Roberto Leonardo Rana, and Caterina Tricase

1 Background

Searching of excellence in energy and environmental achievements is the goal of organizations all over the world, fostering innovative technology to lower pollution level. However, spreading of SARS-CoV-2 virus causing COVID-19 pandemic has threaten this objective as sanitary urgences had to be satisfied sacrificing environmental issues. At the end of 2019 in Wuhan (People's Republic of China), a virus belonging to the coronavirus family, currently known as Sars-CoV-2, responsible for the disease known as COVID19, was identified. Due to the strong viral load of the pathogen, the disease has spread to almost all parts of the world, so much so that it was declared a pandemic by the WHO on March 11, 2020 (WHO, 2020). The virus is highly spread via droplets of saliva emitted from the nose and mouth during coughing, sneezing or simply talking; therefore, protection of the airways is the basis of prevention from the spread of the disease, together with distancing, frequent hand washing. Among sanitary emergencies the use of protective facial masks played a crucial role in defending people from diffusion of the virus, and supply of huge amounts of facepiece respirators, generated both economic and environmental issues. The intensive production of the protective face masks entails a significant environmental impact, considering the entire life cycle of each individual mask (from the cradle to the grave). The from "cradle to cradle" approach has not been followed due to sanitary constrains and the real danger of virus spreading due to the

P. Giungato (✉)
Department of Chemistry, University of Bari Aldo Moro, Bari, Italy
e-mail: pasquale.giungato@uniba.it

R. L. Rana · C. Tricase
Department of Economics, University of Foggia, Foggia, Italy
e-mail: roberto.rana@unifg.it; caterina.tricase@unifg.it

S. S. Muthu (ed.), *Carbon Footprint Assessments*, Environmental Footprints and Eco-design of Products and Processes,
https://doi.org/10.1007/978-3-031-70262-4_6

reuse/recycle of the used masks. Moreover, the production of the non-woven fabrics with the desired filtration efficiency and pandemic containment requires polymers of high purity difficult to obtain from recycled materials. More than 380 and 402 billion units were sold in 2020 and 2021, respectively, compared to about 12 billion in 2019. Although sales returned gradually to the pre-pandemic levels, registering about 28 billion units in 2023, the environmental issues remain a crucial problem considering the goals of sustainability. In the light of the above premises, the current chapter describes the structure of the surgery and FFP masks, the production processes, the environmental impacts in the life cycle perspective, focusing on the studies on carbon footprint. The effect of a proper selection of materials and of sanitization techniques on the carbon footprint as an environmental sustainability indicator to enhance the environmental performance of the face masks supply chain has been extensively reported. The chapter aims to provide a useful tool for supporting policymakers to select the best climate change mitigation strategies in the face masks supply chain.

2 Structure of Face Masks: From Surgery to FFP2/FFP3 Devices

Although cotton masks have been used by people and studied by scholars the variability in the number of layers (and of mass), size (and area), nose clips (and material used), renders the results of the evaluations unreliable for these reasons the present study is limited to only surgical and FFP2 masks, as they comply requirements of recognized standards and results as comparisons (in comparative life cycle assessment studies) will be reliable. Around the 60s disposable single-use face masks, composed of nonwoven synthetic fiber materials, layered gauze face mask, Jel mask with gauze and filter, cellulose-based face mask incorporating cellulose acetate plastics sheets and cotton swabs, were available to public (Matuschek et al., 2020; Spooner, 1967; Strasser & Schlich, 2020). Surgical face masks were typically composed of polypropylene non-woven fabrics but polyesters, fiberglass, polycarbonate, and polyethylene were also employed (Belkin, 1997; Hogan & Samaranayake, 1990). The surgical mask is a rectangular device with two elastic laces fixed to the corners of the shorter sides of the rectangle. Indeed, the surface of the mask is wider, square in shape but pleating allows the mask to adapt to the shape of the operator's nose and mouth (Fig. 1, left).

Surgical masks are composed of three layers of non-woven fabrics, including PP spun-bond and meltblown. In Italy as in main European countries, the so-called surgical masks were used in sanitary environments whereas the facial facepiece respirators (FFPs) were diffused in workplaces as personal protective equipment's. COVID-19 pandemic has partly revolutionized these customary practices introducing the maximum level of protection by extending the use of FFPs in all the workplaces including sanitary environments and limiting the use of cotton in the

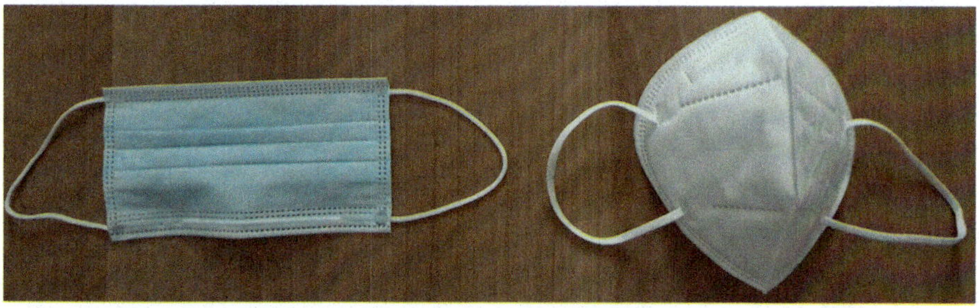

Fig. 1 Surgical (left) and FFP2 (right) face masks. (Source: Authors)

man-made "community masks" which did not guarantee the necessary protection against the spread of the virus. The community masks were useful during the emergency phase, when the shortage of masks was very strong. Measurements of differential pressure and bacterial filtration efficiency of masks, made with different materials and designs, in Italy collected over 3 months during the pandemic, demonstrated that more than 60% of the masks satisfied the standards provided there were at least three layers of the nonwoven fabrics (spunbonded–meltblown–spunbonded) which shown the best results, whereas majority of the community masks made with alternative materials and designs, did not comply with standard requirements (Tessarolo et al., 2021). For these reasons, we talk about the transition from the surgery to the FFPs masks as to underline the change in production, distribution, use and disposal of devices and, consequently, the change in the related environmental impacts. Initially, in the first phases of the pandemic disease, in Italy were distributed and used the surgical masks, widely used in hospitals and retirement homes for the elderly. The surgical mask is a device worn to avoid the spread of pathogens, such as bacteria or viruses, dispersed in respiratory droplets and aerosols that can escape from the mouth and nose of the person using it. Surgical masks comply with the official European Standard (EN 14683) which declare three types of effectiveness, as type 1, type 2 with a bacterial filtration efficiency of over 95% and 98% respectively, and Type R (further distinguished into IR, IIR), in which type IIR is the most resistant to projection of aerosol particles and for all the types described it is recommended to replace it every 4 h of use. The surgery masks were usually used in hospitals and to carry out their task correctly, they must have a very specific structure and the materials they are made of must be suitable for containing droplets. The surgical masks that were sold, although there were various models on the market, were made with three or four overlapping layers of non-woven fabrics composed of polypropylene fibers. Non-woven fabrics nowadays, can be produced with "spunbond" or with "meltblown" technologies: meltblown technology produces non-woven fabrics characterized by poor mechanical resistance, small fiber dimensions which give high filtering power but generate high pressure drop and therefore difficulty in breathing; spunbond technology, on the other hand, allows you to obtain non-woven fabrics characterized by larger fibers, lower pressure drop, greater mechanical resistance but lower filtering power (Fig. 2).

Fig. 2 Final product of the Meltblown process. (Source: Authors)

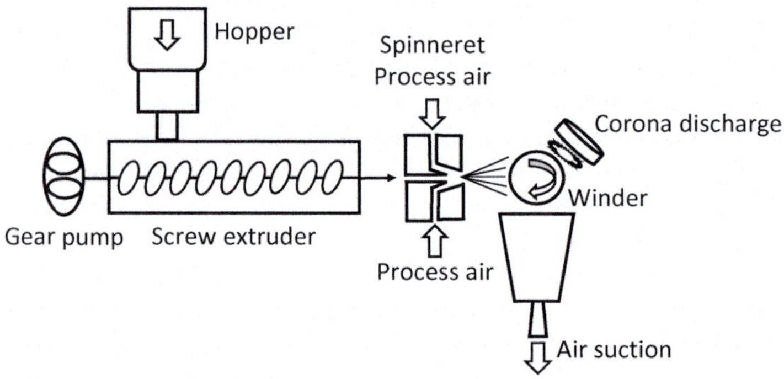

Fig. 3 Schematic of the apparatus for production of non-woven meltblown fabric. (Personal elaboration by the authors, adapted from Cho et al., 2023; Pratumpong et al., 2023)

Meltblown non-woven fabric is produced by heating chips, beads, pellets, or granules of PP over 200 °C in an extruder in which a gear pump loads the melted polymer in the melt-blown die where fine nozzle holes of around 0.3 mm in diameter, sprayed a high-speed air of over 200 °C producing a web of fine fibers randomly oriented on a conveyor-type collector equipped with air suction. The final fabric is wound in a calender which moves alternately on the rotation axis to make the thickness of the fabric uniform where a corona discharge provide electret charging of the fibers network. Properties of non-woven meltblown fabric depend on process variables like material used, its flux, air temperature and pressure, nozzle diameter, conveyor speed, and die to collector distance. The optimization of these parameters has been objective of deep learning model studies as the theoretical basis of the process which connect process variables and final properties has not been yet understood (Fig. 3) (Cho et al., 2023; Pratumpong et al., 2023).

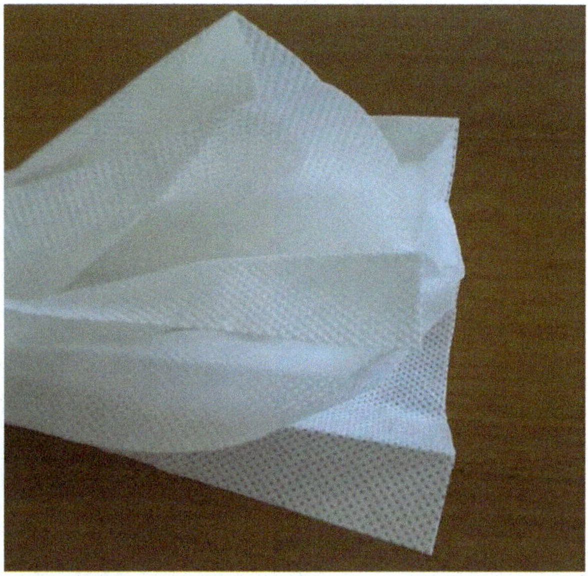

Fig. 4 Finished product after Spunbond processing with diamond calendaring. (Source: Authors)

The largest microfibers in spinning spunbond non-woven fabrics make these layers unsuitable for obtaining high filtration values, as required for surgical and FFP2 masks, as spunbond technology gives an economical, light, robust non-woven fabric, not suitable for fine filtrations (Fig. 4).

The PP fibers, in the form of filaments, are spun and then dispersed directly onto a quenching area, using air jets, where with high speeds microfibers were bonded by resin or thermally on the web forming (Fig. 5). PP spunbonds are produced with even higher speed and a decidedly lower temperatures than non-woven fabrics of different polymers, taking advantage of the low melting point of PP; moreover, polypropylene non-woven fabrics have an intrinsic water-repellent characteristic and are very light, which makes them particularly suitable for disposable products (Dutton, 2008; Lim, 2020; Midha & Dakuri, 2017). Midha and Dakuri in 2017, disclosed a comprehensive review of the spunbonding technique, in terms of fabric characteristics, and process parameters, aimed at optimizing the process parameters to produce the desired properties in the final fabric. They showed that non-woven fabrics made through the spunbond process are used in various applications such as disposable diapers, hygiene and healthcare fabrics, filtration devices, in the automotive industry, civil engineering, packaging applications, carpet backings. There is also a continued growth in the consumption of spunbond fabrics in durable and disposable products, with the development of new products based on this technology expected to further increase the market share (Midha & Dakuri, 2017).

The layers that make up the surgical masks were alternated non-woven fabrics produced with the two different technologies to obtain a final product that enjoys the advantages of both types: layered structures with excellent filtration capacities, low pressure drop and adequate mechanical robustness. To enhance the filtering

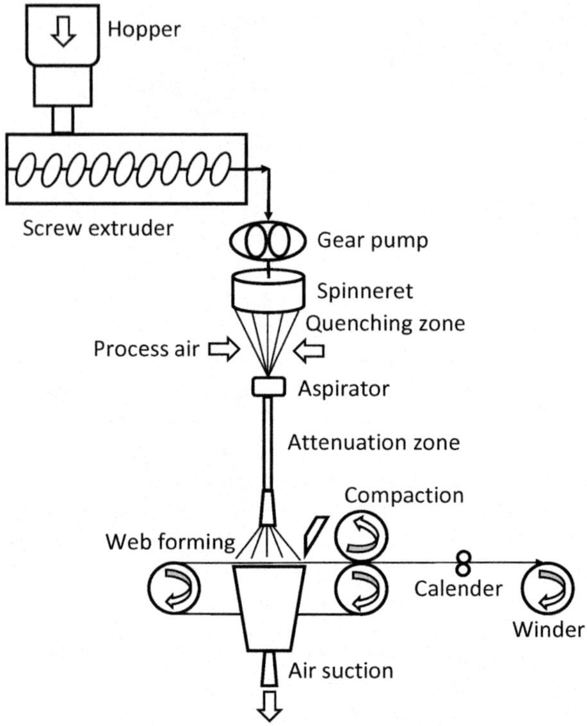

Fig. 5 Schematic of the apparatus for production of non-woven spunbond fabric. (Personal elaboration by the authors, adapted from Leucker, 2020)

capacity, in the case of particulate matter, some technological patents were released over the life cycle of such devices. The virus spread in the same mode as every other influenza virus, via small droplets of short-range aerosols. Fine aerosol particles (with aerodynamic diameter smaller than 5 micrometers) remain suspended and are generally withheld in the upper respiratory tract, and with diameters smaller than 1 micrometer do not settle down but can reach lower respiratory tract, whereas those with aerodynamic diameter greater than 5 micrometers, settle in about 1 h those with aerodynamic diameter greater than 10 micrometers settle suddenly and are few inhaled (Lindsley et al., 2010). As far as the emergency issues raised, both the United States Centers for Disease Control and Prevention (CDC) recommended the usage of N95 filtering facepiece respirators and Italian government made mandatory the equivalent (in terms of filtering capacity) FFP2 filtering facepiece masks as personal protective equipment for healthcare professionals (Fig. 1, right). In fact, Italy was the second country (after the People's Republic of China) to be severely affected by the pandemic in March 2020 and the entire country was placed in lockdown to contain spreading of the virus. Shifting on FFP2 masks, instead of surgical ones, was made necessary by the need to contain the uncontrolled spread of the virus and FFP2 masks were mandatory in in theatres, cinemas, concert, sports halls, and stadiums, in all the shows open to the public that take place indoors (as an example in theatrical or concert halls), in schools, for access in commercial passenger transportation services, as to ferries, ships, aircrafts, trains, comprising those from rental services with driver. Wearing of FFP2 masks was mandatory also in health, social-health and social-assistance facilities, hospitality, and long-term care facilities, assisted healthcare residences, hospices, rehabilitation facilities and residential facilities for older adults for workers, guests, and visitors (President of the Republic of Italy, 2021). FFP which stands for Filtering FacePiece, is a European standard that certified protective masks against droplets, dust particles, and aerosols according to (EN 149:2001) which distinguishes three filter efficiencies: FFP1, with corresponding minimum filtration efficiencies of 80%, FFP2 with corresponding minimum filtration efficiencies of 94%, and FFP3 with corresponding minimum filtration efficiencies of 99%, for 0.3 μm particles (EN, 2001). The FFP1 mask is specific for dust and is used by bricklayers, builders, carpenters but it is only valid with the biggest particles but the protection level with respect to particulate is better than surgical masks, although it lacks the desired protection against aerosols containing viruses. The N95 (U.S.), FFP2 (E.U.), KN95 (People's Republic of China), DS/DL2 (Japan), and KF94 (South Korea) masks comply the filtration capacity determined by the CDC's National Institute of Occupational Safety and Health, which define a minimum filtration efficiency of 95% for 0.3 micrometers of aerodynamic mass mean diameter of sodium chloride aerosols made by a specific apparatus (NIOSH, 1995). FFP2, FFP3, N95, and KN95 masks are widely used in the pharmaceutical industry as protection against spreading of respiratory diseases as tuberculosis bacteria and bird flu (Tcharkhtchi et al., 2021). The FFP3 masks are approximately equivalent to N99 standard and protect against finer particles but not gases, and all the filtering facepiece respirator masks should be changed every 8 h, and reuse is not allowed. N95 masks are made of multiple layers (generally five layers), in which some of them are made of polypropylene fibers with built-in electrostatic charge. They are composed of five layers of non-woven fabrics, including PP spun-bond non-woven (the innermost and outermost layers), PP melt-blown non-woven (the second and fourth layers), and cotton (the middle layer). Filtration efficiency is obtained both from the network structure of the polypropylene meltblown filter layer, by mechanical sieving mechanism, and from the electrostatic charge, due to the presence of electrects. Electrostatic charging can increase the efficiency of mechanical filtration up to 10–20 times. As previously reported, the meltblown non-woven fabric responsible of the filtering capacity is composed of polypropylene randomly oriented microfibers with diameters in the range of ~1–10 μm characterized by high porosity and permeability (Liao et al., 2020). To enhance the filtration efficiency the permeability can be reduced, reducing mean diameter of the voids in the microfibers network, adding mass and density to the fiber network, or by increasing the fiber diameter, but with detrimental pressure drops giving difficulties in breathing. To enhance the filtration capacity without sacrificing the breathability, the fabric is "charged" with a sort of quasi-permanent electrical charges by means of technological charging processes making use of triboelectric technologies or corona discharges.

3 Production and Distribution

The surgical masks are generally composed of at least three or four layers of non-woven fabrics: the external layer of non-woven fabric, is produced with spunbond technology, away from the user's face, has the function of giving the mask mechanical resistance and water repellency; the central layer of non-woven fabric, produced with meltblown technology, increases the filtering power of the multilayer; the internal layer, in contact with the face of the user, produced with spunbond technology, has a protective function for the skin as it avoids direct contact with the filtering layer. Chemically, PP is a thermoplastic polymer, formed by linear or slightly branched chains, obtained by polymerizing propylene, produced from crude oil. Petrochemical production starts with the refining of crude oil in petrochemical plants, with topping towers of a different structure from those used to produce fuels and optimized for the maximum production of a particular fraction: a mixture of light hydrocarbons known as virgin naphtha and, as by-products, fuel oils. Subsequently, the virgin naphtha is subjected to the cracking process which, in the presence of steam (which does not take part of the process) and using catalysts, splits and transforms the molecules of heavy hydrocarbons into smaller molecules including propylene, the starting point to produce plastics, fibers, and rubbers synthetic and other industrial products. Polypropylene is obtained by radical polymerization of propylene, and depending on the type of catalysis, it is possible to obtain three forms of polypropylene. The isotactic and syndiotactic forms are obtained by Ziegler-Natta catalysis. For market reasons, the most used polypropylene is the isotactic one which is found in the semi-crystalline form characterized by a high tensile strength, a low density and good thermal and abrasion resistance. These mechanical properties are due to the regularity in the structure of the isotactic and syndiotactic form, which generates "islands of crystallinity" in a matrix in the final amorphous polymer. This unique structure is the basis of the mechanical properties of isotactic and syndiotactic polypropylene, compared to atactic polypropylene. The polymeric structure lends itself to being extruded to produce continuous yarns for weaving. When air is used, we substantially have the spunbond and meltblown non-woven fabrics production technologies. An advantage of polypropylene is its recyclability as each object produced in PP can be recycled several times while the disposal costs, which were initially very high, have been reduced in recent years thanks to mass awareness among the entire population which has allowed an increase in recycling activity and the conversion of waste disposal companies into companies producing new objects. Nonetheless face masks are often discarded before the end of the technical shelf life, like most of textiles (EEA, 2021a, b) for various reasons: one reason relies in the difficulties in collecting for recycling masks used in urban environments, as it is discarded as an urban waste whereas the masks employed in hospitals and workplaces (classified as special wastes) are collected for incineration for sanitary reasons following the principle of maximum prudence. Increased consumption and production of masks will generate more landfilled and incinerated materials and moreover fostered increased risks of soil (Spennemann,

2022a) and marine (Fadare & Okoffo, 2020) pollution from microplastics generated from the flaking of the material due to atmospheric agents. Municipalities should hurry up in collecting and disposal of stray face masks from the urban environment as the velocity of micro-fiber pollution depends on technology used to their fabrication: spunbonded non-woven fabrics cracks and breaks down much faster due to thinner fibers of shorter length and the lack of thermal bonding points, and photo-oxidation increase the rate of degradation process in fabrics with a light blue dye (Spennemann, 2022b). An LCA study considering landfilling as reference scenario compared with recycled plastics proved that recycling has lower environmental effect on acidification potential, abiotic depletion, human toxicity, and global warming impact categories (Al-Ma'adeed et al., 2011).

4 Environmental Impacts from a Life Cycle Perspective

During the COVID-19 pandemic, there was scarce attention paid to the environmental aspects related to the life cycle of the protective measures to limit the spreading of the virus, but after the initial emergently step, the necessity of a life thinking approach to the environmental impacts' issues emerged strongly. Although there was a literature on environmental impacts from a life cycle perspective on protective equipment in general, the facial masks were not an argument of discussion because too specific in a non-pandemic contest. But with the explosion of the pandemic and the importance raised in the devices for the containment of the respiratory systems, the facial masks assumed a pivotal role and the environmental issues emerged as well. In a first approach of the research, environmental issue of facial masks was supposed to be similar to that of the materials it were composed of. About the disposal of textiles, in the case of facial masks, incineration was supposed to be largely viable and widely used as it is deemed to be safe, simple, and effective as high temperatures completely kill microorganisms like viruses, converting organic matter into inorganic ones. In 2018, Sandin and Peters demonstrate that reusing textile materials is more advantageous than recycling, where reuse is considered reusing the textile material as it is and for the same purpose, not subject to any treatment before the reuse, while recycling is considered the pre- or post-consumer reprocessing of textile waste, for use in new textile or non-textile products. Forty-one studies were examined, of which 85% concern recycling and 41% reuse (27% concern both reuse and recycling). The recycling of fibers appears to be the most studied type of recycling (57%), followed by that of polymers/oligomers (37%), monomers (29%), fabrics (14%). Cotton (76%) and polyester (63%) were the most studied materials. The publications reviewed provide strong support for the claim that reuse and recycling of textiles generally reduce environmental impacts compared to incineration and landfill disposal, and moreover reuse is less impacting than recycling, but there were scenarios in which both reuse and recycling were not beneficial for certain impact categories. For example, as benefits arise primarily from avoided impacts due to production of new devices, benefits lowered in cases where substitution rates were low or the avoided production processes are relatively clean. Raise in customer-induced transportation impacts, in reuse strategies, may cause an environmental impact outweighing the benefits of avoided production, unless the use phase is sufficiently prolonged. Most often it is assumed that recycled textiles are waste with no environmental burdens and that reused products or made from recycled materials replace completely that made from virgin fibers, but in the LCA, as in the CF methodologies, those methodological assumptions are subjected to criticism, due to the different quality of the replaced virgin materials with respect the recycled/reused ones (Sandin & Peters, 2018). Evidences of recycling in the sanitary system were also known as recycling of waste syringes in electric arc furnace of a steelmaking, replacing coke, was proved to significantly reduce environmental issues in most of impact categories of a life cycle assessment (Ghodrat et al., 2017). Reuse of waste masks in production of melt-blown fabric modified bitumen in road engineering has been explored. It was shown that melt-blown fabric modified bitumen has improved resistance of the asphalt mixture to a high-temperature deformation and water damage but at the expense of low-temperature crack resistance (Cheng et al., 2024). As it is known, isolation gowns play a vital role in infection control, protecting visitors, patients, healthcare workers, from transmission of pathogens in controlled environments. These garments and other medical fabrics are available on the market as both reusable and disposable. In the medical sector, opting for a reusable fabric or a disposable material is a decision that depends on a series of factors including the effectiveness of the barrier exerted by the textile material, cost, comfort, and sustainability. Standards to quantify the liquid barrier performance of medical fabrics among which there are isolation gowns were established by both The American National Standards Institute (ANSI) and the Association for the Advancement of Medical Instrumentation (AAMI). A reusable gown and a disposable gown show comparable effectiveness in terms of microbiological barrier. From a sustainability point of view, however, previous case studies had already shown that the adoption of reusable isolation gowns led to a 30% cost reduction compared to disposable gown systems (Kressel et al., 2011), as well as reusable linen or surgical kits and towels offer significant cost savings over disposable ones (Di Giacomo et al., 1992; Siu et al., 2017; Zins & Howard, 2011). In 2018, Vozzola et al. carried out a life cycle study aimed at comparing four environmental impacts (impact categories of energy, global warming potential, water use, and solid waste consumption) exerted by 1000 reuses of insulating polyester clothing, and by 1000 disposable polypropylene clothing, highlighting which phases of the life cycle were particularly impacting. In particular, it emerged that reusing polyester isolation gowns consumed 28% less energy and emitted 30% less greenhouse gases compared to the scenario consisting of 1000 disposable isolation gowns, and that the practice of reuse, involves less water consumption equal to approximately half compared to the disposable. This latter evidence is contrary to the common perception, according to which reusable clothing requires more water due to the periodic washing operation, since public perception often does not take into account the water consumed in the production phase of textiles. There would therefore appear to be greater environmental benefits for all reusable textile items compared to disposable ones. Therefore, opting for reusable items within healthcare facilities, such as gloves, wipes, or masks, would further strengthen the resulting environmental benefits (Vozzola et al., 2018). Consistent with what has already mentioned, Rizan et al. in 2021, in calculating the CF of gloves, aprons, visors, and type IIR and IIIR surgical masks made in polypropylene, highlighted reuse firstly, a local production secondly, as strategies for reducing environmental impacts (Rizan et al., 2021). Even redesigning the face mask from a modeling point of view ("shape reengineering"), perhaps minimizing its area without compromising the filtration capabilities, can be a valid strategy to minimize impacts, or as suggested by Salman et al., redesigning the shape of the laces and the proportion of materials used to manufacture them, although in this study, it emerges that the effect of the laces on the overall impacts is negligible (Salman et al., 2022). Do et al. in 2021 demonstrated that disposable masks provide higher performance from a virus-protection perspective, while worst results in environmental performances. In contrast, reusable masks enable an 85% waste reduction, impact climate change 3.39 times less, and reduce costs by 3.7 times, compared to single-use disposable masks (Do et al., 2021). Recycling of face masks is a challenge because of both the presence of different materials (polymers in the non-woven fabrics, elastomers in the laces, metals, and other polymers in the nose clip) and the possible presence of pathogens, for these reasons are generally incinerated or landfilled. With recycled polypropylene, you can manufacture other objects requiring poor mechanical properties, but no other masks because the latter are devices contacting mouth and by law have to be made of virgin materials and secondly cannot have defects that could compromise the spunbond or meltblown apparatus lowering the filtering capability of the final non-woven fabrics (Fant, 2022).

5 Carbon Footprint of Protective Face Masks

Comparing the CFs of face masks in literature is an interesting challenge as the functional unit used in LCA studies may vary significantly from one paper to another, as also the system boundaries and assumption about the contribution of transport and disposal to the final environmental assessment. In this brief review of the most important studies on this subject, some limitations in the final computed values have been made and highlighted. In the case of disposable single-use masks, the functional unit is the production of a mask while in the case of reuse, the functional unit becomes one use of the mask, only in this case the comparative assessment is reliable. Starting from this assumption, the studies examined in the literature were compared based on a functional unit that was as equal as possible for everyone and, in this case, the single use of the mask was chosen as a comparison. In 2018 Vozzola et al., before the COVID-19 pandemic, made a LCA on polyester-made isolation gowns reusable by laundry operations in hospitals and compared the results with single-use PP-made ones, with a functional unit of 1000 uses of facial masks and a conservative estimate of 60 uses before disposal; the contribution

Table 1 Carbon footprint of face masks reported in literature, with type, reference, and specified single use or reuse (by sanitization)

Use	g CO_2 -eq	Type	References
Single-use	59	Surgical	Klemes et al. (2020)
Single-use	50	N95 - PP	Klemes et al. (2020)
Single-use	20	Surgical IIR	Rizan et al. (2021)
Single-use	13	Surgical II	Rizan et al. (2021)
Single-use	65.5	FFP2 - PP	van Straten et al. (2021)
Reuse	27.7	FFP2 - PP	van Straten et al. (2021)
Single-use	7.52	Surgical	Do et al. (2021)
Single-use	22.38	FFP2 - PP	Do et al. (2021)
Single-use	18.7	Surgical	Lee et al. (2021)
Reuse	10.9	Surgical	Lee et al. (2021)
Single-use	32.7	Surgical	Giungato et al. (2021)
Single-use	32.05	FFP2 - PP	Giungato et al. (2024)
Reuse[a]	6.29	FFP2 - PP	Giungato et al. (2024)
Reuse[b]	5.69	FFP2 - PP	Giungato et al. (2024)
Reuse[c]	3.23	FFP2 - PP	Giungato et al. (2024)
Single-use	51.4	Surgical	Luo et al. (2023)
Single-use	213.7	FFP2 - PP	Luo et al. (2023)

Legend: *PP* polypropylene, *FFP2* Facial FacePiece respirator type 2 with filtration efficiency over 98%, *Surgical* surgical mask undefined, *surgical type II* surgical mask with highest level of filtration for 3 μm particles, surgical type IIR as type II with liquid-repellent layer offering additional protection against particulate matter and infectious splashes

[a]Reuse after laboratory oven dry heating

[b]Reuse after home oven dry heating

[c]Reuse after UV irradiation

analysis highlighted as the production phase of the multilayered structure of the non-woven textiles was the most impacting, and that reuse has been advised as the best choice to lower environmental impact of single-use devices, with a 30% reduction in CF. Although there was no possibility to obtain the CF of the single use face mask, the importance of the paper relies in its original pre-pandemic study on this subject (Vozzola et al., 2018). Rizan et al. estimated a CF of type IIR surgical mask as 20 g CO_2 -eq, and Type II surgical mask 13 g CO_2 -eq per mask (see also Table 1). The contribution of transportation from manufacturing place to use place (UK) was lower than the manufacturing and disposal phases as a reduction of only 12% can be achieved by using local and not imported devices (Rizan et al., 2021). Van Straten et al. computed on FFP2 PP-made mask, a CF of about 6.55 kg CO_2 -eq for single-use face masks and 2.77 kg CO_2 -eq for reprocessed face masks, in this case the functional unit was the protection of 100 health care workers against airborne viruses, using one FFP2 certified face mask in a hospital (The Netherlands). In practice 100 single-use masks including packaging, transportation, use, and disposal and 100 times use of reprocessed masks (considering batch that cannot be reprocessed and a face mask which is sterilized and reused for an additional five times in a medical autoclave). From calculations a 65.5 g CO_2 -eq apiece in the case of single-use and 27.7 g CO_2 -eq per use in the case of reused devices (van Straten et al., 2021). In the work of Do et al., the functional unit was the face masks used in 306 days from March 2020 to December 2020, in Italy by population and produced in PRC. The results gave a computed CF of 7.52 g CO_2 -eq apiece in the case of single-use surgical mask and 22.38 g CO_2 -eq apiece in the case of single-use FFP2 masks (Do et al., 2021). (Lee et al., 2021) compared a single-use PP made surgical mask (composed of an aluminum nose piece, a polyurethane earloop, a poly-laminated of melt-blown PP between two spunbond PP made layers with 25 and 20 g/m^2, respectively) with an embedded filtration layer reusable face mask (composed of polyurethane ear loop laces, a poly-laminated of polyester fabric with a hydrophobic coating, a melt-blown PP made layer, a polyester fabric with hydrophobic coating composite, and a layer of polyester fabric), that can be washed manually with a liquid detergent and water. The functional unit was the consumption of 31 days of face masks by a person in a month. For the single-use surgical PP face mask, was equal to the usage of 31 face masks, whereas for the reusable face mask, was the use of only one face mask washed 30 times (practically, the functional unit in both cases were the use of masks covering a person in a month with a duration of less than 12 h). The use of single-use surgical face mask contributed to climate change with 0.58 kg CO_2-eq whereas the reusable face mask with 0.338 kg CO_2-eq per functional unit. From computations we have a CF of 18.7 and 10.9 g CO_2-eq apiece for the single-use and reusable mask, respectively. Raw material acquisition was the most impacting life cycle step in the case of the single use masks whereas water, detergent, and polyester were predominantly in the case of reusable device. CF assessment of disposable surgical masks and KN95 respirators (like FFP2 for filtering capacity) was 154.328 kg CO_2- eq/FU and 641.249 kg CO_2-eq/FU, respectively with a functional unit the masks used by 100 individuals over a period of 1 month, discarded after each use assuming that each person uses one disposable mask per day (a total of 3000 masks and a cradle-to-grave approach typical of a carbon footprint). These data give values of 51.4 g CO_2-eq apiece in the case of disposable surgical masks and 213.7 g CO_2-eq apiece in the case of KN95 respirators (like FFP2 for filtering capacity) with polymer production the most impacting life cycle phase (Luo et al., 2023). In previous paper of the authors, values of 32.7 g CO_2-eq apiece were computer for surgical masks totally made in Italy (both raw materials and incineration, with energy recovery, occurred in Italy) and values of 32.05 g CO_2-eq apiece were obtained for FFP2-PP made masks, (made in PRC, transported to Italy, and incinerated with energy recovery). Reuse of FFP2-PP made devices was modelled with different types of sanitization dry techniques (to avoid electrets neutralization), computing the CF of the single use of the masks, giving values of 6.29, 5.69, and 3.23 g CO_2-eq apiece depending on the type of sanitization, respectively for reuse after laboratory oven dry heating, reuse after home oven dry heating, and reuse after UV irradiation (Table 1) (Giungato et al., 2024). Some global consideration can be done after reading the papers in literature limiting our study to only surgical and FFP2 masks and considering the CF of a single use of the face mask. Although different assumptions have been made in the literature, as the choice of functional unit, origin of raw materials and production places of the final devices, with additional environmental costs of transportation, use and reuse (with different methods of sanitization) with disposal, some global considerations can be made. CF of FFP2 (or the equivalent N95/KN95 type of face masks) is higher than that of surgical masks in the comparative assessment, in the context of the single published paper (except in the work of Klemes that, however, did not consider transportation in the case of N95). This is mainly due to the higher filtration requirements of FFP2 (and similar as KN95 and N95), which increases the number of fabric layers, resulting in more raw materials and resource consumption, as well as higher environmental burdens arising from disposal. The disinfection changes the role of the most impacting contributes, with shifts from raw materials and disposal to sanitization (the most contributing process of the life cycle was the production of the electricity in the case of ovens, UV, and of all sanitization techniques requiring energy in the dry sanitization, whereas detergents, and water with subsequent disposal, in the case of wet washing). It is quite well established the marginal role of transportation also in the case of devices produces in faraway countries, due to low weight of the devices and their small volume. Although an important role could be attributed to the production of materials in the case of single-use devices, on the contrary the sanitization techniques provide additional environmental issue due to energy and chemicals (solvents) used but it allowed reuse and minor materials consumption, for this reason production of non-woven fabrics is not the principal contributor to the CF in the case of reuse by sanitization. For these reasons, the aim of research in this field is to find the best compromise between the materials used and the disinfection techniques, and life cycle assessment (LCA) is the best choice to afford such an interesting challenge. From study to study, depending on the assumption made and the functional unit choices, results obtained can vary but the literature shows there are not so wide oscillation among results, especially in the most studied CF impacts. But although the results of the CF for a single face mask returns very small quantities, with a restricted oscillation range, such range of values may generate huge variations in the global climate change impacts, when multiplied by the enormous quantity of masks used globally. Cornelio et al. in 2022 estimated the impact of FFP2 face masks composed of PP, calculating the number of masks needed for each person in the country (Italy) considering the following hypotheses: the assumption that masks should be changed every 8 h, the use of two masks a day, 5 days a week, and one mask a day during the weekend was considered for workers. For non-workers (both unemployed and inactive such as pensioners and people, such as students, who are not looking for work), four masks per week were considered. Since children under six are not required to wear masks, they were not included in the study. Normally, the total number of masks consumed in a year is around 21.96 billion, but when masks became mandatory, i.e., in the period March 2020—end of December 2021, equal to 95 weeks, the masks consumed could be estimated at 40.12 billion. Based on these estimates, the total CO_2 emissions associated with surgical masks in the period under examination were about 240 kton, whereas fabric masks (composed in cotton) contributed with 7 kton. If the transportation is considered, the contribution was 350 kton of CO_2 by surgical masks use (mostly imported and not made in Italy). However, this used approach was a simplified LCA analysis, as the carbon footprint for each material composing the masks was evaluated through the database Ces Selector 2019 and refers to the production of the fabric (PP or cotton) only, not to the entire life cycle, but among conclusions, it should be worth of mention that surgical masks have a higher environmental impact due to the huge number of masks needed to meet the sanitary and hospital demand and are specifically imported from abroad (in particular PRC, Germany, France, Spain in the order) including the emissions due to their production and transportation. Cotton masks are composed of cotton, degradable and reusable, produced locally with minor contribution of the transportations on to the CF. The authors suggest that, for the future, it could be possible to evaluate the recovery of surgical masks with disinfection techniques to minimize the CF. We should consider that globally more than 380 and 402 billion face masks were sold in 2020 and 2021, respectively, compared to about 12 billion in 2019, in the pre-pandemic. Sales returned gradually to the pre-pandemic levels, registering about 28 billion units in 2023 (Cornelio et al., 2022).

6 Strategies to Reduce the Carbon Footprint of Protective Face Masks: Effect of Materials

In the first months of the pandemic, masks shortage fostered emergency use of man-made cotton masks (the so called cloth masks) in Italy called also "community masks." Bhattacharjee et al. (2021) studied the performance of cloth masks against violent respiratory events, such as human sneezing, in the aim to provide guidance for the optimal design of a device able to outperform a traditional three-layer surgical mask. It was shown that a minimum of three layers is recommended to ensure the performance of surgical masks composed of a combination of cotton/linen for the inner layer, blends for the middle layer and polyester/nylon for the outer layer. The authors tested different combinations of fabrics and measured their effectiveness in blocking particles of different sizes generated by a simulated sneeze, and evaluated the breathability, comfort, and durability of the cloth masks, showing that cloth masks can be an alternative to medical masks during pandemics, if correctly designed and realized. Possible innovations for cloth masks may be the incorporation of nanoparticles with antimicrobial and multifunctional properties. The advantages of cloth masks rely in the capacity to resist to machine washing at 60 °C without significantly impacting the filtering performance. Conventional face masks do not impeded survivability of viruses on its surface as shown by experiments conducted on respirator filters and surgical masks loaded with bacteria (Mycobacterium abscessus, Staphylococcus epidermidis, and Bacillus subtilis subsp. Niger) (Brosseau et al., 1997). Survivability depends also on levels of dust-loaded polypropylene filtering nonwovens and type and the amount of dust, as demonstrated on dust-loaded polypropylene nonwovens (melt-blown, spun-bonded, and needle-punched) using microbiological quantitative method (Majchrzycka et al., 2018). SARS-CoV-2 virus reduced on plastic materials from 103.7 to 100.6 TCID50 (50% tissue culture infectious dose) per milliliter after 72 h with an estimated median half-life of approximately 6.8 h (Liao et al., 2020). Inner and outer layers of face masks retain SARS-CoV-2 viruses for a period of 4 and 7 days, respectively, confirming the stability of the virus in the environment of touching surfaces (Aboubakr et al., 2020; Chin et al., 2020). Another issue related to face masks is safe management and disposal of used devices, to overcome that, research and development has fostered the attention on antimicrobial fabrics, equipped with surface killing properties occurring by contact, in addition to the required compulsory particle filtration properties. Generally, the most used materials to make non-woven fabrics useful for face masks are polypropylene (PP) and polyethylene terephthalate (PET). As already emerges in Dutton (2008) and highlighted years later by Lim (2020), the spunbond and meltblown technologies, used to manufacture the non-woven fabrics utilized in the assembly of multilayer surgical and FFP2 masks, allow the use of a large variety of thermoplastic materials, such as isotactic PP, PET, polyethylene (PE), polyamide (Nylon 6 and 6.6), polystyrene (PS), polycarbonate (PC), polyurethane (PUR), polyvinylidene fluoride (PVDF), polyacrylonitrile (PAN), and polybutylene succinate (PBS). Normally, PP is the most used nowadays by producers, as it is cheaper and provide a high fiber/kg yield, obtainable from the raw granulate. PET exhibits higher tensile strength, modulus, and heat stability than PP but it is more expensive and difficult to process. PE is characterized by good electrical insulation, chemical resistance, and hydrophobicity. Polyamide (including nylon 6 and 6.6) requires more energy for its production and is used normally for packaging materials. PUR has the elastic properties required for the aforementioned applications, is used basically to make diapers, medical plasters, and elastic materials. PVDF is gaining interest in non-woven fabric for face masks, as its nanofibers can reach 50 nm mean diameter giving large surface area/volume ratio, good breathability. Due to conductive nature of the native polymer, the final fabric has good electret filter properties useful for the capture of nanoparticles in workplaces (Sanyal & Sinha-Ray, 2021). Monolayer fibrous structures of PAN-electrospun nanofibers deposited onto non-woven PP fabrics proven to reach FFP2-type filtration efficiencies to the final device. The balance between efficiency and breathability can be reached by variation in process parameters whereas ZnO particles added into the PAN solution gave strong bactericidal activity to the final PAN-ZnO_3, (3 wt.% of ZnO_3 content) without increase of breathing resistance (Pardo-Figuerez et al., 2021). When research moved on to determine whether an alternative material to PP could minimize the CF, PE appears to be the least impactful material, although it is currently little used for the manufacture of face masks since the technology does not guarantee highly performing products. PP appears to be the most used material and less impactful than many others, followed, in ascending order of environmental impact, by polystyrene PS, polyurethane PUR, and polylactic acid PLA. The latter, although it is a material obtained from biomass, capable of fixing CO_2 during photosynthesis, has a much greater impact than PP and PE, which instead are of fossil origin. This happens because the PLA production process is complex and

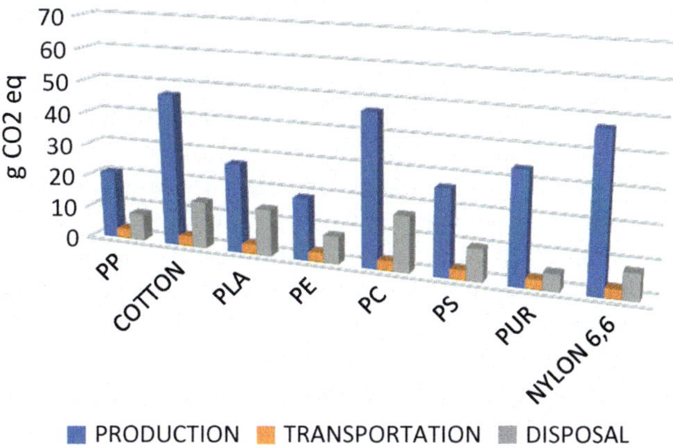

Fig. 6 CF expressed as CO_2 -eq apiece at each life cycle stage with the functional unit 1000 FFP2 single-use face masks with the same weight made in: PP, cotton, PLA, PE, PC, PS, PUR and Nylon 6,6. (Adapted from: Giungato et al., 2024)

requires several steps and the use of many raw materials, some of which are not incorporated into the final product, which leads to a high ratio between the waste mass and the mass of desired product. Furthermore, energy consumption for this manufacture is also higher than for other productions. The CF of Nylon 6.6 is even higher, but slightly lower than that of cotton and PC which is, by far, the most impactful material (Giungato et al., 2024) (Fig. 6).

In the case of cotton, although it is a natural cellulose fiber or a renewable raw material, its cultivation raises many environmental concerns, as it requires considerable quantities of water for its cultivation (Chapagain et al., 2006), in addition to the massive use of pesticides and insecticides to prevent the presence of parasites and the onset of plant diseases (Indhu Kavi et al., 2018). Moreover, the production of PC face masks of the same weight as the PP masks showed the absolute highest CFs, due to the production process and the disposal worst environmental score, which includes, in the polymer production process step as in the polymer disposal, the presence of highly impactful monomers, such as bisphenol A (BPA). Another interesting innovation that can be applied relies on the use of bioplastics. However, there is a significant risk of virus infiltration into improvised face coverings due to humidity, liquid diffusion, and virus retention (Rowan & Lafey, 2021). As an alternative to petroleum-based plastics, the bioplastics are rocketing materials useful when biodegradation and carbon neutrality are required in industrial applications (Emadian et al., 2016; Thakuv et al., 2018). Derived partly or wholly from polymers of biological origin like potato, starch, cellulose, cane, cotton, straw, sugar (Thakuv et al., 2018) and, recently, seaweed (Thiruchelvi et al. (2020) and microalgae (Rizman et al., 2018), the bioplastics are degraded in composting plants or simply by air exposure, by fungi and bacteria using their enzymes. Also, Polybutylene Succinate (PBS) which is biodegradable like PLA has been used in melt-blown fabric production with few limitations in the quality of the final product (Pratumpong et al., 2023). Polylactic acid (PLA) has also proven to be a material capable of guaranteeing the production of high-performance masks, especially if functionalized, as shown by Nicosia et al., who prepared polylactide/polyhydroxy butyrate (PLA/PHB) nanofibers by electrospinning. Functionalization was achieved by using ammonium-based ionic liquid belonging to quaternary ammonium compounds, using different quantities, and obtaining a non-woven fabric capable of filtering aerosol particles containing pathogens. The filtration performance of these devices revealed that the quality requirements were retained with the addition of ionic liquids and main differences depend on the time of the electrospinning production. Quantitative analysis of antimicrobial activity showed that renewable materials, such as PLA mixed with PHB, can be useful for production of bioplastics filters, according to the market needs of developing eco-friendly products. The addition of quaternary ammonium compounds salt to the biopolymer mixture allowed to fabricate a multilayered, device in which both the biodegradation and antimicrobial properties are integrated (Nicosia et al., 2015).

7 Strategies to Reduce the Carbon Footprint of Protective Face Masks: Effect of Sanitization Techniques

The urgent demand for disposable surgical face masks and filtering facepiece respirators surpassed the manufacturing capacities and, resulted in significant shortages in pandemic period partly compensated by single use devices and community masks (improvised face coverings by the public that carries a lower transmission risk and made using cotton, silk, or non-woven paper fabrics). Already in 2020 Klemes et al. demonstrated that reuse is a valid alternative to reduce energy consumption and the environmental impact of measures to fight COVID-19, including face masks, provided that their effectiveness remains unchanged (Klemes et al., 2020). Recycling of face masks is difficult as the presence of different materials (polymers in the multilayer of the non-woven fabrics and in the nose clips, elastomers in the laces, metals) and above all the possible presence of residual pathogens, guides the choices of political decision makers towards the safest solutions such as incineration or landfilling. For these reasons, the only possible solution will be the implementation of reuse strategies. Recycling of materials, as well as being difficult to implement at municipal level, for technical reasons due to the spunbond and meltblown processes, does not allow the manufacturing of new masks. The only viable strategy to lower the environmental impacts remain the reuse, by means of disinfection, and in literature are reported numerous papers in which essentially are used vaporized hydrogen peroxide (30–35% volume of H_2O_2), ozone, ultraviolet light at 254 nm, dry and moist heat, ethylene oxide, use of bleach (Rowan & Lafey, 2021), antimicrobial photodynamic inactivation using methylene blue as photosensitizer (Ludwig-Begall & Heyne, 2024). Vapor of hydrogen peroxide, alone or with ozone, has been used for sanitization of N95 (like FFP2) and has been certified for personal protective equipment decontamination under emergency use authorization by the U.S. FDA to address shortages in supplies during pandemic. The same occurred in many countries in which sanitization with reuse of face masks is allowed but the final decision is leaved to health service managers. Ultraviolet (UV) light using UVC wavelength is a valid cheaper alternative for sanitization but, unfortunately, SARS-CoV-2 can penetrate in deep layers of non-woven fabrics resulting hidden and unreachable by the irradiation. In most cases, UV dose (expressed in mW/s cm^2 or mJ/cm^2) plays a crucial role in sanitization efficiency. Moist heating (60–70 °C) with high humidity for 60 min is also a valid sanitization technique easily scalable and with high throughput, provided the filtration efficiency is retained due to electret inactivation caused by water neutralization. Microwave generated steam, physical irradiation (both gamma and electron beam), and ethylene oxide were not applicable for masks reuse (Rowan & Lafey, 2021). Ethylene oxide is harmful to the wearer, even at very low concentrations, with carcinogenic and teratogenic risks although there is no lowering of filtration efficiency (Bergman et al., 2010; Viscusi et al., 2009, 2011). Sinkko et al. (2023) compared two scenarios: the first one a single use disposable FFP2 mask usually used in hospitals, whereas the second one, the same mask with the same >94% filtration efficiency and used in hospitals, decontaminated by a low temperature atmospheric plasma discharge, and re-used before disposal. The work demonstrated honestly that decontaminated mask impacts more than the single use disposable one in all the examined impact categories, due to huge consumes of He needed to cool the electrodes between which the discharge is produced, and that replacement of He with alternatives does not revert the results: only the reduction of at least 95% of used cooling He gas (by means of recirculation) can make the reuse scenario more environmental friendly than the single-use device. Vaporized hydrogen peroxide (H_2O_2) was able to remove viruses without affecting filtering capacity, leaving residues of decontaminant within safe limits, but no data about the effect of various cycle of sanitization with vaporized hydrogen peroxide were provided. On the disinfection with other wet chemicals like ethanol, isopropyl alcohol, and sodium hypochlorite data were not favorable due to filtration capacity failure and safety concerns (O'Hearn et al., 2020). Another issue arises as masks reuse should be an accepted practice by healthcare workers. Efficacy of filtering power of facial masks depends on the appropriate wearing to fit nose and mouth covering, type of material used, number of layers, duration of wearing, and durability of ties over ear loops instead of laces, use of nose clips helping to fit of the device to the face of the user. The CF of a single use of a FFP2 mask after sanitization with dry heating (ten times in home oven, laboratory oven and UV irradiation in a UV cabinet) is significantly lower than that of a disposable mask: 6.29 g of CO_2 -eq. apiece if sanitized in a laboratory oven (an 80.4% reduction), 5.69 g of CO_2 -eq. apiece for a mask sanitized in a home oven, 3.23 g of CO_2 -eq. apiece in the case of UV irradiation (Table 1) (Giungato et al., 2024). These results are in agreement with those of (van Straten et al., 2021) who reported a 58% CF reduction of autoclave sanitized masks (five times) with respect to the single use disposable ones. Contribution analysis reveals that the CF was due basically to the energy consumption of the static ovens used, except for the UV cabinet with less electricity consumed with respect to ovens and consequently, the production of the materials remained the most impacting life cycle phase. Considering as functional unit the single use, the sanitizing procedure give significant reduction in the CF of the reused masks, compared with the disposable single use ones, but such results strictly depend to the load capacity of the sanitizing apparatus.

8 Conclusions

The achievement of the objectives of containment of climate change has been put to the test by the health emergency due to the COVID-19 pandemic. One of the fundamental objectives in stopping the diffusion of the virus was to contain its spread through the use of personal protective equipment that could stop the emission of saliva particles into the air. In Italy as in most countries, to achieve this purpose, face masks made from various materials were used in the first months during the pandemic, due to the shortage of face masks sufficient to meet the needs of the population. The use of improvised masks also called "community masks" made with easily available materials such as cotton was allowed. Once the emergency phase was over, the need arose to procure as many face masks as possible for the purpose and during this phase, surgical masks have played a fundamental role. However, the needs to contain the spread of the virus have led the Italian authorities to adopt more reliable personal protective equipment, the FFP2 face masks. Both surgical and FFP2 masks meet specific quality standards in their construction and their filtration efficiency. Modern tools for comparing environmental aspects regarding products or processes such as life cycle assessment require the use of products that have the same qualities and technical characteristics in order to provide credible data. For this reason, in our discussion, community masks made with easily available materials such as cotton were neglected while attention was focused on standard products such as type II and IIR surgical masks and FFP2, N95 type masks, which have been widely used in Italy to contain the spread of the virus although in different periods. Once the pandemic emergency was over, the researchers' attention was focused on the production, but above all on the evaluation of the environmental impacts linked to the use of these masks. In evaluating environmental impacts, the best possible choice is to evaluate them along the entire life cycle of the product, starting from the production of raw materials through the production of the finished product, its distribution, use and finally, disposal. Certainly, one of the most studied impact categories in the literature is the carbon footprint which constitutes one of the environmental impacts treated in life cycle assessment studies. However, a reliable assessment of the carbon footprint as a life cycle assessment study requires a series of fundamental assumptions that greatly influence the results, which are the choice of the functional unit, of the system boundaries, of the databases from which to obtain the data inventory, the local transport and disposal systems. Due to a precautionary principle, most public institutions have recommended incineration with or without energy recovery as the primary disposal option, while reuse and recycling have been discouraged for health reasons. The technologies available for the manufacturing of the filter elements of these masks require the use of virgin raw materials and can hardly operate with recycled raw materials. Taking into account such limitations the literature revealed that the CF of an FFP2 mask is in most of the studies analyzed, higher than that of a surgical mask, the reason is that the first have a greater mass and require a greater use of materials in their production, the filtration efficiency is superior and the FFP2 face mask also have the advantage of adapting better to the user's face and therefore to contain the uncontrolled spread of viral particles in the ambient air by fugitive emissions. Researchers' attention, therefore, with limitation in recycling options, shifted to reuse through disinfection to contain the environmental impacts linked to the use of these masks, and among the reuse techniques, some have proven valid while others have not. Another possibility to reduce the carbon footprint of masks is to use alternative materials to those usually used in their production. However, this option is difficult to apply industrially and sometimes lead to results that are worse than the traditional materials. Nonetheless the variability induced by the crucial choices typical of every CF study, the contribution's analysis shows that in the case of single-use masks, the main impact derives from the production of the material constituting the filtering membranes while transport does not significantly affect the generation of the carbon footprint. The situation changes drastically in the case in which a reuse of the masks through sanitization is modeled; in this case, if the sanitization takes place through systems that use electricity (dry heating, UV), the greater environmental impact derives from the production of electricity while in the case of reuse systems through wet washing, it comes from the production of water and the detergents used for washing, obviously including their disposal. It should be noted that none of the disinfection methods are particularly effective in maintaining the filtering performance of the devices. In fact, some wet washing techniques cause a reduction in the filtering power of the masks due to the neutralization of the electrets present on the polymer; furthermore, the number of washes that can be carried out is not unlimited but responds to precise numerical limitations as demonstrated by studies on the quantity of virus present on the masks at the end of a disinfection process. Some disinfection processes also have severe limitations due to the presence of residues of toxic substances on the masks at the end of the disinfection process, still others are completely ineffective in containing the proliferation of the virus. Recently, research has shifted to the use of traditional materials containing substances that increase their antiviral and antibacterial properties; however, the environmental compatibility of these masks remains to be demonstrated given that many of these chemical substances need a lot of energy to be produced, generate significant environmental impacts when disposed of and may have harmful effects on the users. A policy to reduce the carbon footprint of face masks to contain the spread of viruses and bacteria and, in general of airborne diseases, must take into account numerous factors that act sometimes in a synergistic manner, sometimes in an antagonistic manner, the choice depends greatly by local variables such as the availability of low-cost energy and raw materials, the impact of transportation in the production (that may happen locally or the devices can be imported), distribution, and disposal process. Disposal of face masks varies greatly from country to country, so the results are strongly influenced by local variables; energy recovery incineration systems are still preferable to those without energy recovery, but such combined heat and electricity multifunctional incinerating systems are not used in all countries, or the mix between them is locally different. The fundamental aspect that emerges from these studies is the need to combine research on materials and reuse/recycle strategies with research on the environmental impacts linked to the life cycle of devices, to contain the spread of the airborne diseases, and probably the best approach is that of safe and sustainability by design of products as depicted by the recent EU document (Caldeira et al., 2022). The approach would support the design, development, production, and use of both chemicals and materials taking into account not only the environmental aspects but also the harmful impacts to human health. The integration of these two aspects, the domain of safety, and that of circularity of chemicals and materials, in the view of sustainability considerations throughout the lifecycle, will foster the transition to safe, zero pollution, climate-neutral, and resource-efficient economy in the EU.

CRediT Authorship Contribution Statement *Pasquale Giungato*: background, structure of face masks, production, carbon footprint of protective face masks, strategies reducing carbon footprint: effect of materials and of sanitization, conclusions. *Roberto Leonardo Rana*: background, production and distribution, carbon footprint of protective face masks, strategies reducing carbon footprint: effect of materials. *Caterina Tricase*: background, production and distribution, carbon footprint of protective face masks, strategies reducing carbon footprint: effect of sanitization.

References

Aboubakr, H. A., Sharafeldin, T. A., & Goyal, S. M. (2020). Stability of SARS-CoV-2 and other coronaviruses in the environment and on common touch surfaces and the influence of climatic conditions: A review. In *Transboundary and emerging diseases* (pp. 1–17).

Al-Ma'adeed, M., Ozerkan, G., Kahraman, R., Rajendran, S., & Hodzic, A. (2011). Life cycle assessment of particulate recycled low density polyethylene and recycled polypropylene reinforced with talc and fiberglass. *KEM, 471–472*, 999–1004. https://doi.org/10.4028/www.scientific.net/kem.471-472.999

Belkin, N. L. (1997). The evolution of the surgical mask: Filtering efficiency versus effectiveness. *Infection Control and Hospital Epidemiology, 18*, 49–57.

Bergman, M. S., Viscusi, D. J., Heimbuch, B. K., Wander, J. D., Sambol, A. R., & Shaffer, R. E. (2010). Evaluation of multiple (3-cycle) decontamination processing for filtering facepiece respirators. *Journal of Engineered Fibers and Fabrics, 5*(4), 33–41.

Bhattacharjee, S., Bahl, P., de Silva, C., Doolan, C., Chughtai, A. A., Heslop, D., & Mac Intyre, C. R. (2021). *ACS Biomaterials Science & Engineering, 7*(6), 2791–2802. https://doi.org/10.1021/acsbiomaterials.1c00368

Brosseau, L. M., McCullough, N. V., & Vesley, D. (1997). Bacterial survival on respirator filters and surgical masks. *Journal of the American Biological Safety Association, 2*, 32–43.

Caldeira, C., Farcal, R., Garmendia, A. I., Mancini, L., Tosches, D., Amelio, A., Rasmussen, K., Rauscher, H., Riego, S. J., & Sala, S. (2022). *Safe and sustainable by design chemicals and materials – Framework for the definition of criteria and evaluation procedure for chemicals and materials, EUR 31100 EN*. Publications Office of the European Union. https://doi.org/10.2760/487955, JRC128591. ISBN 978-92-76-53264-4.

Chapagain, A. K., Hoekstra, A. Y., Savenjie, H. H. G., & Gautam, R. (2006). The water footprint of cotton consumption: An assessment of the impact of worldwide consumption of cotton products on the water resources in the cotton producing countries. *Ecological Economics, 60*, 186–203. https://doi.org/10.1016/j.ecolecon.2005.11.027

Cheng, P., Zheng, C., Zhang, Z., Li, Y., Huang, K., Yu, D., & Ji, Y. (2024). Preparation and performance of bitumen modified by melt-blown fabric of waste mask based on Grey relational and radar chart analysis. *Polymers, 16*, 153. https://doi.org/10.3390/polym16010153

Chin, A. W. H., Chu, J. T. S., Perera, M. R. A., Hui, K. P. Y., Yen, H.-L., Chan, M. C. W., et al. (2020). Stability of SARS-CoV-2 in different environmental conditions. *Lancet Microbe, 1*, e10.

Cho, K.-C., Park, S.-W., Lee, I., & Shim, J. (2023). Process prediction and feature visualization of Meltblown nonwoven fabrics using scanning electron microscopic (SEM) image-based deep neural network algorithms. *Processes, 11*, 3388. https://doi.org/10.3390/pr11123388

Cornelio, A., Zanoletti, A., Federici, S., Ciacci, L., Depero, L. E., & Bontempi, E. (2022). Environmental impact of surgical masks consumption in Italy due to COVID-19 pandemic. *Materials, 15*, 2046. https://doi.org/10.3390/ma15062046

Di Giacomo, J., Odom, J., Ritota, P., & Swan, K. (1992). Cost containment in the operating room: Use of reusable versus disposable clothing. *The American Surgeon, 58*, 654–656.

Do, T., Huyen, T., Mizsey, P., & Andras, J. T. (2021). Applicability of membranes in protective face masks and comparison of reusable and disposable face masks with life cycle assessment. *Sustainability, 13*(22), 12574. https://doi.org/10.3390/su132212574

Dutton, K. C. (2008). Overview and analysis of the meltblown process and parameters. *Journal of Textile and Apparel, Technology and Management, 6*(1), 1–24.

EEA. (2021a). *Progress towards preventing waste in Europe — The case of textile waste prevention* (EEA report no 15/2021). European Environment Agency. Available at: https://www.eea.europa.eu/publications/progressing-towards-waste-prevention-in. Accessed 9 July 2022.

EEA. (2021b). *Diversion of waste from landfill in Europe*. European Environment Agency. Available at: https://www.eea.europa.eu/ims/diversion-of-wastefrom-landfill. Accessed 9 July 2022.

Emadian, S. M., Onay, T. T., & Demirel, B. (2016). Biodegradation and bioplastics in natural environments. *Waste Management, 59*, 526–536.

EN. (2001). *Respiratory protective devices. Filtering half masks to protect against particles.* Requirements, testing, marking. European standard EN 149, 2001+A1: 2009.

Fadare, O. O., & Okoffo, E. D. (2020). Covid-19 face masks: A potential source of microplastic fibers in the environment. *Science of Total Environment, 737*, 140279. https://doi.org/10.1016/j.scitotenv.140279

Fant, S., (2022). *Recycling COVID masks: Here are those who attempt it around the world.* Renewable Matter. https://www.renewablematter.eu/articles/article/recycling-covid-masks-here-are-those-who-attempt-it-around-the-world. Accessed 16 Jan 2024.

Ghodrat, M., Rashidi, M., & Samali, B. (2017). Life cycle assessments of incineration treatment for sharp medical waste. *Energy Technology*, 131–143.

Giungato, P., Rana, R. L., Nitti, N., Cavallari, C., & Tricase, C. (2021). Carbon footprint of surgical masks made in Taranto to prevent SARS-CoV-2 diffusion: A preliminary assessment. *Sustainability, 13*, 6296. https://doi.org/10.3390/su13116296

Giungato, P., Moramarco, B., Rana, R. L., & Tricase, C. (2024). Carbon footprint of FFP2 protective facial masks against SARS-CoV-2 used in the food sector: Effect of materials and dry sanitisation. *British Food Journal, 126*(1), 33–47. https://doi.org/10.1108/BFJ-09-2022-0773

Hogan, B., & Samaranayake, L. P. (1990). The surgical mask unmasked: A review. *Oral Surgery, Oral Medicine, and Oral Pathology, 70*, 34–36.

Indhu Kavi, P., Amarnath, J. S., & Sivasankari, B. (2018). Economic and environmental impact of pesticide use in conventional cotton and bt cotton. *Economic Affairs, 63*(2), 425–431. https://doi.org/10.30954/0424-2513.2.2018.19

Klemes, J. J., Fan, Y. V., & Jiang, P. (2020). The energy and environmental footprints of COVID-19 fighting measures-PPE, disinfection, supply chains. *Energy, 211*, 118701. https://doi.org/10.1016/j.energy.2020.118701

Kressel, A. B., McVey, J. L., Miller, J. M., & Fish, L. L. (2011). Hospitals learn their collective power: An isolation gown success story. *American Journal of Infection Control, 39*, 76–78.

Lee, A. W. L., Neo, E. R. K., Khoo, Z.-Y., Yeo, Z., Tan, Y. S., Chng, S., Yan, W., Lok, B. K., & Low, J. S. C. (2021). Life cycle assessment of single-use surgical and embedded filtration layer (EFL) reusable face mask, resources. *Conservation and Recycling, 170*, 105580. https://doi.org/10.1016/j.resconrec.2021.105580

Leucker, K. (2020). *Influence of the calender pattern on the mechanical properties of polypropylene spunbond nonwovens*. PhD dissertation, Advisor: Dirk W. Schubert, Faculty of Technology, Friedrich Alexander University Erlangen-Nuremberg.

Liao, L., Xiao, W., Zhao, M., Yu, X., Wang, H., Wang, Q., Chu, S., & Can, Y. (2020). N95 respirators be reused after disinfection? How many times? *ACS Nano, 14*, 6348–6356. https://doi.org/10.1021/acsnano.0c03597

Lim, H. (2020). A review of spun bond process. *Journal of Textile and Apparel, Technology and Management, 6*(3), 1–13.

Lindsley, W. G., Blachere, F. M., Thewlis, R. E., Vishnu, A., Davis, K. A., Cao, G., Palmer, J. E., Clark, K. E., Fisher, M. A., Khakoo, R., & Beezhold, D. H. (2010). Measurements of airborne influenza virus in aerosol particles from human coughs. *PLoS One, 5*(5), e15100.

Ludwig-Begall, L. F., & Heyne, B. (2024). aPDI meets PPE: Photochemical decontamination in healthcare using methylene blue—Where are we now, where will we go? *Photochemical & Photobiological Sciences*. https://doi.org/10.1007/s43630-023-00514-1

Luo, Y., Yu, M., Wu, X., Ding, X., & Wang, L. (2023). Carbon footprint assessment of face masks in the context of the COVID-19 pandemic: Based on different protective performance and applicable scenarios. *Journal of Cleaner Production, 387*, 135854. https://doi.org/10.1016/j.jclepro.2023.135854

Majchrzycka, K., Okrasa, M., Jachowicz, A., Szulc, J., & Gutarowska, B. (2018). Microbial growth on dust-loaded filtering materials used for the protection of respiratory tract as a factor affecting filtration efficiency. *International Journal of Environmental Research and Public Health, 15*, 1902. https://doi.org/10.3390/ijerph15091902

Matuschek, C., Moll, F., Fangerau, H., Fischer, J. C., Zänker, K., van Griensven, M., et al. (2020). The history and value of face masks. *European Journal of Medical Research, 25*, 1–6.

Midha, V., & Dakuri, A. (2017). Spun bonding technology and fabric properties: a review. *Journal of Textile Engineering and Fashion Technology, 1*(4), 126–133. https://doi.org/10.15406/jteft.2017.01.00023

Nicosia, A., Gieparda, W., Foksowicz-Flaczyk, J., Walentowska, J., Wesolek, D., Vazquez, B., Prodi, F., & Belosi, F. (2015). Air filtration and antimicrobial capabilities of electrospun PLA/PHB containing ionic liquid. *Separation and Purification Technology, 154*, 154–160. https://doi.org/10.1016/j.seppur.2015.09.037

NIOSH. (1995). Public Health Service, Department of Health and Human Services, 42 CFR Part 84, 30336 Federal Register, Vol. 60, No. 110, June 8, 1995, Rules and Regulations, available at: https://www.govinfo.gov/content/pkg/CFR-2022-title42-vol1/pdf/CFR-2022-title42-vol1-part84.pdf. Accessed 8 Jan 2024.

O'Hearn, K., Gertsman, S., Webster, R., Tsampalieros, A., Ng, R., Gibson, J., Sampson, M., Sikora, L., & McNally, J. D. (2020). Efficacy and safety of disinfectants for decontamination of N95 and SN95 filtering facepiece respirators: A systematic review. *Journal of Hospital Infection, 106*(3), 504–521. https://doi.org/10.1016/j.jhin.2020.08.005

Pardo-Figuerez, M., Chiva-Flor, A., Figueroa-Lopez, K., Prieto, C., & Lagaron, J. M. (2021). Antimicrobial nanofiber based filters for high filtration efficiency respirators. *Nanomaterials, 11*, 900. https://doi.org/10.3390/nano11040900

Pratumpong, P., Cholprecha, T., Roungpaisan, N., Srisawat, N., Toommee, S., Pechyen, C., & Parcharoen, Y. (2023). Effects of melt-blown processing conditions on nonwoven Polylactic acid and polybutylene succinate. *Polymers, 15*, 4189. https://doi.org/10.3390/polym15204189

President of the Republic of Italy. (2021). Decreto-Legge 24 dicembre 2021, n. 221 Proroga dello stato di emergenza nazionale e ulteriori misure per il contenimento della diffusione dell'epidemia da COVID-19. (21G00244) (GU Serie Generale n.305 del 24-12-2021). Available at: https://www.gazzettaufficiale.it/eli/id/2021/12/24/21G00244/sg. Accessed 8 Jan 2024.

Rizan, C., Reed, M., & Bhutta, M. F. (2021). Environmental impact of Personal Protective Equipment distributed for use by health and social care services in England in the first six months of the COVID-19 pandemic. *MedRxiv*. https://doi.org/10.1101/2020.09.21.20198911

Rizman, M., Mujtaba, G., Memon, S. S., Lee, K., & Rashid, N. (2018). Exploring the potential of microalgae for new biotechnology applications and beyond: A review. *Renewable and Sustainable Energy Reviews, 92*, 394–404.

Rowan, N. J., & Lafey, J. G. (2021). Unlocking the surge in demand for personal and protective equipment (PPE) and improvised face coverings arising from coronavirus disease (COVID19) pandemic—Implications for efficacy, re-use and sustainable waste management. *Science of the Total Environment, 752*, 142259. https://doi.org/10.1016/j.scitotenv.2020.142259

Salman, A., Sholihah, M., Mitake, Y., Tsutsui, Y., Wang, H., & Shimomura, Y. (2022). A sustainable approach towards disposable face mask production amidst pandemic outbreaks. *Sustainability, 14*(7), 3849. https://doi.org/10.3390/su14073849

Sandin, G., & Peters, G. M. (2018). Environmental impact of textile reuse and recycling-a review. *Journal of Cleaner Production, 184*, 353–365. https://doi.org/10.1016/j.jclepro.2018.02.266

Sanyal, A., & Sinha-Ray, S. (2021). Ultrafine PVDF nanofibers for filtration of air-borne particulate matters: A comprehensive review. *Polymers, 13*, 1864. https://doi.org/10.3390/polym13111864

Sinkko, T., Ardente, F., Scaccabarozzi, D., & Fumagalli, F. (2023). Life cycle assessment of face mask decontamination via atmospheric pressure plasma. *Journal of Cleaner Production, 422*, 138308. https://doi.org/10.1016/j.jclepro.2023.138308

Siu, J., Hill, A. G., & MacCormick, A. D. (2017). Systematic review of reusable versus disposable laparoscopic instruments: Costs and safety. *ANZ Journal of Surgery, 87*(1–2), 28–33. https://doi.org/10.1111/ans.13856

Spennemann, D. H. R. (2022a). COVID-19 face masks as a long-term source of microplastics in recycled urban green waste. *Sustainability, 14*(1), 207. https://doi.org/10.3390/su14010207

Spennemann, D. H. R. (2022b). Environmental decay of single use surgical face masks as an agent of plastic micro-fiber pollution. *Environments, 9*, 94. https://doi.org/10.3390/environments9070094

Spooner, J. L. (1967). History of Surgical Face Masks: The myths, the masks, and the men and women behind them. *AORN Journal, 5*, 76–80.

Strasser, B. J., & Schlich, T. (2020). A history of the medical mask and the rise of throwaway culture. *Lancet, 396*, 19–20.

Tcharkhtchi, A., Abbasnezhad, N., Zarbini Seydani, M., Zirak, N., Farzaneh, S., & Shirinbayan, M. (2021). An overview of filtration efficiency through the masks: Mechanisms of the aerosols penetration. *Bioactive Materials, 6*, 106–122.

Tessarolo, F., Nollo, G., Maniglio, D., Rigoni, M., Benedetti, L., Helfer, F., Corradi, I., Rovati, L., Ferrari, A., Piccini, M., Accorsi, L., Veronesi, E., Cuoghi, A., Baglio, S., Tuccitto, N., Stefani, S., Stracquadanio, S., Caraci, F., Terrasi, A., Tricomi, A., Musumeci, M., Miraglia, A., Cuttone, G., Cosentino, S., Muscas, C., Vitali, L. A., Petrelli, D., Angrisani, L., Colicchio, R., D'Anna, A., Iavicoli, I., De Falco, G., Di Natale, F., Di Maio, E., Salvatore, P., Quaglia, F., Mingoia, M., Castellini, P., Chiariotti, P., Simoni, S., Montalto, L., Baleani, A., & Paone, N. (2021). Testing surgical face masks in an emergency context: The experience of Italian laboratories during the COVID-19 pandemic crisis. *International Journal of Environmental Research and Public Health, 18*(4), 1462. https://doi.org/10.3390/ijerph18041462. PMID: 33557403; PMCID: PMC7915703.

Thakuv, S., Chaudhary, J., Sharma, B., Verma, A., Tamulevinus, S., & Thakur, V. K. (2018). Sustainability of bioplastics – Opportunities and challenges. *Current Opinion in Green, Sustainable Chemistry, 13*, 68–75.

Thiruchelvi, R., Kavitha, K., & Shankari, K. (2020). New biotechnological routes for greener bioplastics from seaweed. *Research Journal of Pharmacy and Technology, 13*(5). https://doi.org/10.5958/0974-360X.2020.00444.8

van Straten, B., Ligtelijn, S., Droog, L., Putman, E., Dankelman, J., Sperna Weiland, N. H., & Horeman, T. (2021). A life cycle assessment of reprocessing face masks during the Covid-19 pandemic. *Scientific Reports, 11*, 17680. https://doi.org/10.1038/s41598-021-97188-5

Viscusi, D. J., Bergman, M. S., Sinkule, E., & Shaffer, R. E. (2009). Evaluation of the filtration performance of 21 N95 filtering face piece respirators after prolonged storage. *American Journal of Infection Control, 37*, 381–386.

Viscusi, D. J., et al. (2011). Impact of three biological decontamination methods on filtering facepiece respirator fit, odor, comfort, and donning ease. *Journal of Occupational and Environmental Hygiene, 8*(7), 426–436.

Vozzola, E., Overcash, M., & Griffing, E. (2018). Environmental considerations in the selection of isolation gowns: A life cycle assessment of reusable and disposable alternatives. *American Journal of Infection Control, 46*, 881–886. https://doi.org/10.1016/j.ajic.2018.02.002

WHO. (2020). Director-General's opening remarks at the media briefing on COVID-19—11 March 2020. Retrieved Gen 6, 2024, from: https://www.who.int/director-general/speeches/detail/who-director-general-s-opening-remarks-at-the-media-briefing-on-covid-19%2D%2D-11-march-2020.

Zins, H. M., & Howard, M. (2011). Reusable medical textiles. In V. Bartels (Ed.), *Handbook of medical textiles* (pp. 80–105). Woodhead Publishing Limited.

Part II
Innovative & Best Practices in Carbon Footprint Assessments

Nature-Based Solutions to Reduce Carbon Footprint and Increase Resilience of Road Transportation Infrastructure

Victor Hugo Souza de Abreu, Rafael Ferraz dos Santos, Sandra Oda, and Andrea Souza Santos

1 Introduction

Road transportation, like other critical infrastructures, is profoundly impacted by the effects of climate change in the short, medium, and long term (Moretti & Loprencipe, 2018). This sector will continue to face significant challenges, including reduced efficiency in transportation operations and the limited capacity of infrastructure to withstand extreme weather events beyond the usual patterns (De Abreu et al., 2022b, c, 2023b; Evans et al., 2009). Transport networks play a crucial role in the economy and society (Santos et al., 2021), and adapting these infrastructures to climate change is imperative (de Abreu et al., 2021a, b; Santos et al., 2020).

In this context, Nature-Based Solutions (NbSs) are crucial for building resilience in road transportation infrastructure and other critical infrastructures, in addition to sequestering carbon and enhancing biodiversity, offering a range of other benefits for people, such as improving health and well-being (Baskin, 2023; Plovnick & Filosa, 2023). NbS refers to approaches that utilize ecosystems and their services to address challenges related to climate change and other environmental issues (Seddon et al., 2020). This includes measures such as ecosystem restoration, the creation of green corridors, and the implementation of sustainable soil and water management practices (IPCC, 2022). These strategies expand the capacity for mitigation and adaptation to climate change in urban environments (Lafortezza & Sanesi, 2019; Osaka et al., 2021). NbS addresses global challenges directly linked to the Sustainable Development Goals (SDGs) (Somarakis et al., 2019).

V. H. S. de Abreu (✉) · R. F. dos Santos · S. Oda · A. S. Santos
Federal University of Rio de Janeiro (Universidade Federal do Rio de Janeiro – UFRJ), Transport Engineering Programme (Programa de Engenharia de Transportes – PET), COPPE-UFRJ, Rio de Janeiro, Brazil
e-mail: victor@pet.coppe.ufrj.br

S. S. Muthu (ed.), *Carbon Footprint Assessments*, Environmental Footprints and Eco-design of Products and Processes,
https://doi.org/10.1007/978-3-031-70262-4_7

Despite the proven benefits of NbS for climate adaptation, it is concerning to note that less than 10% of public funding allocated to climate-related issues specifically targets NbS for adaptation. This situation worsens when considering the transportation sector, where NbS receives even less attention (Baskin, 2023). This lack of investment and attention to NbS in the transportation sector is alarming, given that the impacts of climate change pose a significant threat to road infrastructure and the safety of transportation systems. Rising temperatures, the occurrence of more intense and prolonged extreme weather events, such as storms, and sea-level rise jeopardize the efficient operation of highways, bridges, and drainage systems (De Abreu et al., 2021b, c).

Within this context, the aim is to understand the importance of NbS for reducing carbon footprint and increasing the resilience of critical infrastructures. In this perspective, the research seeks to map the current NbS available to strengthen the resilience of the road transportation sector in the face of climate change impacts. Additionally, the goal is to identify the main challenges and offer potential solutions associated with promoting more sustainable and adaptive development.

2 Methodological Approach

Faced with the difficulty of obtaining a solid database through direct searches in databases using search terms, particularly when it comes to introducing NbS and Ecosystem-Based Adaptation (EbA) in the road transportation sector, it became necessary to adopt a broader document review approach, seeking directly applicable materials to the theme.

To provide an in-depth review, this study utilized a systematic review with a bibliometric approach. This involved conducting direct searches in the Web of Science and Scopus databases, which are known for their wide reach and comprehensive coverage. However, when using the search terms *TS = ("nature-based solution"AND "road transport*") OR TS = ("ecosystem-based adaptation" AND "road transport*")*, it was observed that there was a lack of relevant results for the research. This limitation suggests a literature gap in the field of studies related to NbS applied to road transportation, thus representing a research opportunity for understanding and applying these solutions in this specific context.

In addition to academic databases, a diverse range of materials, including technical reports, government documents, and gray literature, were explored to conduct a comprehensive review on the integration of NbS in the road transportation sector. This inclusive approach led to the selection of 59 studies, with insights drawn from various domains. Notably, the latest Intergovernmental Panel on Climate Change (IPCC) Sixth Assessment Report (AR6) (IPCC, 2022) played a central role. It is essential to highlight that around 71% of these studies addressed critical infrastructure and climate change solutions more broadly, emphasizing the necessity for an integrated perspective that considers the interdependencies among different transportation modes and overall infrastructure.

Another relevant aspect that deserves emphasis is the importance of the studies incorporated into the database. To ensure the quality and relevance of the selected scientific knowledge, priority was given to research conducted by renowned international institutions, as well as papers published in reputable journals or presented at prestigious international conferences, all published in recent years. This meticulous selection aimed to ensure the reliability and currency of the information used as the foundation for the study.

3 The Importance of NbS for Carbon Footprint Reduction and Enhancing the Resilience of Critical Infrastructures

Traditionally, urban planners and land and resource management professionals have relied on conventional engineering solutions to adapt to climate change. However, this approach is not always economical, sufficient, and/or sustainable. NbS presents a valuable alternative, addressing the social challenges resulting from climate change and urbanization in a sustainable manner (Somarakis et al., 2019). Additionally, NbS offers various additional benefits to urban residents, including improvements in health, well-being, and the promotion of the local green economy (Kabisch et al., 2017).

Thus, the increasing use of NbS is observed in guiding the design of resilient landscapes and cities (Lafortezza et al., 2018). By capitalizing on ecosystem services, these innovative solutions use natural elements to achieve environmental and social objectives, representing significant potential to provide efficient responses in terms of energy and resources to climate change while enhancing our natural capital (Kabisch et al., 2017; Lafortezza et al., 2018).

According to Keesstra et al. (2018), NbS can be categorized into two main sets of strategies: (i) soil solutions and (ii) landscape solutions. Soil solutions aim to enhance soil health and functions by promoting the maintenance or restoration of local ecosystem services. On the other hand, landscape solutions primarily focus on the concept of connectivity. By making the landscape less connected, the transformation of rainfall into runoff is facilitated, reducing the risk of flooding. Moreover, this approach increases soil moisture, decreasing the likelihood of droughts and erosion. These measures directly contribute to promoting sustainability and enhancing ecosystem services.

It is important to emphasize that EbA is a subset of NbS that has been widely employed as an effective strategy to assist people in adapting to climate change. EbA aims to utilize natural ecosystems and their services to strengthen the resilience of communities and social systems in the face of challenges posed by environmental changes. EbA involves the conservation, restoration, and sustainable management of natural ecosystems such as forests, wetlands, mangroves, coral reefs, and other habitats to provide adaptive benefits to human populations. These ecosystems play a fundamental role in mitigating the negative climate change impacts, offering valuable ecosystem services such as local climate regulation, protection against storms and floods, water purification, and maintenance of biodiversity (IPCC, 2022).

NbS plays a vital role in achieving carbon-neutral cities. However, to thoroughly grasp the climate neutrality potential of NbS, it is imperative to comprehensively estimate and quantify their impacts, encompassing both direct and indirect effects on social and economic systems (Chausson et al., 2020; Pan et al., 2023). Additionally, realistic implementations of NbS to achieve carbon neutrality must be effective on a global scale while adapting to specific socio-economic and physical conditions prevailing at the local level (Cong et al., 2023).

In this context, a study conducted by Pan et al. (2023) assessed the carbon mitigation potential of NbS in European cities. The analysis focused especially on often overlooked indirect pathways, such as human behavioral interventions and resource savings. The results indicate that considering theoretical maximum implementation, NbSs in the residential, transportation, and industrial sectors have the potential to reduce urban carbon emissions by up to 25%. The spatial prioritization of different types of NbSs in 54 major cities in the European Union could result in an average reduction of 17.4% in anthropogenic carbon emissions.

This suggests that NbS provides essential environmental benefits, such as enhancing biodiversity, increasing carbon storage, reducing extreme temperatures, and controlling rainwater flooding (Epelde et al., 2022). However, to re-naturalize cities through the implementation of NbS, the European Commission (2015) proposed a set of actions as presented in Fig. 1.

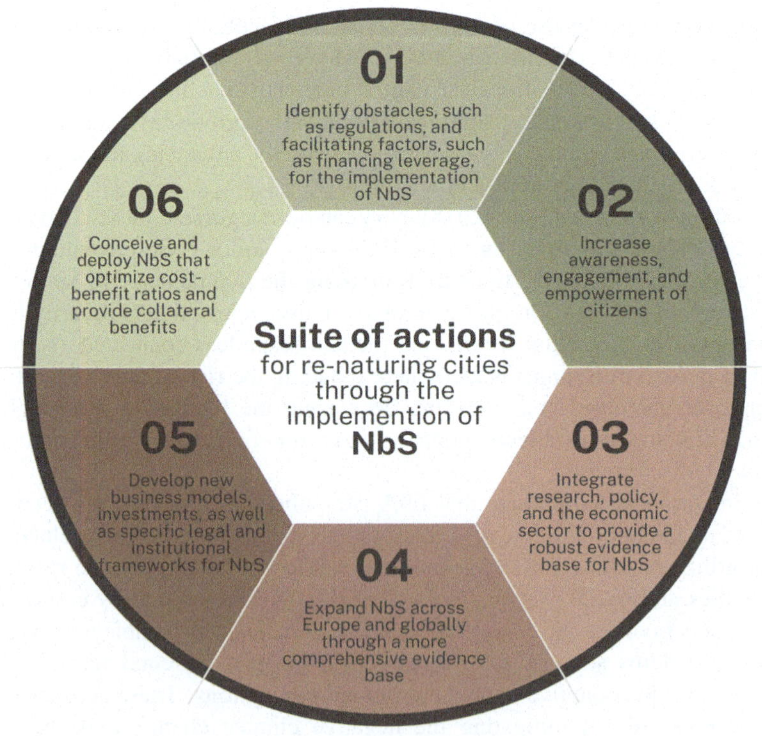

Fig. 1 Actions in favor of the implementation of NbS

It is worth noting that despite substantial evidence of the benefits associated with NbS, especially in highly dense cities, Lafortezza et al. (2018) indicate that additional application studies are needed to address the following challenges: (i) understanding the connections between NBS and the associated ecosystem services in the four main categories of provisioning, regulating, cultural, and supporting, at different scales (e.g., from the "central" urban area to the broader peri-urban area); (ii) assessing NBS using various sensors and data sources, including high-resolution satellite sensor images, field sensors, aerial LiDAR, and field data.

Additionally, according to Bush and Doyon (2019), urban planning plays a crucial role in supporting the implementation of NbS and in managing trade-offs and conflicts, including considerations of social equity dimensions. Urban planning contributes to processes and mechanisms involving the definition of goals and strategic objectives, the application of participatory approaches in knowledge and data collection, as well as decision-making processes related to land use. If these urban planning processes are applied to the design and implementation of NbS, they have the potential to support and promote wider implementation. Furthermore, integrating NbS into land use planning provisions and regulations can also encourage increased implementation.

According to Kabisch et al. (2022), aiming to strengthen the development of resilient future cities, it is imperative that the design, planning, policy formulation, and governance of NbS incorporate the aspects highlighted in Fig. 2.

Each of the aspects mentioned in Table 2 can be understood as (Kabisch et al., 2022):

- Systemic understanding: In addition to considering the individual elements of NbS, it is crucial to adopt a holistic perspective, understanding the interconnections between different components of the urban system. This implies analyzing not only immediate benefits but also long-term impacts throughout the urban ecosystem.
- Community well-being and biodiversity: The contribution of NbS should go beyond mere mitigation of urban challenges; it should be a source of human well-being improvement. This includes creating accessible green spaces, promoting sustainable agricultural practices, and preserving natural habitats. Thus, NbS not only benefits people but also contributes to the health and diversity of local ecosystems.
- Inclusive long-term solutions: To ensure long-term sustainability, NbS should be conceived as inclusive solutions that address the needs of all layers of society. This requires considering social, economic, and cultural aspects, ensuring that NbS not only addresses immediate challenges but also promotes equity and resilience over time.

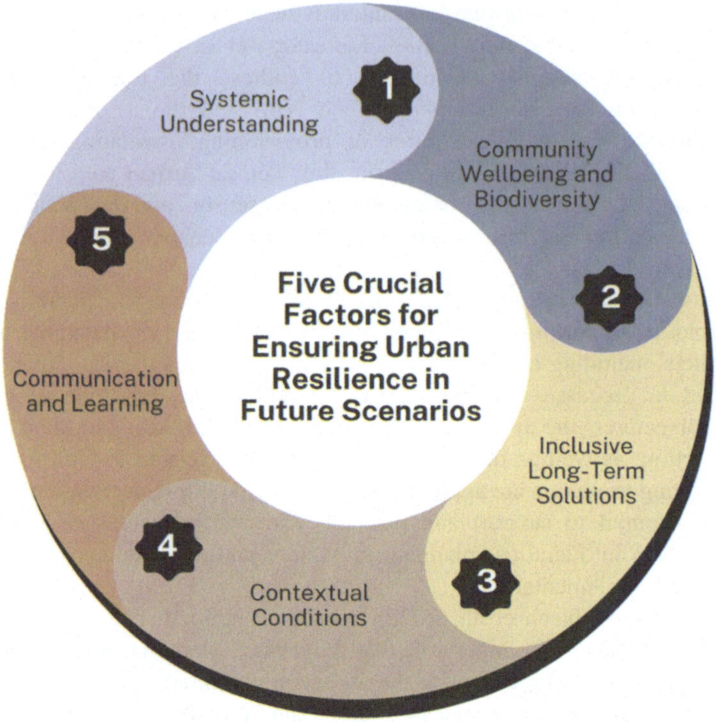

Fig. 2 Actions to strengthen the development of resilient future cities through NbS

- Contextual conditions: Each urban area has unique characteristics, and NbS should be adapted to the specific conditions of the context. This involves considering factors such as climate, topography, population density, and existing infrastructure. Customizing NbS according to local conditions enhances their effectiveness and community acceptance.
- Communication and learning: The effectiveness of NbS is intrinsically linked to transparent communication and continuous promotion of learning. Actively involving the community in the design and implementation of NbS, providing clear information about the benefits, and engaging in ongoing educational programs are fundamental elements for the long-term success of NbS in resilient urban environments.

4 The Impacts of Climate Change on the Road Transport Sector and the Need for Adaptation

From an infrastructure perspective, the story of climate change can be told through the increasing number of occurrences of extreme weather events and their impacts (Guy, 2015). Climate change will amplify existing risks related to weather and may create new ones (IPCC, 2014b). A key factor in determining a society's resilience to climate change is the vulnerability of national critical infrastructure systems to current and future natural disasters and extreme events (Hawchar et al., 2020), which can lead to cascading effects across different sectors (Santos et al., 2020).

Those responsible for the development and improvement of infrastructure in the transportation sector face a new fundamental challenge considering the growing concern about climate change (de Abreu et al., 2022a). This issue is considered one of the planet's most significant global challenges today (Santos et al., 2020). The climate change impacts on the transportation network can result in the loss of infrastructure assets and hinder the recovery and resilience of the entire system (Trindade et al., 2019).

Climate risks directly affect physical infrastructure assets and have a significant impact on the country's economy (Alhjouj et al., 2022). Therefore, qualifying and quantifying climate-related risks provide a greater margin for maneuvering to reduce disasters and losses at the national level. According to the IPCC (2014a), the risk of impact is the emerging result of the interaction between its dimensions (climate threat, exposure, and vulnerability—sensitivity and adaptive capacity), which are intrinsically associated with changes in the indicators to which they are subjected. Climate Risk Assessments (CRA) are therefore carried out considering these three dimensions of impact risk.

The climate change impacts on road infrastructure can be direct or indirect, with direct impacts having an immediate effect on physical infrastructure and, consequently, on existing transportation systems and networks (TRB, 2008). The direct impacts of extreme weather conditions on transportation, such as damage to pavements, bridges, and drainage systems (requiring infrastructure repair in terms of replacing damaged components) and landslides, are more easily assessed than indirect impacts or potential indirect effects between systems, such as indirect costs of delays, detours, and trip cancellations, which can also be substantial (Forzieri et al., 2018).

The threats posed by climate change can result in serious damage to road transportation infrastructure and, consequently, impact the safety of its operations, potentially leading to a significant transformation of the system, whether it be gradual or sudden (de Abreu et al., 2022b). Therefore, it becomes essential to recognize which threats exist and to what extent each threat interferes with the risk to the infrastructure. The study of threat indicators constitutes a fundamental step in this process (de Abreu et al., 2022c). Additionally, it is important to identify the road infrastructure or specific road sections that are at high risk of being affected by certain climatic events, giving priority to critical infrastructure (de Abreu et al., 2023b).

Furthermore, since the climate change impacts manifest locally, individual assets have different exposures to threats depending on their geographical location (Forzieri et al., 2018). Thus, the assessment of road infrastructure exposure to extreme weather events is of great importance for highway authorities, governmental institutions, and safety researchers worldwide (Schlögl & Laaha, 2017).

Due to their exposure to environmental conditions, transportation infrastructure systems, including highway pavement, bridges, and drainage systems, are particularly sensitive to climate change (Santos, 2014), threatening economic growth, development, and social well-being. This necessitates current adaptive capacity and timely future adaptation integrated into decision-making (Bles et al., 2016). Moderate climate changes may have minimal impacts on transportation. However, climate change and extreme weather events can generate significant impacts (PBMC, 2014).

Climate change can create a link between academic research and political requirements by bringing together intricate situations like the vulnerability of different regions, families, or countries into one vulnerability indicator. This indicator can be used easily in public policies (Hinkel, 2011). In this context, it becomes crucial for decision-makers (e.g., infrastructure owners, state agencies, local authorities, the general society, etc.) to gain an understanding of the vulnerability of critical infrastructure to current and future climate-related threats. This is essential for developing effective adaptation strategies (Bles et al., 2016; Hawchar et al., 2020).

Previous studies have shown that during extreme weather events, the integrity and safety of road infrastructure can be compromised, often leading to severe accidents (Michalis et al., 2016). Additionally, there is concrete evidence of the reduced lifespan of pavement related to climate change due to material deterioration, emphasizing the need to consider climate change in regular road infrastructure projects (Mallick et al., 2014; Mndawe et al., 2015).

This raises the question of how road authorities should proceed to integrate the climate variable into their regular asset management processes (World Bank Group, 2017). Therefore, transportation decision-makers (including road engineers, asset managers, climate change adaptation professionals, innovation managers, and project managers) must conduct a thorough CRA. Understanding and quantifying these risks is crucial for planning appropriate adaptation measures to safeguard and ensure the functioning of society (Forzieri et al., 2018). Moreover, by implementing adaptation measures, the risk profile of threats will decrease and should reach an acceptable level of risk (Bles et al., 2016).

Adaptation should be understood as a co-benefit of sustainable transportation and should contribute to increasing the resilience of transportation infrastructure for its long-term sustainability (Valenzuela et al., 2017). Climate change adaptation can be seen as a series of responses to the current and potential impacts of climate change, to minimize possible damages and capitalize on potential opportunities (De Abreu et al., 2023a). These responses can take various forms, ranging from infrastructure through engineering works to non-structural measures through political, educational, and social actions (IPCC, 2014a, b).

It is crucial, however, to initially identify road infrastructure that is at high risk of being affected by specific climatic events (prioritizing critical infrastructure) and, thus, develop adaptation measures aligned with the region under analysis within a limited budget (Hawchar et al., 2020). All of this requires efficient decision-making that prioritizes cost-benefit economic analysis, including social benefits, environmental costs, and the climate change impacts (Schweikert et al., 2014).

5 Nature-Based Solutions Applied to Road Transportation

The NbS demonstrates good performance in both rural and urban contexts, and its effectiveness is enhanced when combined with traditional infrastructure options. The use of hybrid solutions, integrating both green infrastructure and conventional (gray) infrastructure, for example, can result in significant benefits such as lower initial cost, more agile implementation, greater long-term sustainability, and lower maintenance costs. It is essential to involve local communities in the design and implementation of NbS in the road sector (IPCC, 2022).

To develop NbS suitable for the context, as presented in Table 1, it is crucial to conduct a comprehensive analysis of locations susceptible to climate threats and more intense biophysical impacts. The efficient allocation of resources for potential projects based on this information becomes a crucial responsibility of infrastructure planners (Chinowsky et al., 2013). Integrating NbSs at all stages of the project development process is one of the most effective approaches to achieving this goal (Mott Macdonald, 2023).

Table 1 highlights measures to enhance the resilience of road transportation in the face of climate change. Green infrastructure, ecological corridors, soil conservation, permeable materials, ecosystem restoration, natural drainage systems, natural barriers, sustainable stormwater management, urban agriculture, and natural river structures are presented as strategies. These actions aim to address issues ranging from rainwater absorption, urban heat reduction, and air quality improvement to biodiversity preservation and prevention of damage caused by extreme weather conditions.

6 Problems Identified

Despite the significant potential of NbS to minimize environmental impacts and promote social and economic gains in the road transportation sector, there are challenges to overcome for optimizing their effective implementation. Table 2 comprehensively synthesizes these challenges, addressing crucial aspects of applying NbS.

Among the primary obstacles, issues such as the need for more robust public policies and financial incentives to foster widespread adoption of sustainable practices stand out. Additionally, the lack of awareness and understanding of the benefits of NbS, both among managers and the community, emerges as a significant challenge. Efficient integration of NbS into existing infrastructure projects and overcoming regulatory barriers also prove crucial for a successful transition toward more sustainable practices in the road transportation sector. A detailed analysis of these challenges provides valuable insights to guide future strategies for the effective implementation of NbS in the context of road transportation.

Table 1 Mapping of NbS to enhance the resilience of the road transportation sector to the impacts of climate change

Adaptation measure	Description
Green infrastructure	Implementing green infrastructure involves incorporating natural elements such as trees, vegetation, rain gardens, and permeable areas along roadways (Cohen-Shacham et al., 2016). These features assist in the absorption and infiltration of rainwater, reducing the risk of floods and improving water quality. Additionally, vegetation can help mitigate the impact of urban heat and enhance air quality (de Abreu et al., 2023b; Jiricka-Pürrer, 2020; Plovnick & Filosa, 2023)
Ecological/green corridors	Ecological corridors are strips of natural habitat that connect protected areas or fragments of ecosystems, enabling the movement of species and the maintenance of biological diversity. In road transportation, establishing ecological corridors along highways can help mitigate habitat fragmentation and facilitate species dispersal, reducing negative impacts on wildlife and flora (Baskin, 2023)
Soil conservation measures	The adoption of soil conservation practices, such as the use of no-till farming techniques, vegetative cover, and terracing, helps reduce erosion, sedimentation, and soil degradation along roadways. These measures protect soil quality, minimize sediment runoff, and reduce the negative impact on adjacent water resources (Mott Macdonald, 2023)
The use of permeable materials	The use of permeable materials in road and parking lot construction allows water to infiltrate the soil, reducing surface runoff and the risk of floods (De Abreu et al., 2023a). These materials, such as permeable pavements and porous concrete pavements, help recharge groundwater and mitigate the effects of soil impermeability. Their installation can be up to 50% cheaper than traditional pavement, and their maintenance is generally more cost-effective (Baskin, 2023; de Abreu et al., 2022a). It is worth noting, however, that this NbS needs to be aligned with the implementation of efficient drainage systems
Ecosystem restoration	The restoration of natural ecosystems, such as beaches, dunes, wetlands, mangroves, maritime forests, and reefs, along coastlines can provide significant benefits (Plovnick & Filosa, 2023). These ecosystems play a crucial role in reducing water runoff, stabilizing the soil, and protecting against soil and rockslides, which can disrupt highway operations and even cause significant damage to road infrastructure. Additionally, they provide habitats for native species (de Abreu, 2023)
Natural drainage systems	The implementation of natural drainage systems, such as infiltration trenches and retention basins, can help control water runoff and filter pollutants before they reach adjacent water bodies. These measures assist in reducing floods, which significantly impact roadway infrastructure and improve water quality (de Abreu et al., 2023b)
Natural barriers	The use of natural barriers, such as dense vegetation belts, can help protect highways from strong winds and storms, reducing the risk of damage caused by extreme weather conditions (Baskin, 2023)
Sustainable stormwater management	The implementation of sustainable stormwater management practices along roads, such as the use of rain gardens and infiltration trenches, can help reduce water runoff, improve water quality, and mitigate the risk of floods (de Abreu et al., 2023b)

(continued)

Table 1 (continued)

Adaptation measure	Description
Urban agriculture	Promoting urban agriculture around roadways can contribute to reducing urban heat, improving air quality, and enhancing local food security. Additionally, this practice can aid in stormwater management and promote biodiversity (Artmann & Sartison, 2018)
Natural river structures	The creation or restoration of natural river structures, such as meanders and sandbanks, along rivers and streams near roads, can help improve water absorption capacity, reduce erosion, and provide habitats for local fauna and flora (IPCC, 2022)

The results presented in Table 2 provide a consolidated view of the complexities associated with the successful adoption of NbS, thus contributing to a deeper understanding of the factors influencing the effectiveness of these innovative approaches. Mapping NbS for the road transportation sector should consider the specific geographical, climatic, and ecological characteristics of each region to identify the most suitable and effective solutions. It is important to engage multidisciplinary experts and conduct technical-economic feasibility studies to ensure correct implementation and maximize the benefits of NbS in strengthening the resilience of the road transportation sector against impacts. NbSs are more likely to be effective and equitable when aligned with the scale of the challenge, involve diverse voices, and are adapted to specific social, cultural, ecological, and technological contexts (Hobbie & Grimm, 2020).

Highway authorities need to incorporate Disaster Risk Management into transport planning. Moreover, integrating adaptation interventions, such as NbSs, is important to enhance resilience in both existing and future projects. Policy management that encompasses these solutions should adopt a more comprehensive approach and coordinate with various stakeholders, including water management sectors, highway agencies, environmental experts, maintenance engineers, and local populations. Additionally, the selection and implementation of NbSs should consider technical, economic, and environmental feasibility. Each region may present specific challenges and opportunities, so it is crucial to involve experts, local communities, and relevant stakeholders to identify the most suitable solutions and maximize their benefits. NbSs do not imply quick fixes, and to achieve their full potential, they require a fundamental reshaping of society's relationship with nature (Nelson et al., 2020).

7 Final Considerations

Analyzing the evidence and experiences documented in the literature allowed for the identification of NbSs that have proven effective in enhancing the resilience of the road transportation sector. These measures range from incorporating natural

Table 2 Main challenges encountered

Challenge	Description	How to minimize it?
Cost-benefit ratio compared to projected alternatives	It refers to the comparison between the costs associated with a particular solution and the benefits that this solution provides. The cost-benefit ratio is calculated by dividing the total cost by the total benefit, providing a quantitative measure of return on investment. Many NbSs may involve significant initial costs. Transportation managers may hesitate to invest in these solutions due to uncertainty about long-term financial returns (Seddon et al., 2020)	Although there are claims that NbS can outperform conventional interventions, especially considering co-benefits, it is imperative to conduct comprehensive economic and/or social assessments (Albert et al., 2019). These assessments are crucial to support arguments about the effectiveness of NbS and to inform data-driven decisions that address concerns related to the cost-benefit ratio (Nelson et al., 2020)
Implementation time	Some NbSs, such as tree planting to create windbreaks, may take time to fully develop. This can be challenging when there is a need for quick solutions to immediate problems (De Abreu et al., 2023a)	It is important to continue focusing on lessons learned from case studies and building an evidence base for NbS to support future replication and expansion (Cohen-Shacham et al., 2016)
Adaptation to different climatic and geographic contexts	The success of NbS often depends on specific environmental characteristics, such as climate and geography. Adapting these solutions to different contexts can be a challenge (Chausson et al., 2020; Croeser et al., 2021)	Infrastructure engineering professionals must incorporate the complexity and regenerative properties of natural and nature-based systems for NbS to fully realize their potential (Nelson et al., 2020). Additionally, decision support tools, such as Multicriteria Decision Analysis (MCDA), can help navigate complex decisions related to the best choice among NbS (Croeser et al., 2021)
Trade-offs and co-benefits of various types of NbSs	NbS plays a crucial role in the pursuit of sustainable approaches to environmental and social challenges (De Abreu et al., 2023a). However, when implementing these solutions, it is essential to consider the trade-offs and co-benefits associated with different types of NbSs (Hutchins et al., 2021)	Multifunctional assessment tools should be developed to identify the optimal placement of NbS across a diverse set of socio-environmental indicators, incorporating considerations of the relative location of implementation areas and the accumulation of benefits (Hutchins et al., 2021)
Lack of knowledge and awareness	Lack of understanding about NbSs and their benefits can be an obstacle. It is important to educate stakeholders about the long-term benefits of these approaches (Nelson et al., 2020)	It is essential to quantify the benefits of NbS through easily measurable outcomes (Nelson et al., 2020). These results should be readily understood by professionals, municipal policymakers, and members of community organizations, considering specific geographical and climatic contexts (Bayulken et al., 2021)

(continued)

Table 2 (continued)

Challenge	Description	How to minimize it?
Coordination among different stakeholders	Effective implementation often requires coordination among various stakeholders, including government authorities, private companies, and local communities. Lack of coordination can hinder successful implementation (Hobbie & Grimm, 2020)	The public and private sectors should engage with policymakers to ensure that NbS can realize its potential to address the climate and biodiversity crisis while also contributing to sustainable development (Cohen-Shacham et al., 2016; Seddon et al., 2020)
Continuous maintenance	Some NbSs require continuous maintenance to ensure their effectiveness. Lack of resources or commitment can lead to the degradation of these solutions over time	Adequate resource planning should be undertaken before implementing NbSs, evaluating the necessary resources for regular maintenance and incorporating them into the budget (Bush & Doyon, 2019). Additionally, involving the local community in the maintenance of NbSs can increase commitment and ensure a sustainable approach, including the implementation of training programs to empower them to perform simple maintenance tasks
Legal and regulatory challenges	Legal and regulatory issues can arise when implementing NbS. For example, alterations to natural areas may face resistance due to environmental concerns or the need to comply with local regulations. Additionally, NbS often involves various actions occurring in broad landscapes and marine environments, crossing jurisdictional boundaries (Seddon et al., 2020)	The effectiveness of the governance of NbS depends on active cooperation and coordinated action among stakeholders (Nelson et al., 2020). The priorities, interests, and values of these parties must be aligned, requiring the resolution of any existing conflicts to achieve success in the implementation of NbS (Seddon et al., 2020)
Limited funding	The scarcity of funding emerges as a significant obstacle, potentially compromising the successful implementation of NbS. Securing financial resources for these innovative projects not only proves challenging but also underscores the urgent need for awareness and support from various sectors (Seddon et al., 2020)	Robust funding proposals highlighting the environmental, social, and economic benefits of NbS should be developed, presenting concrete data on costs, returns, and expected impacts. Additionally, specific government funding programs for environmental projects should be explored, as many governments provide grants and incentives for initiatives aimed at conservation and environmental sustainability

elements into road infrastructure to the restoration of adjacent ecosystems, aiming to promote environmental sustainability and adaptation to changing climatic conditions.

However, the research faced challenges in identifying specific studies on NbSs applied to road transportation. This lack of direct information indicates an opportunity to fill the existing literature gap in this area. By adopting a more comprehensive approach, relevant studies addressing critical infrastructure in general, including road transportation as an illustrative element, were identified. The inclusion of research from reputable institutions and well-regarded publications provided a solid foundation of scientific knowledge to support the study and contribute to the advancement of this emerging research area.

For future research, several proposals can be explored to further enhance the resilience of road transportation. An interesting research direction would be to investigate the application of innovative technologies, such as the use of advanced permeable pavement materials or the development of smart drainage systems capable of adapting automatically to weather conditions. Additionally, conducting case studies in different geographical regions would be valuable to assess the effectiveness of NbSs in varied contexts and identify best practices for implementation.

Another promising research area would be the analysis of the impact of NbSs in mitigating the effects of climate change on the road transportation sector. This would include assessing quantitative benefits such as carbon emissions reduction and decreasing costs related to natural disasters. Furthermore, studies could be conducted to investigate the positive effects of NbSs on air quality, public health, and the well-being of local communities.

By mapping the impacts of climate change and identifying promising NbSs, this study aimed to provide valuable information for decision-makers in the road transportation sector. Based on these insights, they can take proactive measures to adapt and strengthen infrastructure and operations, thereby reducing risks and enhancing resilience in the face of evolving climate challenges. However, there is still ample room for future research to further enhance knowledge about NbSs and their application in the context of road transportation.

Understanding the effectiveness, benefits, and limitations of NbSs in the context of road transportation is essential to guide planners, engineers, and decision-makers in seeking more sustainable and resilient NbSs. Through this approach, it is possible to promote concrete and effective actions to address the challenges posed by climate change in the transportation sector, contributing to environmental protection, infrastructure safety, and the well-being of communities that depend on this vital system.

References

Albert, C., Schröter, B., Haase, D., Brillinger, M., Henze, J., Herrmann, S., et al. (2019). Addressing societal challenges through nature-based solutions: How can landscape planning and governance research contribute? *Landscape and Urban Planning, 182*, 12–21. https://doi.org/10.1016/j.landurbplan.2018.10.003

Alhjouj, A., Bonoli, A., & Zamorano, M. (2022). A critical perspective and inclusive analysis of sustainable road infrastructure literature. *Applied Sciences, 12*(24), 12996. https://doi.org/10.3390/app122412996

Artmann, M., & Sartison, K. (2018). The role of urban agriculture as a nature-based solution: A review for developing a systemic assessment framework. *Sustainability, 10*(6), 1937. https://doi.org/10.3390/su10061937

Baskin, A. (2023). *Nature based solutions (NBS) in the transport sector*. Disponível em: https://transformative-mobility.org/wp-content/uploads/2023/04/NBS-Article_Ariadne-Baskin.pdf

Bayulken, B., Huisingh, D., & Fisher, P. M. (2021). How are nature based solutions helping in the greening of cities in the context of crises such as climate change and pandemics? A comprehensive review. *Journal of Cleaner Production, 288*, 125569. https://doi.org/10.1016/j.jclepro.2020.125569

Bles, T., Bessembinder, J., Chevreuil, M., Danielsson, P., Falemo, S., Venmans, A., et al. (2016). Climate change risk assessments and adaptation for roads–results of the ROADAPT project. *Transportation Research Procedia, 14*, 58–67. https://doi.org/10.1016/j.trpro.2016.05.041

Bush, J., & Doyon, A. (2019). Building urban resilience with nature-based solutions: How can urban planning contribute? *Cities, 95*, 102483. https://doi.org/10.1016/j.cities.2019.102483

Chausson, A., Turner, B., Seddon, D., Chabaneix, N., Girardin, C. A., Kapos, V., et al. (2020). Mapping the effectiveness of nature-based solutions for climate change adaptation. *Global Change Biology, 26*(11), 6134–6155. https://doi.org/10.1111/gcb.15310

Chinowsky, P. S., Price, J. C., & Neumann, J. E. (2013). Assessment of climate change adaptation costs for the US road network. *Global Environmental Change, 23*(4), 764–773. https://doi.org/10.1016/j.gloenvcha.2013.03.004

Cohen-Shacham, E., Walters, G., Janzen, C., & Maginnis, S. (2016). *Nature-based solutions to address global societal challenges* (Vol. 97, pp. 2016–2036). IUCN. https://doi.org/10.2305/IUCN.CH.2016.13.en

Cong, C., Pan, H., Page, J., Barthel, S., & Kalantari, Z. (2023). Modeling place-based nature-based solutions to promote urban carbon neutrality. *Ambio, 52*(8), 1297–1313. https://doi.org/10.1007/s13280-023-01872-x

Croeser, T., Garrard, G., Sharma, R., Ossola, A., & Bekessy, S. (2021). Choosing the right nature-based solutions to meet diverse urban challenges. *Urban Forestry & Urban Greening, 65*, 127337. https://doi.org/10.1016/j.ufug.2021.127337

de Abreu, V. H. S. (2023). *Diretrizes Para Um Plano Setorial De Adaptação Da Infraestrutura Rodoviária Brasileira Frente Aos Possíveis Impactos Da Mudança Do Clima*. Tese Para Concessão Do Grau De Doutor Em Ciências Em Engenharia De Transportes.

de Abreu, V. H. S., Ribeiro, F. B., & Santos, A. S. (2021a). Impactos Da Mudança Climática Na Infraestrutura De Transporte Rodoviário À Base De Revisão Bibliográfica. In *35° ANPET – Congresso de Pesquisa e Ensino em Transportes*.

de Abreu, V. H. S., Ribeiro, F. B., & Santos, A. S. (2021b). Impactos da Mudança Climática na Infraestrutura de Transporte Terrestre à Luz de Revisão Bibliográfica com Abordagem Bibliométrica. In *Via Viva 2021*. Ministério da Infraestrutura. https://doi.org/10.54507/viaviva.2021.01

de Abreu, V. H. S., Santos, A. S., & Monteiro, T. G. M. (2022a). Climate change impacts on the road transport infrastructure: A systematic review on adaptation measures. *Sustainability, 14*(14), 8864. https://doi.org/10.3390/su14148864

de Abreu, V. H. S., Ribeiro, F. B., & Santos, A. S. (2022b). Identificação de Indicadores de Vulnerabilidade para Análise de Risco Climático da Infraestrutura Rodoviária. In *Via Viva 2022*. Ministério da Infraestrutura.

de Abreu, V. H. S., Monteiro, T. G. M., Ribeiro, F. B., & Santos, A. S. (2022c). Identificação De Ameaças E Impactos Da Mudança Climática Na Infraestrutura De Transporte Rodoviário. *MIX Sustentável, 8*(3), 142–156. https://doi.org/10.29183/2447-3073.MIX2022.v8.n3.142-156

De Abreu, V. H. S., de Lima, G. B. C., Monteiro, T. G. M., & Santos, A. S. (2023a). Implications of climate change for the Brazilian road infrastructure. In *COVID-19 and climate change in BRICS nations* (pp. 80–99). Routledge.

de Abreu, V. H. S., Monteiro, T. G. M., de Oliveira Vasconcelos, A., & Santos, A. S. (2023b). Climate change adaptation strategies for road transportation infrastructure: A systematic review on flooding events. In *Transportation systems technology and integrated management* (pp. 5–30). https://doi.org/10.1007/978-981-99-1517-0_2

Epelde, L., Mendizabal, M., Gutiérrez, L., Artetxe, A., Garbisu, C., & Feliu, E. (2022). Quantification of the environmental effectiveness of nature-based solutions for increasing the resilience of cities under climate change. *Urban Forestry & Urban Greening, 67*, 127433. https://doi.org/10.1016/j.ufug.2021.127433

European Commission (EC). (2015). *Nature-Based Solutions & Re-Naturing Cities. Final Report of the Horizon 2020 Expert Group on 'Nature-Based Solutions and Re-Naturing Cities'.* Directorate-General for Research and Innovation–Climate Action, Environment, Resource Efficiency and Raw Materials (p. 74). https://doi.org/10.2777/765301.

Evans, C., Tsolakis, D., & Naudé, C. (2009, September). Framework to address the climate change impacts on road infrastructure assets and operations. In *Proceedings of the Australasian transport research forum (ATRF), 32nd*. Disponível em: https://www.australasiantransportresearchforum.org.au/sites/default/files/2009_Evans_Tsolakis_Naude.pdf. Acesso em: 01 maio 2023.

Forzieri, G., Bianchi, A., e Silva, F. B., Herrera, M. A. M., Leblois, A., Lavalle, C., et al. (2018). Escalating impacts of climate extremes on critical infrastructures in Europe. *Global Environmental Change, 48*, 97–107. https://doi.org/10.1016/j.gloenvcha.2017.11.007

Guy, F. (2015). Vulnerability and Adaptation of Transportation Infrastructure to Climate Change. In *The 2015 Conference of the transportation Association of Canada Charlottetown, PEI, 2015.*

Hawchar, L., Naughton, O., Nolan, P., Stewart, M. G., & Ryan, P. C. (2020). A GIS-based framework for high-level climate change risk assessment of critical infrastructure. *Climate Risk Management, 29*, 100235. https://doi.org/10.1016/j.crm.2020.100235

Hinkel, J. (2011). "Indicators of vulnerability and adaptive capacity": Towards a clarification of the science–policy interface. *Global Environmental Change, 21*(1), 198–208. https://doi.org/10.1016/j.gloenvcha.2010.08.002

Hobbie, S. E., & Grimm, N. B. (2020). Nature-based approaches to managing climate change impacts in cities. *Philosophical Transactions of the Royal Society B, 375*(1794), 20190124. https://doi.org/10.1098/rstb.2019.0124

Hutchins, M. G., Fletcher, D., Hagen-Zanker, A., Jia, H., Jones, L., Li, H., et al. (2021). Why scale is vital to plan optimal nature-based solutions for resilient cities. *Environmental Research Letters, 16*(4), 044008. https://doi.org/10.1088/1748-9326/abd9f4

Intergovernmental Panel On Climate Change – IPCC. (2014a). *Climate change 2014: Impacts, adaptation, and vulnerability. Part A: Global and sectoral aspects. Contribution of working group II to the fifth assessment report of the intergovernmental panel on climate change.* Cambridge University Press; INBS 9781107641655.

Intergovernmental Panel On Climate Change – IPCC. (2014b). *Summary for policymakers. Mitigation of climate change. Contribution of working group III to the fifth assessment report of the intergovernmental panel on climate change.* Cambridge University Press.

Intergovernmental Panel On Climate Change – IPCC. (2022). *Climate change 2022. Impacts, vulnerability and adaptation. Summary for policymakers.* Disponível em: https://www.ipcc.ch/report/ar6/wg2/

Jiricka-Pürrer, A. (2020). *Nature based solutions for climate proofing in transport planning and relevant EU regulations.* Disponível em: https://static1.squarespace.com/static/5e249ec8b3dbd6661a1d5306/t/5f6b570f5143276f761c40c2/1600870177504/Naturebased+Solutions+for+Climateproofing%281%29.pdf

Kabisch, N., Korn, H., Stadler, J., & Bonn, A. (2017). *Nature-based solutions to climate change adaptation in urban areas: Linkages between science, policy and practice*. Springer Nature. https://doi.org/10.1007/978-3-319-56091-5
Kabisch, N., Frantzeskaki, N., & Hansen, R. (2022). Principles for urban nature-based solutions. *Ambio, 51*(6), 1388–1401. https://doi.org/10.1007/s13280-021-01685-w
Keesstra, S., Nunes, J., Novara, A., Finger, D., Avelar, D., Kalantari, Z., & Cerdà, A. (2018). The superior effect of nature based solutions in land management for enhancing ecosystem services. *Science of the Total Environment, 610*, 997–1009. https://doi.org/10.1016/j.scitotenv.2017.08.077
Lafortezza, R., & Sanesi, G. (2019). Nature-based solutions: Settling the issue of sustainable urbanization. *Environmental Research, 172*, 394–398. https://doi.org/10.1016/j.envres.2018.12.063
Lafortezza, R., Chen, J., Van Den Bosch, C. K., & Randrup, T. B. (2018). Nature-based solutions for resilient landscapes and cities. *Environmental Research, 165*, 431–441. https://doi.org/10.1016/j.envres.2017.11.038
Mallick, R. B., Radzicki, M. J., Daniel, J. S., & Jacobs, J. M. (2014). Use of system dynamics to understand long-term impact of climate change on pavement performance and maintenance cost. *Transportation Research Record, 2455*(1), 1–9. https://doi.org/10.3141/2455
Michalis, D., Efthymis, L., Iraklis, S., & Evangelos, M. (2016). Vulnerability of transport infrastructure to extreme weather events in small rural catchments. *European Journal of Transport & Infrastructure Research, 16*(1). https://doi.org/10.18757/ejtir.2016.16.1.3117
Mndawe, M. B., Ndambuki, J. M., Kupolati, W. K., & Badejo, A. A. (2015). Assessment of the effects of climate change on the performance of pavement subgrade. *African Journal of Science, Technology, Innovation and Development, 7*(2), 111–115.
Moretti, L., & Loprencipe, G. (2018). Climate change and transport infrastructures: State of the art. *Sustainability, 10*(11), 4098. https://doi.org/10.3390/su10114098
Mott MacDonald. (2023). *Highway to nature-based solutions*. Disponível em: https://www.mottmac.com/views/highway-to-nature-based-solutions
Nelson, D. R., Bledsoe, B. P., Ferreira, S., & Nibbelink, N. P. (2020). Challenges to realizing the potential of nature-based solutions. *Current Opinion in Environmental Sustainability, 45*, 49–55. https://doi.org/10.1016/j.cosust.2020.09.001
Osaka, S., Bellamy, R., & Castree, N. (2021). Framing "nature-based" solutions to climate change. *Wiley Interdisciplinary Reviews: Climate Change, 12*(5), e729. https://doi.org/10.1002/wcc.729
Painel Brasileiro De Mudanças Climáticas – PBMC. (2014). Impactos, vulnerabilidades e adaptação às mudanças climáticas. Contribuição do Grupo de Trabalho 2 do Painel Brasileiro de Mudanças Climáticas ao Primeiro Relatório da Avaliação Nacional sobre Mudanças Climáticas (Assad, E.D., Magalhães, A. R. (Eds.)). COPPE. Universidade Federal do Rio de Janeiro, Rio de Janeiro, RJ, Brasil, 414 pp.
Pan, H., Page, J., Shi, R., Cong, C., Cai, Z., Barthel, S., et al. (2023). Contribution of prioritized urban nature-based solutions allocation to carbon neutrality. *Nature Climate Change, 13*(8), 862–870. https://doi.org/10.1038/s41558-023-01737-x
Plovnick, A., & Filosa, G. (2023). Transportation resilience in the United States and The Netherlands: Summary of collaboration on nature-based solutions and application of infrastructure resilience tools, 2016-2022 (No. DOT-VNTSC-FHWA-23-10). United States. Department of Transportation. Federal Highway Administration. Office of Natural Environment.
Santos, A. S. (2014). A Importância do Setor de Transportes para o Aumento de Resiliência das Cidades frente à Mudança Climática: Uma Proposta de Plano de Ação para a Cidade do Rio de Janeiro (Doctoral dissertation, Tese (Doutorado em Ciências)-Universidade Federal do Rio de Janeiro, Programa de Pós-Graduação e Pesquisa de Engenharia (COPPE), Engenharia de Transportes. Disponível em: https://www.pet.coppe.ufrj.br/index.php/pt/producao-academica/teses/2014/143-a-importancia-do-setor-de-transporte-para-o-aumento-de-resiliencia-das-cidades-frente-a-mudanca-climatica-uma-proposta-de-plano-de-acao-para-a-cidade-do-rio-de-janeiro

Santos, A. S., Ribeiro, S. K., & de Abreu, V. H. S. (2020). Addressing climate change in Brazil: Is Rio de Janeiro City acting on adaptation strategies? In *2020 international conference and utility exhibition on energy, environment and climate change (ICUE)* (pp. 1–11). IEEE. https://doi.org/10.1109/ICUE49301.202

Santos, A. S., de Abreu, V. H. S., de Assis, T. F., Ribeiro, S. K., & Ribeiro, G. M. (2021). An overview on costs of shifting to sustainable road transport: A challenge for cities worldwide. In *Carbon footprint case studies: Municipal solid waste management, sustainable road transport and carbon sequestration* (pp. 93–121). https://doi.org/10.1007/978-981-15-9577-6_4

Schlögl, M., & Laaha, G. (2017). Extreme weather exposure identification for road networks–a comparative assessment of statistical methods. *Natural Hazards and Earth System Sciences, 17*(4), 515–531. https://doi.org/10.5194/nhess-17-515-2017

Schweikert, A., Chinowsky, P., Espinet, X., & Tarbert, M. (2014). Climate change and infrastructure impacts: Comparing the impact on roads in ten countries through 2100. *Procedia Engineering, 78*, 306–316. https://doi.org/10.1016/j.proeng.2014.07.072

Seddon, N., Chausson, A., Berry, P., Girardin, C. A., Smith, A., & Turner, B. (2020). Understanding the value and limits of nature-based solutions to climate change and other global challenges. *Philosophical Transactions of the Royal Society B, 375*(1794), 20190120. https://doi.org/10.1098/rstb.2019.0120

Somarakis, G., Stagakis, S., Chrysoulakis, N., Mesimäki, M., & Lehvävirta, S. (2019). *ThinkNature nature-based solutions handbook.* European Union.

Transportation Research Board – TRB. (2008). *Potential impacts of climate change on U.-S. transportation* (Special report no. 290). TRB, Washington, DC. https://doi.org/10.26225/jerv-w202

Trindade, D. Q. L., Abreu, V. H. S., & Santos, A. S. (2019). Medidas de Adaptação frente aos Possíveis Impactos das Mudanças Climáticas no Sistema de Transportes: Uma Revisão da Literatura. In *XXXIII Congresso Nacional de Pesquisa e Ensino em Transportes.*

Valenzuela, Y. B., Rosas, R. S., Mazari, M., Risse, M., & Rodriguez-Nikl, T. (2017). Resilience of road infrastructure in response to extreme weather events. In *International conference on sustainable infrastructure 2017* (pp. 349–360).

World Bank. (2017). *Integrating climate change into road asset management.* Disponível em: http://documents1.worldbank.org/curated/en/981831493278252684/pdf/114641-WP-ClimateAdaptationandAMSSFinal-PUBLIC.pdf

Carbon Footprint as a Tool to Improve Sustainability: Evidence from the Agri-Food Supply Chain

Caterina Tricase, Roberto Leonardo Rana, Mariarosaria Lombardi, Paola Geatti, Pasquale Giungato, Vera Amicarelli, and Christian Bux

1 Introduction

In 1987, in the document entitled "Our Common Future" (also known as "Brundtland Report") and distributed by the United Nations through the Oxford University Press, the notion of sustainable development appeared for the first time. It defines sustainability as "meeting the needs of the present without compromising the ability of future generations to meet their own needs" (Brundtland, 1987) and underlines the different paradigms of sustainable development, which must consider environmental, economic and social factors. The reasons why it appears essential to deal with sustainability lie in the significant growth of the population, which has reached eight billion people (Dodson et al., 2020; Ritchie et al., 2023), and in the limitation and scarcity of resources available to satisfy the growing needs of human beings (Pearse, 1991). For a long time, the question among researchers was: Can the world population grow infinitely or is there an objective limit? Some years later, in

C. Tricase · R. L. Rana · M. Lombardi · C. Bux (✉)
Department of Economics, University of Foggia, Foggia, Italy
e-mail: caterina.tricase@unifg.it; roberto.rana@unifg.it; mariarosaria.lombardi@unifg.it; christian.bux@unifg.it

P. Geatti
Department of Economics and Statistics, University of Udine, Udine, Italy
e-mail: paola.geatti@uniud.it

P. Giungato
Department of Chemistry, University of Bari Aldo Moro, Bari, Italy
e-mail: pasquale.giungato@uniba.it

V. Amicarelli
Department of Economics, Management and Business Law, University of Bari Aldo Moro, Bari, Italy
e-mail: vera.amicarelli@uniba.it

S. S. Muthu (ed.), *Carbon Footprint Assessments*, Environmental Footprints and Eco-design of Products and Processes,
https://doi.org/10.1007/978-3-031-70262-4_8

1991, the World Conservation Union developed an alternative definition for sustainable development, namely "improving the quality of human life while living within the carrying capacity of supporting ecosystems" (World Conservation Union, 1991), and such a definition was soon implemented by the first ecological economist, Herman Daly, who defined "a steady-state economy as one where material and energy use are stabilized and kept within ecological limit" (O'Neill, 2023).

Four pillars must be considered when speaking about sustainability, as follows: (i) the weight of the anthropogenic impact on the natural systems; (ii) the usage rate of non-renewable resources and their speed of regeneration; (iii) the release of polluting and climate-altering substances and waste into the biosphere; and (iv) the substitution of non-renewable with renewable sources. The sustainable development, therefore, consists of the rational use of the resources present on the planet, also allowing the exploitation and benefit for future generations (Rana, 2020).

Nowadays, sustainability has become a competitive factor for companies, especially for those that intend to respond and invest in the growing demand for Corporate Social Responsibility (CSR) to penetrate local and international market (Tarnovskaya, 2023). Regarding the environmental aspects, organizations have to necessarily pay attention to the consumption of fossil fuels and energy and the related greenhouse gas (GHG) emissions, as well as the use of water for the production of goods and the generation of waste. Additionally, they have to consider the other impacts of human activities, which are also responsible for the release of wastewater into water basins, the loss of biodiversity and the harm to soil quality.

International policies, starting from the 17 Sustainable Development Goals (SDGs) in the UN 2030 Agenda (United Nations, 2015) and considering the Circular Economy Action Plan (European Commission, 2020) or the European Green Deal (European Commission, 2019a, b) in the European Union (EU), have underlined the sustainability issue at an entrepreneurial level. In the EU, in the last 20 years, agriculture with reduced environmental impact was promoted. The Common Agricultural Policy, in recent years, introduced the use of sustainable agricultural practices with a reduced environmental impact (European Commission, 2023). Specifically, many agri-food businesses invested to adapt to the new needs of consumers who pay attention to the environment and demand a lower consumption of natural resources. The phenomenon of consumer awareness of environmental sustainability issues has very important effects on a business level, since it requires a modification of the production cycles of goods in order to reduce, among other things, the GHG emissions into the environment. In regulatory terms, the request for more sustainable production of goods has pushed the legislator to adopt appropriate tools capable of guiding companies towards a renewal of production processes and making products with low environmental impact recognizable (Calabro & Vieri, 2024).

In this context, the carbon footprint (CF), based on the measurement of GHG emissions associated with the life cycle of a product, has assumed great importance as an environmental sustainability indicator. Its concept dates back to 1996, when Wackernagel and Rees (1996) developed the "ecological footprint" as "a biologically productive land and sea area required to sustain a given human population

Table 1 GWP_{100} values of GHGs[a] (2021)

Greenhouse gas	Formula	GWP_{100} value per kg of emission
Carbon dioxide	CO_2	1
Methane (fossil origin)	CH_4	29.8
Methane (non-fossil origin)	CH_4	27.2
Nitrous oxide	N_2O	273
Hydrofluorocarbons	HFCs	4.84–14 600
Per-fluorocarbons	PFCs	0.004–12 400
Sulphur hexafluoride	SF_6	25 200

Source: International Panel for Climate Change (2021)
[a]By convention, a time span of 100 years is taken into account in CF valuations

expressed in global hectares". Today, one of the most comprehensive definitions is "the quantity of GHGs expressed in terms of CO_2eq, emitted into the atmosphere by an individual, organization, process, product, or event from within a specified boundary", which underlines the importance of the methodological approach adopted (Pandey et al., 2011). The GHGs are climate-altering substances responsible for the so-called "artificial greenhouse effect" that causes the global warming of Earth. These gases differ from each other in their ability to absorb energy (their "radiative efficiency") and how long they stay in the atmosphere (also known as their "lifetime"), that is the global warming potential (GWP) index. Specifically, it estimates the quantity of energy that the emission of 1 t of GHG will absorb over a given period of time (generally 20, 100 or 500 years), with reference to the emissions of 1 tCO_2 (IPCC, 1994; UNEP, 2024). Table 1 illustrates the Kyoto Protocol GHGs and the respective GWP_{100} values, as revised in 2021 by the International Panel for Climate Change (2021).

In the light of above, this chapter describes the importance of applying the CF as an environmental sustainability indicator to enhance the performance of the agri-food supply chain. By describing the main international standards and guidelines, and by presenting the main findings from two case studies conducted in Italy, the chapter aims to highlight the advantages for the entire agri-food sector and to provide a useful tool for supporting the climate change mitigation strategies requested at the global level.

2 Environmental Sustainability of the Agri-Food Supply Chain

Nowadays, consumers are oriented towards healthy and environmentally friendly products and ask for goods and services with low consumption of natural resources (Rana, 2020), and this trend has led to the development of the ecological product industry (Polonsky et al., 1997). So, to adapt to consumer needs, companies have started to adopt green marketing strategies (Rahman & Nguyen-Viet, 2023). Recent surveys demonstrated that consumers are willing to accept a higher price for sustainable products (Chen, 2010), and a momentous study highlighted that 89% of consumers are willing to pay to reduce air pollution, whereas 11% trust that dumpers and governments should pay for environmental defence (Shahzad et al., 2022).

Sustainability, intended in the field of economy, society and the environment, could help companies in improving their competitiveness in the market through some business incentives, such as: (i) the improvement in the financial performance; (ii) the increase in the stakeholder engagement; (iii) the improvement in risk management; (iv) the enhancement in workplace reputation and talent acquisition; and (v) the boost of innovation (Bonini & Swartz, 2014; Capozucca, 2012; Eccles et al., 2012). In the field of the financial performance, environmental and social sustainability programmes are strongly correlated with good financial performance, namely a lower cost of debt and equity (Roche Diagnostics, 2024). Moreover, from the stakeholder (and consumer) side, a sustainable business is more likely to create value for a wider range of stakeholders, climate change being a phenomenon that affects everyone, regardless of their specific economic or social needs. In terms of workplace reputation and talent acquisition, recent studies (Carballo-Penela et al., 2023) highlight that the value a company places on sustainability is important for job seekers and for those who want to work for socially and ethically responsible companies (Jenkin, 2015).

In a highly competitive market, among several environmental sources of supply differentiation (Galletto & Barisan, 2019), the assessment of the CF by companies represents a clear and comprehensive strategy for communicating to consumers an interest in environmental protection and climate change mitigation.

The competitive advantage is defined as "the conditions that make a business more successful than the businesses it is competing with, or a particular thing that makes it more successful" (Cambridge Dictionary, 2024), and it requires companies to differentiate their products or services in order to gain competitiveness in the market. In this context, the success of competitive strategies mainly depends on the value that consumers perceive, both for goods and services (Galletto & Barisan, 2019), and it is essential for companies to communicate in a transparent and clear manner their activities through advertising, branding, labels or social media, for instance (Flint & Golicic, 2009).

Specifically, the CF assessment could help companies achieve several organizational benefits, such as the increase in the credibility, consistency and transparency of the GHGs quantification; monitoring and reporting, in order to enhance consumers' trust and tackle the consumers' uncertainty (Chen & Chang, 2013); the simplification in the development and implementation of sustainable strategies and plans by companies; and the development of carbon-offset programmes, intending the attribution of credits and their trading among companies (van Kooten & de Vries, 2013). In terms of competitive benefits, the CF assessment at the organizational level helps: (i) the benchmarking between similar products or services across competing companies; (ii) the evaluation of alternative product design, procurement and production methods; (iii) the selection of more sustainable suppliers on the basis of their estimated GHGs; (iv) the identification of concrete opportunities to save energy costs; (v) the demonstration to customers and stakeholders of the real environmental and social commitments, in terms of corporate social responsibility; (v) the opportunity to participate to grants and public calls (i.e., Green Public Procurement); and (vi) the adaptation to eventual requirements imposed by local and international laws.

But a question arises spontaneously: How can companies demonstrate that they have conducted a CF analysis? The communication of sustainability is essential for companies that want to gain a competitive advantage, and for this it is necessary to resort to the CF labels, so-called carbon labelling (Cohen & Vandenbergh, 2012). Nowadays, various labels have been used worldwide that certify the GHGs of a product or service, but most of them are issued without taking into account an international standard, rather they are based on procedures mostly identified at national or individual company level (Rana, 2020). For instance, the carbon labelling programme in the United Kingdom was launched in 2007 (i.e., Carbon Reduction Label), whereas the Product Carbon Footprint Project in Germany (i.e., Blaue Engel) and in Switzerland (i.e., ClimaTop) were adopted in 2008, as well as the CF programme in Japan. The carbon labelling system in South Korea started in 2009 (Rana, 2020; Zhao et al., 2020).

3 Methods and Data for Carrying Out the Carbon Footprint

In recent times, the development of a globalized market has required the adoption of common rules among States, with the aim of eliminating information asymmetries and minimizing the margins of discretion in environmental impact assessment studies (FAO, 2011). Hence, international reference standards were introduced, capable of replacing those present at a local level. Consequently, the estimation of the CF starts from the analysis of the life cycle stages included in the production of a product or a service or in a specific system, such as agricultural production, industrial transformation, distribution and logistics, transportation and end-of-life stages. From the practical perspective, different standards and guidance have been developed on a global scale, such as: (a) the GHG Protocol of WSI/SBCSD, namely the World Resource Institute and the World Business Council on Sustainable Development; (b) the ISO 14025; (c) the ISO 14064; (d) the ISO 14067; (e) the PAS2050 of the British Standard Institution; (f) the 2006 IPCC guidelines for National Greenhouse Gas Inventories. In addition to internationally accepted tools, some countries have developed local guidelines, such as those promoted by the Department of Food and Rural Affair (DEFRA) and the Carbon Trust in the United Kingdom and those adopted by the Environmental Protection Agency (EPA) in the United States (Pandey et al., 2011).

In the case of the CF of a product, one of the first methodologies was introduced in Great Britain in 2008, defined as Publicly Available Specifications-2050 (PAS 2050-1:2012). This latter was developed by the British Standard Institute, the

Carbon Trust and the Department of Environment, Food and Rural Affairs (British Standard Institute, 2012). Subsequently, additional standards were introduced by the International Organization for Standardization (ISO), namely the ISO 14067:2018 on "Greenhouse gases—Carbon footprint of products—Requirements and guidelines for quantification" for the assessment of GHG emissions at the product level (ISO, 2018a) and the ISO 14064:2018 for the evaluation of the CF at the organization level (ISO, 2018b).

3.1 Software for the Assessment of the Carbon Footprint

Companies are asked to minimize their environmental impact and, therefore, it is essential to measure and monitor the CO2eq emissions over time. Recently, several software tools have been developed to enhance the environmental management and accounting at the organization level, but it should be highlighted that not all companies own the suitable expertise to use these environmental accounting tools within their boundaries (Ormazabal et al., 2014). Companies that have not yet reached the level of maturity that allows for the internal assessment of the environmental impacts can rely on external consultancies, which can be made by private companies or research centres, such as universities (Finnveden & Schneider, 2023), and green soft skills are required to boost sustainability and competitiveness among companies (Farao et al., 2023).

In recent years, companies reached a certain level of environmental maturity by implementing the ISO 14000 guidelines on the environmental management (ISO, 2024), which cover a set of instruments for improving sustainability such as audits, communications, labelling and life cycle analysis, including the assessment of the CF at the process (ISO 14064) and the product (ISO 14067) level. Specifically, the ISO 14001 on environmental management systems (EMS) is the only standard that can be certified and is recognized in over 180 countries worldwide (ISO, 2024). Practically, several software tools have been developed to help companies estimate their environmental impact, but they differ in terms of implementation, data and objectives. These software tools can focus on single organizations, specific projects undertaken by organizations or final users, activities or products along their entire lifecycle (Ormazabal et al., 2014).

Generally, software tools for estimating the CF are independent of the type of organization or the sector in which it operates and are therefore considered transversal. Among them, there are the Sphera (formerly GaBi) and the SimaPro software, then Autodesk or Intelex (Ormazabal et al., 2014). However, there are some sector-specific software, which have been developed for estimating GHG emissions for s specific sector, such as the Cement CO2 and Energy Protocol (World Business Council for Sustainable Development, 2011) or, more specifically, there is software for estimating the F in the waste management at the country level, namely the CO2ZW (Sevigné Itoiz et al., 2013), and another software that estimates the total

CF of a data centre over its lifecycle operations, considering IT, mechanical, electrical and public security infrastructures, developed by Schneider Electric (2023).

The question regards where to collect data for the environmental impact assessment, since data collection represents one of the most critical steps for developing CF studies for either researchers or practitioners. By definition, "finding appropriate and suitable life cycle assessment data has been a research job itself", and for this reason, the openLCA Nexus was developed with the aim of making the data collection as simple as possible, and fast as well. According to the openLCA Nexus (2024), practitioners willing to estimate the CF of their products and services can rely on both free and for-purchase databases. Among the most common free databases, it is possible to consult the Environmental Footprints, which is under the guidance of the European Commission and its Joint Research Centre and was updated in June 2023, or the Agrybalise database (version 3.1.1.), that is specific for the agriculture and the food sector and contains the lifecycle inventory of about 2600 agricultural and food products. Moreover, companies could also consider using the Emission Factor Database (EFDB) developed by the International Panel on Climate Change (IPCC), which provides emission coefficients for energy, industrial processes and product use, agriculture forestry and other land use, waste and residual categories, and is site-specific (IPCC, 2021).

In the field of the for-purchase databases, companies can rely on the ecoinvent database, which is one of the most famous ones worldwide and is used by more than 5000 organizations (Ecoinvent, 2024), and which can be completed by the EN15804 add-on, which is a database for Environmental Product Declarations (EPDs) developed by GreenDelta GmbH according to the EN15802 norm.

3.2 Guidelines Proposed by the ISO 14067:2018

One of the most common options to evaluate the CF, as outlined by the ISO 14067: 2018, follows the methodological framework proposed by the ISO 14040:2006 on the life cycle assessment (LCA). Specifically, an analysis of CF should respect four steps, namely: (i) definition of the goals and scope, which regards the choice of the functional unit and the system boundaries; (ii) the life cycle inventory analysis; (iii) the life cycle impacts assessment; and (iv) the interpretation of results. Figure 1 illustrates the four steps for conducting a CF.

3.2.1 Definition of the Goal and Scope

The first step to calculate the CF of a product, a service or a system is to identify the emission sources, which can be either direct or indirect. Direct emission may appear as the easiest to recognize, since they are generated by the combustion of fossil fuel or associated with other processes that lead to the GHG emissions, such as the

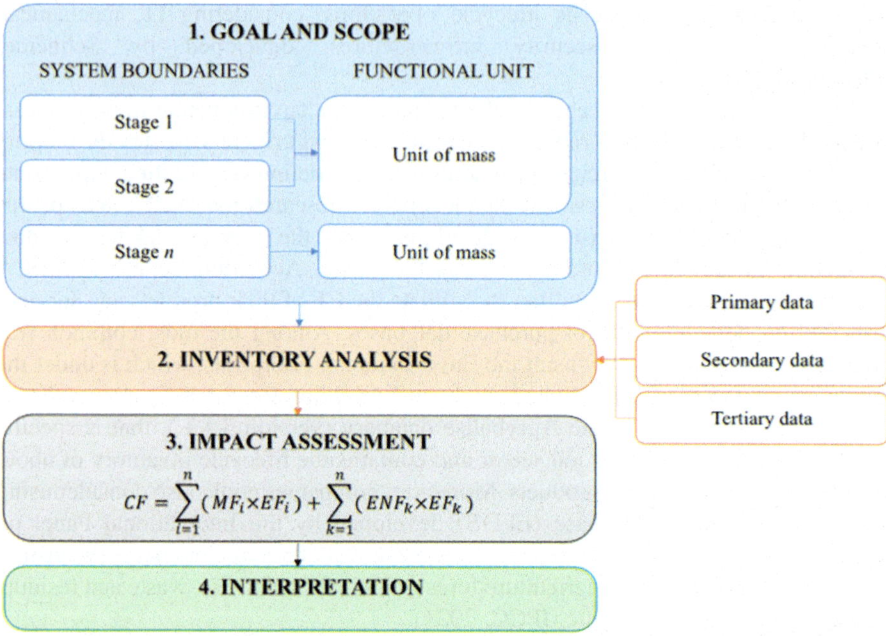

Fig. 1 Steps for conducting a CF according to the ISO 14040:2006. (Note: CF carbon footprint, MF mass flows, EF emission factors, ENF energy flows, *i* material input, *k* energy input)

anaerobic decomposition of organic matter that generates the CH_4, for instance. Indirect emissions are rather complex to be depicted, being caused by the demands that the subject of the CF exerts upon the wider economy in which it operates (United States Environmental Protection Agency, 2023). Specifically, these types of emissions must be found in the phases preceding those being analysed, for example the manufacturing phase of a specific product, or the mining extraction phase for the production of electricity necessary for the production process. In the field of the source categories, the International Panel for Climate Change (2006) has identified some categories, as follows: (i) stationary combustion; (ii) mobile combustion; (iii) GHG emissions from chemical processes; (iv) agriculture; (v) land use, land use change and forestry; and (vi) stock.

In the definition of the goals and scope, researchers must identify the system boundaries for the CF assessment, so as to develop a system which is clear, accurate and easy to understand (Rubinstein & Kroese, 2016). Selecting specific system boundaries can facilitate the determination of the flows of mass and energy that should be included in the assessment. The process of evaluating the CF requires costs in terms of time and resources, essential to collect and analyse data, and it is not practical to consider all plausible emission sources for a certain study. In the light of Williams et al. (2012), two different system boundaries should be considered, namely "geographical boundaries" and "temporal boundaries". Last, the CF assessment requires the identification of the functional unit, which is described as a quantity of a product or product system on the basis of the performance it delivers in its end-use application. The functional unit is essential, as it enables objective comparisons across different products or systems that serve the same final functions, and for this reason, the functional unit must be a convenient and physically meaningful quantity.

3.2.2 Inventory Analysis

The data collection phase is crucial to guarantee a reliable result. The inventory analysis is a quantitative analysis of resource, energy consumption, or emissions to the environment throughout all the life cycle stages of product, service or processes (Chen & Wang, 2017). To do this, it is necessary to identify of all input and output included in the selected system boundaries and expressed in units of measure (e.g., mass, volume). Specifically, according to the ISO 14040 guidelines, the inventory analysis is defined as the "compilation and quantification of inputs and outputs for a product throughout its life cycle" (Muthu, 2020).

Considering the ISO 11820:2022, on the "Measuring circularity—Methods and indicators for measuring circular processes in organizations" (Ente Italiano di Normazione, 2022), prior to the data acquisition phase, the perimeter of the evaluation being measured (i.e., system boundaries) and the data quality requirements, not only for the data themselves but also for any metadata, must be defined. The evaluation reliability is strongly influenced by the quality, reliability and the verifiability of the data collected. To provide flexibility in measuring and evaluating the CF, it is recommended to identify, measure and track individual input and output resource flows instead of consolidating them. The maintenance of separate resource flows during the measurement simplifies resource-related calculations. All flows that have an impact should be included, measured and recorded separately.

For compiling a clear and comprehensive life cycle inventory, the system under evaluation is first divided into necessary elementary components. This research activity can take the form of a graphical diagram, such as a resource flow diagram (through the use of the material flow analysis). In data collection, however, it is possible to encounter some different situations. Researchers can find readily available data, which can be used immediately for calculating the CF (i.e., primary data), or some data not immediately available but that can be acquired with reasonable resources, such as data from literature (i.e., secondary data). Ultimately, it can happen that data are not accessible, and in this case, the consequences must be evaluated. If missing data have a critical impact on quality requirements, scholars can use modelling, some basic data for the sector or proxy data, for instance data adapted from another country (i.e., tertiary data).

3.2.3 Impact Assessment

For simplicity of calculation, the CF evaluation requires the estimate of the GHG emissions associated with each individual material and energy flow. A simplified Eq. (1) for calculating the CF is as follows:

$$CF = \sum_{i=1}^{n} (MF_i \times EF_i) + \sum_{k=1}^{n} (ENF_k \times EF_k)$$

where CF is carbon footprint, MF identifies the mass flows, EF the emission factors and ENF the energy flows, whereas *i* is related to material inputs and *k* to the energy ones.

3.2.4 Interpretation of the Results

The calculation of the CF can be useful for guiding policymakers, entrepreneurs or consumers towards the implementation of virtuous actions and behaviours that reduce GHG emissions. However, the CF is not just a tool for controlling and reducing emissions, since it can be used as a tool for communicating a company's sensitivity to environmental sustainability issues. This indicator can be an important element in making products more competitive on local and international markets. Furthermore, companies can benefit not only in terms of consumers, increasing their reputation, but also by intervening to reduce energy consumption and rationalize raw materials.

Nonetheless, the CF represents only a starting point in quantifying environmental sustainability, since it is necessary to integrate the results with sustainability communication tools, including environmental impact labels and brands.

4 Case Studies on the CF in the Agri-Food Sector

Companies should reduce their carbon emissions to enhance the sustainability of the agri-food supply chain, which currently represents one of the largest polluters to the environment. Latest statistics (Ritchie & Roser, 2021) reveal that quite half of the habitable land is addressed to agriculture and crop production, for about 48 million km^2 (Rana et al., 2023), and that about 17.9 $GtCO_2eq$ are generated in the food system. In Europe, Italy represents the fifth largest polluter soon after Germany and France, accounting for about 98 Mt of CO_2eq per year, whereas China is the major polluter worldwide (2.41 $GtCO_2eq$) (Ritchie & Roser, 2021).

Nowadays, a much-debated issue is how to measure the environmental sustainability of the agri-food supply chain. In recent years, various methodologies have been developed, all relating to life cycle thinking (Mazzi, 2020), including the environmental life cycle assessment (E-LCA) developed in 1999 by the Society of Environmental Toxicology and Chemistry (2024). Studies on the environmental sustainability of the agri-food supply chain can concern individual production phases, such as the agricultural or transformation one, or consider an integrated approach from cradle to grave. The approaches in the agri-food supply chain can be micro, meso or macro (Molthan-Hill et al., 2020), depending on the selected system boundaries.

The chapter presents the results of two case studies conducted in Southern Italy on two different agri-food commodities, namely the globe artichoke [*Cynara cardunculus* L. var. *scolymus* (L.) Fiori] (Rana et al., 2023) and the sweet cherry (*Prunus avium* L.), which represent an herbaceous plant and a domesticated plant, respectively. The globe artichoke represents one of the most cultivated vegetables worldwide, with a yield of 1.58 Mt in 2022, and Italy is the top producer (0.37 Mt) (FAOstat, 2023). As regards sweet cherries, their production was 2.7 Mt in 2022, and Italy ranks sixth on the global scale, with a yield of 0.10 Mt per year (FAOstat, 2023).

4.1 The CF of the Globe Artichoke

The first case study considers the so-called "rustic artichoke" produced in Italy, in the province of Foggia. Its goal is to evaluate the GWP100 associated with the agricultural production and the industrial processing of the globe artichoke. The functional unit is 1 high-density-polyethylene (HDPE) drum of 299 kg of rustic artichoke, composed as follows: (a) 150 kg of final product; (b) 70 kg of preserving liquid; (c) 9 kg of HDPE drum. Specifically, the case study evaluates two intermediate functional units, namely 1 ha of cultivated land and 1 kg of HDPE for the packaging production. In terms of system boundaries, the agricultural production encompasses seven phases, namely: (i) carduccio planting; (ii) mineral fertilization; (iii) organic fertilization; (iv) pest management; (v) irrigation; (vi) farm management; and (vii) harvesting. During the industrial processing stage, five phases are considered, as follows: (i) calibration; (ii) washing and trimming; (iii) blanching; (iv) preserving liquid production; (v) barrels filling. Last, in the packaging production stage, the case study considers the production of virgin HPDE, the material recovery of HDPE waste and the HDPE drum production. Transport occurs from agricultural production and packaging production to industrial processing.

Table 2 illustrates the inventory. For the agricultural stage, the functional unit is 1 ha of cultivated land, whereas for the HDPE drum production, the functional unit is 1 kg of HDPE drum. The ultimate functional unit in the industrial processing is 229 kg of HDPE drum of rustic artichoke.

In terms of agricultural production outputs, the primary one is the fresh artichoke, specifically 12,500 kg per 1 ha of arable land, whereas the primary output for the HDPE drum production is 1 kg of HDPE drum. Overall, the entire supply chain accounts for several outputs, such as the rustic artichoke, but also leaves and by-products, second calibre artichokes, wastewater and of course GHG emissions.

Table 2 Input inventory

Stage	Input	Quantity	Source
AG	Manure	25,000 kg	PD
	N fertilizer	265 kg	PD
	P_2O_5 fertilizer	100 kg	PD
	K_2O fertilizer	110 kg	PD
	Fungicides/insecticides	10 kg	PD
	Herbicides	3 kg	PD
	Water	6500 m^3	PD
	Electricity	8665 kWh	PD
	Diesel (farm activities)	875 kg	PD
	Diesel (reservoir operations)	53 kg	Martin-Gorriz et al. (2020)
	Diesel (field operations)	319 kg	Martin-Gorriz et al. (2020)
	Diesel (fertilizers transportation)	5.8 kg	EEA (2021)
HDPE	Virgin HDPE	0.94 kg	Istrate et al. (2021)
	Recycled HDPE	0.09 kg	Istrate et al. (2021)
	Electricity	0.58 kWh	Bassi et al. (2020)
	HDPE drums	0.03 kg	EEA (2021)
IP	Fresh artichokes	370 kg	PD
	Water (industrial operations)	410 kg	PD
	Water (preserving liquid)	59 kg	PD
	Vinegar (preserving liquid)	7 kg	PD
	Salt (preserving liquid)	4 kg	PD
	HDPE drum	9 kg	PD
	LPG (cooking)	4.05 kg	PD
	Electricity (industrial activities)	4.56 kWh	PD
	Diesel (transportation)	20.02 kg	EEA (2021)

Note: *AG* agricultural stage, *HDPE* drum production, *IP* industrial processing, *EEA* European Environment Agency, *PD* primary data

In the light of the CF analysis, the CF for the functional unit was estimated at about 336 kgCO_2eq, distinguishing between 82% due to the agricultural operations and 18% due to the industrial processing and HDPE drum production. Considering 1 kg of fresh artichokes, its CF was evaluated at about 0.75 kgCO_2eq.

Figure 2 illustrates the results of the CF conducted in the artichoke supply chain, from agricultural production to industrial transformation.

The results of a CF provide not only a quantitative but also a qualitative overview of the process under observation. In the case of the globe artichoke, it is clear that the majority of GHG emissions derive from the use of chemical fertilizers, followed by the use of natural fertilizers and diesel, both for agricultural operations and for the transport of food from field to the industrial processing plant, as well as for the transport of HDPE drum from abroad (import from Spain). Therefore, the CF is not just a tool for measuring and quantifying environmental impacts aimed at obtaining environmental certifications, but useful for penetrating markets that are increasingly attentive to environmental issues and for winning over customers interested in

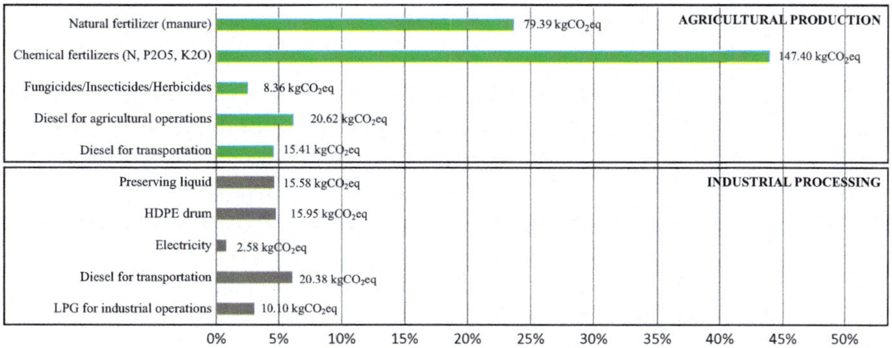

Fig. 2 CF of 229 kg of HDPE drum of rustic artichoke

products with a low environmental impact. The CF is also used to identify hotspots during the production chain and intervene for the programming of plans and strategies aimed at reducing GHG emissions. These latter, which are currently increasingly useful for structuring sustainability communication by companies (e.g., sustainability reports), can place agri-food companies in a position of great commercial visibility, allowing the acquisition of new consumers who, over time, can concretely see the result of sustainability strategies.

In the specific case of the globe artichoke, the CF calculation underlines the need to substitute chemical fertilizers, such as N fertilizers, with manure, so as to reduce the environmental impacts from 12.43 kgCO2eq to 0.07 kgCO2eq per kg of fertilizers used. Moreover, as outlined by previous studies on the topic (Elsharkawy et al., 2021), chemical fertilizers could also be replaced by PK_3 fertilizers and biostimulants, which are useful to increase the productivity and yield of artichoke crops in the Mediterranean area. Moreover, it is demonstrated that the use of bio-stimulants from seaweed extracts could increase plant height and the number of leaves per plant, allowing for the reduction of agrochemicals.

In concerning the industrial processing stage, it is advisable for agri-food managers to purchase HDPE drums from Italy instead of importing them from Spain, since such an operation can reduce the transportation environmental emissions. Moreover, from the energetic perspective, the operations of the sector can substitute diesel and LPG energy sources with greener ones, such as biomethane generated from the anaerobic digestion of the artichoke by-products themselves. Indeed, in the current perspective, organic waste from the agricultural and industrial stage are addressed to feeding livestock, which seems to be essential to reduce the quantity of CH4 in the animals' effluents. However, biowaste from the artichoke supply chain could be alternatively used to produce bio-methane through anaerobic digestion, as well as digestate (as a co-product of the bio-methane production), compost, biomaterials and inulin (Rana et al., 2023). Considering the circular economy paradigm, the recovery of biomethane and digestate from biowaste can reduce the quantity of non-renewable energy sources and the use of chemical fertilizers during the agricultural stage.

4.2 CF of Processed Sweet Cherries

The second case study regards the CF of 0.50 kg of sweet cherries packaged in a polyethylene terephthalate (PET) clamshell of 0.019 kg, considering a cradle-to-gate approach, in Southern Italy. Specifically, in terms of agricultural production, the research considers all steps from nursery to fruit distribution, including the production of the packaging (i.e., clamshell) on the basis of different intermediate functional units: (i) in the orchard phase, the functional unit is 1 ha; and (ii) in the clamshell production, it is 1 kg of PET. Considering the damage allocation, the case study is based on an economic allocation, namely: (i) 95.05% of fresh cherries and 4.95% of woods in the orchard phase; (ii) 77.76% of conforming and 22.24% of non-conforming cherries in the industrial processing; and (iii) 95.35% of clamshell and 4.65% of PET slops.

In detail, in the orchard phase, the case study considers the mineral (i.e., N, P_2O_5, K_2O) and the organic fertilization (i.e., manure), the pest management, the irrigation and the farm management (i.e., harvesting, mechanized pruning, tillage, dismantling). In the industrial processing stage, the case study considers the energy consumption associated with hydrocooling, calibration, selection, packaging and labelling, as well as cold storage, whereas in the clamshell production, it estimated the carbon emissions due to the extrusion of PET resin and the thermoforming of PET sheets.

In the field of the GHG emissions during the orchard phase, the case study considers the CH_4 emissions associated with manure along the entire orchard lifetime (i.e., 20 years), namely 343.6 $kgCH_4$, as well as the N_2O emission, namely 11.46 kgN_2O. In terms of soil management, the case study estimates the direct emissions of N_2O equal to 59.21 kg/ha and the indirect ones equal to 11.29 kg/ha. As regards the emissions due to farm management and transportation of materials and workers, they are estimated on the basis of the subsequent assumptions: (a) the orchard management consumes 7199 L of diesel; (b) the transportation of materials to orchard consumes 3364 L of diesel; and (c) the transportation of workers to orchard consumes 439 L of diesel. In terms of electricity consumption, the case study considers a quantity of 30,380 kWh per 20 years.

Table 3 illustrates the inventory.

For the production of 1 kg of PET clamshell, the case study considers the subsequent input inventory: (i) 0.17 kg of recycled postconsumer PET pellet; (ii) 0.97 kg of thermoforming; (iii) 0.01 L of diesel for transportation from Brazil to Rotterdam, then 0.03 L for transportation from Rotterdam to Rutigliano (Southern Italy). Last, for the industrial processing stage of 1 t of processed cherry, the subsequent input inventory was considered: (i) 0. 35 kg of sodium hypochlorite; (ii) 1.78 kg of potassium chloride; (iii)1.72 L of diesel for cherries transportation; (iv) 116.21 kWh of electricity; and (v) 12,000 L of water.

Findings of the case study highlight that the CF of 0.5 kg of the packaged sweet cherries is 0.58 $kgCO_2eq$, with the agricultural management as the most burdening

Table 3 Input inventory for the agricultural phase of 1 ha of cherry orchard

Input	Unit	Nursery	1st to 3rd year	4th to 7th year	7th to 20th year
Compost	kg	1400	0	0	0
N fertilizer	kg	6.48	0	61.8	574
P_2O_5	kg	10.8	0	121.8	854
K_2O	kg	4.32	0	250.8	1414
Insecticides	kg	0.576	0	0.9	14
Fungicides	kg	1.68	2.7	10.35	67.2
Diesel for ploughing	L	0	300	300	1400
Diesel for fertilization	L	0	135	135	630
Diesel for compost tr.	L	6.35	0	0	0
Diesel for NPK tr.	L	2.33	0	150.91	849.08
Diesel for manure tr.	L	0	346.37	346.37	1616.37
Diesel for workers tr.	L	0.26	14.02	59.58	359.81
Electricity	kWh	222	1855	3450	24,853

Note: *Tr* transportation. The research considers primary data and data from the EcoInvent database (2013). Data were retrieved from Rana et al. (2019)

phase (0.44 kgCO_2eq), followed by PET clamshell production (0.07 kgCO_2eq) and cherries' processing (0.06 kgCO_2eq). Figure 3 illustrates the CF of the sweet cherries.

The most carbon-intensive phase is the orchard management, specifically the years of the full production (i.e., from 7th to 20th year), since there is a high consumption of chemicals, water and fuels. Specifically, the highest amount of carbon emissions is due to irrigation and fertigation in terms of electricity consumption, which represent 15% (0.09 kgCO_2eq) of the entire CF of the cherries production, followed by manure transportation (6%) and ploughing activities (5%). In addition, as outlined also by the case study related to the artichoke, high quantities of CO_2eq are due to fertilizers' production, which in this case contributes for the 4% to the entire CF. However, although high quantities of GHG are released into the atmosphere, the cherries' production contributes to the so-called carbon sequestration into the soil, for an amount of 9080 kgCO_2eq along the entire 20-year lifespan.

In terms of the industrial processing phase, carbon emissions are mainly associated with electricity consumption by air conditioning in the processing room, for an amount of 35% of the total, and with hydrocooling for 33%. Moreover, the processing phase contributes to the generation of waste, which consists of non-conforming cherries and other by-products (e.g., peduncles, scraps), currently addressed to landfilling, and these residues contribute to the generation of CH_4, CO_2 and SF_6 emissions. Findings suggest the use of organic waste in other industrial processes, for example for the production of flavour or aroma compounds, antioxidants, natural colourants or dietary nutrients, according to a circular economy approach. Moreover, the substation of the current electricity mix by adding more renewables (e.g., photovoltaic energy) could reduce the CO_2eq emissions by 19%, from 0.58 to 0.47 kgCO_2eq.

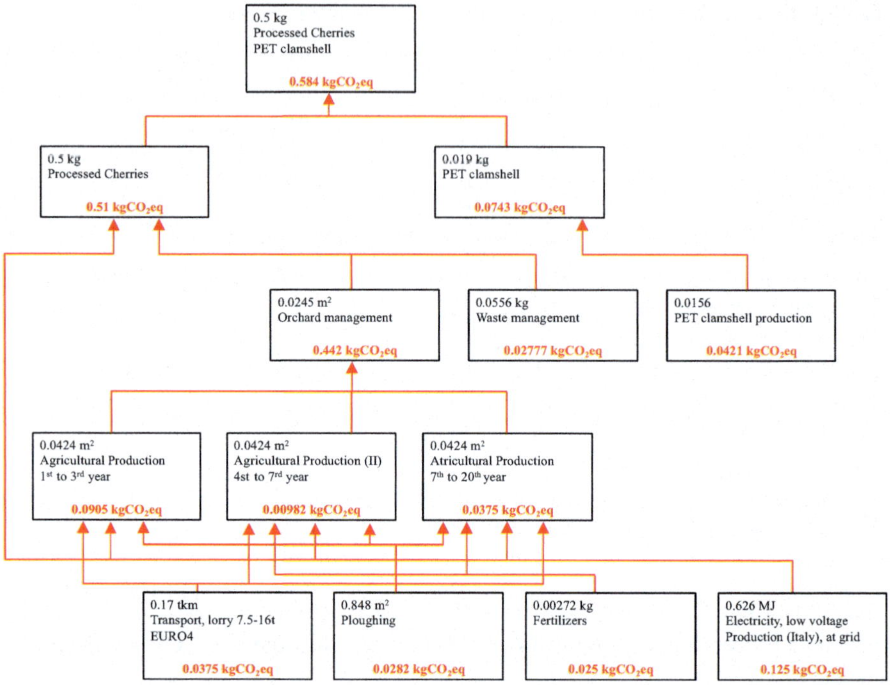

Fig. 3 CF of 0.50 kg of sweet cherries. (Note: The flow diagram represents an alternative manner of graphically representing CF results. One option is related to the bar graphs (Fig. 2), whereas alternative options are the flow diagram or the Sankey diagram)

Last, the PET clamshell production contributes to the entire CF for 7%, specifically in terms of thermoforming for a quantity of 0.014 kgCO_2eq and of extrusion for 0.01 kgCO_2eq. For the managerial perspective, companies could contribute to the sustainability of the agri-food chain by replacing virgin with recycled packaging, or by substituting PET with alternative organic materials, such as edible coating.

5 Carbon Footprint and Carbon Border Adjustment Mechanism

The environmental legislation, together with the standardized measurement of environmental benefits and burdens associated with the production of products and services, is essential to guarantee sustainable development and economic growth. Indeed, the adoption of environmental legislation provides an essential instrument to the government to influence the competitiveness of an economy (as a whole) or the attractiveness of some sectors. In light of Dechezleprêtre and Sato (2017), environmental regulations require a transition (and a transformation) from polluting facilities to more sustainable ones, which may impose costs on businesses sometimes difficult to afford. These differences across firms, sectors or countries can cause changes in the relative production and final product costs, leading to possible inequalities in markets.

Considering the standard theories of international trade (Rauscher, 1994), restrictive policies towards the use of natural resources could cause internal movements from pollution-intensive sectors to cleaner sectors, or international trades from developed economies to developing ones. For instance, the prohibition on disposable plastic items could cause companies producing this product to collapse in favour of industries that produce resistant plastic, or where such companies were not present in a certain country it would import from third-party economies that do not have a specific environmental legislation.

In order to put an obstacle to environmental dumping within the EU's GHG emissions control, also known as "carbon leakage" (Selicato, 2022), and thus ensure that the EU's environmental objectives are not compromised by the relocation of production to Countries with less binding GHG emission policies, the EU has recently established the so-called Carbon Border Adjustment Mechanism (CBAM) governed by EU Regulation 2023/956, the implementing rules of which are laid down in the Commission Implementing Regulation (EU) 1773/2023 (OJEU, 2023a, b).

The aim of this measure is to prevent goods imported into the EU (from non-EU countries) from enjoying an unfair competitive advantage, due to the lack of "carbon" costs in the exporter Countries. With the application of the CBAM (which in fact constitutes an environmental tax, a tax revenue destined for the EU budget), the production cost of European companies, which pay a carbon price for these products based on the EU Emission Trading System (ETS), would be equated to that of products imported from Countries where no such mechanisms exist (OJEU, 2003). A single and shared methodology for the evaluation of the "carbon" levels associated with a particular product would therefore be desirable because various tools currently exist such as CF methodologies, reporting systems, etc. (Sect. 3), which in fact can create imbalances between operators and distortions in the markets.

From this perspective, in addition to being an environmental indicator, the CF plays a key role in the implementation of the CBAM. The standardized CF, specifically applied at the sectorial and the company level, could be the benchmark to provide better results among EU countries (and other governments) that are adopting this mechanism to combine environmental protection and business competitiveness, as well as to ultimately provide correct evaluation of carbon emissions that can allow an effective comparison even in different geographical areas (Gailhofer & Graichen, 2023).

The CBAM instrument is part of the "Fit for 55" proposals that would aim to reduce GHG emissions by at least 55% by 2030 compared to 1990 values, as part of the Green Deal EU plan (European Parliament, 2022). The CBAM instrument was created to complement the ETS, relying on a system of certificates to cover emissions incorporated in products that are imported into the EU. These certificates will have to be purchased by importers of the goods covered by the measure (European

Commission, 2021). The price of the certificates, proportional to the emission content of the imported product, will be calculated on the basis of the average weekly auction price of EU ETS allowances expressed in €/tonne of CO_2 emitted (European Commission, 2021). Differently from ETS allowances, CBAM certificates cannot be traded nor banked (Carbon Market Watch, 2021).

Initially, this measure will only be applied to a selected number of goods from carbon-intensive manufacturing sectors (cement and cement products, electricity, mineral and chemical fertilizers, iron and steel products, aluminium products, hydrogen) and provides for information and reporting obligations for companies importing these types of products into the EU customs territory. The European Commission, however, reserves the right to broaden the categories of goods covered (OJEU, 2023a). A gradual introduction of the CBAM measure is planned. In the first transitional phase (1 October 2023–31 December 2025), information will be collected on the quantities of incoming products subject to the CBAM, but no levy will be applied to imported goods. It will be necessary for importers (or customs representatives) to prepare and submit via the "CBAM Transition Register" (an electronic database) a quarterly report containing information on the quantity of imported goods, the direct and indirect emissions incorporated in those goods and the carbon price due for those emissions, including the carbon price due for emissions incorporated in precursor materials. As of 1 January 2026, the CBAM mechanism will become definitively operational and, operators will have to submit a "CBAM declaration" (OJEU, 2023b).

The EU CBAM measure in its early stage is going to have a direct impact on the agri-food supply chain due to the use of fertilizers in agriculture to improve soil fertility and increase crop productivity. Fertilizers play a very important role in agriculture: it has been estimated that the use of nitrogen (N) fertilizers was responsible for an increase of up to 50% in agricultural yields during the twentieth century, and by the end of the same century, approximately 40% of the world's population was eating food from crops grown with synthetic fertilizers (Erisman et al., 2008).

According to data provided by Fertilizer Europe (an association grouping the main mineral fertilizer producers operating in the European Union), in the year 2021 world fertilizers production was 259.1 Mt, of which 17.3 Mt were produced in Europe, representing 6.7% of global production (Fertilizers Europe, 2023). EU countries are both exporters and importers of fertilizers. In 2022, the total fertilizer consumption in the EU was 16 Mt, of which approximately 10 Mt were N ones. It is well known that N fertilizers production and use are responsible for approximately 5% of global GHG mainly due to their energy-intensive manufacturing processes and accounts for about 2–3% of global energy consumption (Gao & Cabrera Serrenho, 2023; Ouikhalfan et al., 2022). Even though, in many countries, recent environmental regulations and energy costs have prompted fertilizer manufacturers to look for alternative production technologies, deploying energy-efficient processes, materials and technologies for carbon capture, use or storage, alternatives to fossil fuels, etc., to estimate their CF could be a useful tool to allow CBAM evaluation for fertilizer entering in EU agri-food sector. A combined CF and CBAM contribution could improve sustainability in the agri-food supply chain.

6 Conclusions

Moving from the assumption that companies are asked to minimize their environmental impact by measuring and monitoring the CO_2eq emissions over time, the research examined the role of the CF assessment to gain competitive advantage among agri-food companies and improve financial performance, increase stakeholder engagement, boost innovation and reduce carbon emissions, as well.

Technically, there are different standards and guidance to measure the CF, such as the GHG protocol of WSI/SBCSB, the ISO 14025, the ISO 14064, the ISO 14067, the PAS2050 of the British Standard Institution and the 2006 IPPC guidelines for National Greenhouse Gas Inventories, but in addition to internationally accepted tools, there are also local guidelines, implemented within single countries. In order to conduct CF studies, specific software tools were developed, namely the Sphera (formerly GaBi) and the SimaPro software, but also the Autodesk or the Intelex. Of course, one challenge regards the need to collect data, since it represents "a research job itself", and the openLCA Nexus helps practitioners and academics in creating suitable inventories for the CF assessment. Specifically, there are free databases, such as the Environmental Footprints, under the guidance of the European Commission or the Agrybalise, but it is possible to rely on for-purchase databases, for instance EcoInvent.

In detail, the chapter illustrated the guidelines for the CF assessment proposed by the ISO 14067:2018, highlighting the typical four steps, namely: (i) definition of the goals and scope, which includes the identification of the functional unit and the system boundaries; (ii) the life cycle inventory analysis; (iii) the life cycle impacts assessment; and (iv) the interpretation of results. Moreover, to give evidence to the CF evaluation, the study proposed two case studies conducted on agri-food Italian products. On the one hand, the globe artichokes, and on the other hand, the sweet cherries. Considering in both cases a cradle-to-gate approach and including also packaging production, the case studies highlighted the importance of CF results to enhance the sustainability of agri-food supply chains. In particular, findings highlighted the need of switching from non-renewable to renewable energy, as well as the importance of replacing chemical with organic fertilizers. In addition, suitable waste management represents a key towards sustainable development and economic growth.

Last, the chapter contextualized the CF role in the novel Carbon Border Adjustment Mechanism, which constitutes an environmental tax destined to the EU budget and whose aim is to obstacle the environmental dumping ("carbon leakage") and to enhance the sustainability of the agri-food sector as a whole.

CRediT Authorship Contribution Statement *Caterina Tricase*: Carbon footprint and Carbon Border Adjustment Mechanism; *Roberto Leonardo Rana*: Introduction, Case studies on the carbon footprint in the agri-food sector; *Mariarosaria Lombardi*: Introduction, Environmental sustainability of the agri-food supply chain; Case-studies on the carbon footprint in the agri-food sector; Conclusions; *Paola Geatti*: Carbon footprint and Carbon Border Adjustment Mechanism; *Pasquale*

Giungato: Case studies on the carbon footprint in the agri-food sector; *Vera Amicarelli*: Methods and data for carrying out the carbon footprint, Conclusions; *Christian Bux*: Introduction, Case studies on the carbon footprint in the agri-food sector, Methods and data for carrying out the carbon footprint, Case studies on the carbon footprint in the agri-food sector, Conclusions.

References

Bassi, S. A., Boldrin, A., Faraca, G., & Astrup, T. F. (2020). Extended producer responsibility: How to unlock the environmental and economic potential of plastic packaging waste? *Resources, Conservation and Recycling, 162*, 105030. https://doi.org/10.1016/j.resconrec.2020.105030

Bonini, S., & Swartz, S. (2014). Profits with purpose: How organizing for sustainability can benefit the bottom line. In *McKinsey on sustainability & resource productivity*. McKinsey & Company.

British Standard Institute. (2012). PAS 2050-1:2012. Assessment of life cycle greenhouse gas emissions from horticultural products – Supplementary requirements for the cradle to gate stages of GHG assessments of horticultural products undertaken in accordance with PAS 2050.

Brundtland, G. H. (1987). *Our common future: Report of the World Commission on Environment and Development*. UN-Dokument A/42/427.

Calabro, G., & Vieri, S. (2024). Limits and potential of organic farming towards a more sustainable European agri-food system. *British Food Journal, 126*(1), 223–236. https://doi.org/10.1108/BFJ-12-2022-1067

Cambridge Dictionary. (2024). *English Dictionary*. https://dictionary.cambridge.org/dictionary/english/. Accessed 1 Feb 2024.

Capozucca, P. (2012). *Sustainability 2.0. Using sustainability to drive business innovation and growth*. Deloitte Insights. https://www2.deloitte.com/us/en/insights/deloitte-review/issue-10/sustainability-2-0-innovation-and-growth-through-sustainability.html. Accessed 1 Feb 2024.

Carballo-Penela, A., Ruzo-Sanmartín, E., & Sousa, C. M. P. (2023). Does business commitment to sustainability increase job seekers' perceptions of organisational attractiveness? The role of organisational prestige and cultural masculinity. *Business Strategy and the Environment, 32*(8), 5521–5535.

Carbon Market Watch. (2021). *A brief explanation of the Carbon Border Adjustment Mechanism (CBAM)*. https://carbonmarketwatch.org/wp-content/uploads/2022/01/a-brief-explanation-of-the-cbam-proposal-1-1.pdf. Accessed 16 Jan 2024.

Chen, Y. S. (2010). The drivers of green brand equity: Green brand image, green satisfaction, and green trust. *Journal of Business Ethics, 93*, 307–319. https://doi.org/10.1007/s10551-009-0223-9

Chen, Y. S., & Chang, C. H. (2013). Greenwash and green trust: The mediation effects of green consumer confusion and green perceived risk. *Journal of Business Ethics, 114*, 489–500. https://doi.org/10.1007/s10551-012-1360-0

Chen, H., & Wang, L. (2017). Chapter 9: Coproducts generated from biomass conversion processes. In H. Chen & L. Wang (Eds.), *Technologies for biochemical conversion of biomass* (pp. 219–264). Academic. https://doi.org/10.1016/B978-0-12-802417-1.00009-0

Cohen, M., & Vandenbergh, M. P. (2012). The potential role of carbon labeling in a green economy. *Energy Economics, 34*(1), 52–63. https://doi.org/10.1016/j.eneco.2012.08.032

Dechezleprêtre, A., & Sato, M. (2017). The impacts of environmental regulations on competitiveness. *Review of Environmental Economics and Policy, 11*(2), 183–206. https://doi.org/10.1093/reep/rex013

Dodson, J. C., Dérer, P., Cafaro, P., & Götmark, F. (2020). Population growth and climate change: Addressing the overlooked threat multiplier. *Science of the Total Environment, 748*, 141346. https://doi.org/10.1016/j.scitotenv.2020.141346

Eccles, R. G., Ioannou, I., & Serafeim, G. (2012). *The impact of corporate sustainability on organizational processes and performance*. NBER Working Paper No. 17950.

Ecoinvent. (2024). *Database*. https://ecoinvent.org/database-login/. Accessed 2 Feb 2024.

Elsharkawy, G. A., Ibrahim, H. A. H., Salah, A. H., Akrami, M., Ali, H. M., & Abd-Elkader, D. Y. (2021). Early and total yield enhancement of the globe artichoke using an ecofriendly seaweed extract-based biostimulant and PK fertilizer. *Agronomy, 11*(9), 1819. https://doi.org/10.3390/agronomy11091819

Ente Italiano di Normazione. (2022). *UNI/TS 11820; Misurazione Della Circolarità—Metodi ed Indicatori per la Misurazione dei Processi Circolari Nelle Organizzazioni*. Ministero dell'Ambiente e della Sicurezza Energetica, Rome.

Erisman, J. W., Sutton, M. A., Galloway, J., Klimont, Z., & Winiwarter, W. (2008). How a century of ammonia synthesis changed the world. *Nature Geoscience, 1*, 636–639.

European Commission. (2019a). *The European Green Deal sets out how to make Europe the first climate-neutral continent by 2050, boosting the economy, improving people's health and quality of life, caring for nature, and leaving no one behind*. https://ec.europa.eu/commission/presscorner/detail/en/ip_19_6691. Accessed 15 Jan 2024.

European Commission. (2019b). Fertilisers in the EU Prices, trade and use. *EU Agricultural Markets Briefs, 15*, 1–8. http://ec.europa.eu/agriculture/markets-and-prices/market-briefs/index_en.htm. Accessed 30 Jan 2024.

European Commission. (2020). *A new circular economy action plan. For a cleaner and more competitive Europe*. https://eur-lex.europa.eu/legal-content/EN/TXT/HTML/?uri=CELEX:52020DC0098. Accessed 15 Jan 2024.

European Commission. (2021). *Carbon border adjustment mechanism: Questions and answers*. https://ec.europa.eu/commission/presscorner/api/files/document/print/en/qanda_21_3661/QANDA_21_3661_EN.pdf. Accessed 10 Jan 2024.

European Commission. (2023). *The common agricultural policy at a glance. The common agricultural policy supports farmers and ensures Europe's food security*. https://agriculture.ec.europa.eu/common-agricultural-policy/cap-overview/cap-glance_en#documents. Accessed 15 Jan 2024.

European Environment Agency. (2021, October). *EMEP/EEA Air pollutant emission inventory guidebook 2019 – Update*. https://www.eea.europea.eu/publications/emep-eea-guidebook-2019. Accessed 12 Sept 2022.

European Parliament. (2022). Press release. *CBAM: Parliament pushes for higher ambition in new carbon leakage instrument*. https://www.europarl.europa.eu/pdfs/news/expert/2022/6/press_release/20220603IPR32157/20220603IPR32157_en.pdf. Accessed 10 Jan 2024.

FAO. (2011). *Environmental impact assessment. Guidelines for FAO field projects*. FAO.

FAOstat. (2023). *Crops and livestock products*. http://fenix.fao.org/faostat/internal/en/#data/QCL. Accessed 11 Jan 2024.

Farao, C., Bernuzzi, C., & Ronchetti, C. (2023). The crucial role of green soft skills and leadership for sustainability: A case study of an Italian small and medium enterprise operating in the food sector. *Sustainability, 15*, 15841. https://doi.org/10.3390/su152215841

Fertilizers Europe. (2023). *Fertilizer industry facts & figures 2023*. https://www.fertilizerseurope.com/wp-content/uploads/2023/07/Industry-Facts-and-figures-2023.pdf. Accessed 30 Jan 2024.

Finnveden, G., & Schneider, A. (2023). Sustainable development in higher education—What sustainability skills do industry need? *Sustainability, 15*, 4044. https://doi.org/10.3390/su15054044

Flint, D. J., & Golicic, S. L. (2009). Searching for competitive advantage through sustainability: A qualitative study in the New Zealand wine industry. *International Journal of Physical Distribution and Logistics Management, 39*, 841–860.

Gailhofer, P., & Graichen, V. (2023). *Monitoring, reporting and verification in a Carbon Border Adjustment Mechanism (CBAM). Which rules and standards for calculating and certifying product-related emissions?* (Technical Report Texte 154/2023). German Environment Agency. https://doi.org/10.13140/RG.2.2.25874.84161. Accessed 30 Jan 2024.

Galletto, L., & Barisan, L. (2019). Carbon footprint as a lever for sustained competitive strategy in developing a smart oenology: Evidence from an Exploratory Study in Italy. *Sustainability, 11*, 1483. https://doi.org/10.3390/su11051483

Gao, Y., & Cabrera Serrenho, A. (2023). Greenhouse gas emissions from nitrogen fertilizers could be reduced by up to one-fifth of current levels by 2050 with combined interventions. *Nature Food, 4*, 170–178. https://doi.org/10.1038/s43016-023-00698-w

International Panel for Climate Change. (2006). In H. S. Eggleston, L. Buendia, K. Miwa, T. Ngara, & K. Tanabe (Eds.), *IPCC guidelines for national greenhouse gas inventories*. Institute for Global Environmental Strategies, Hyama.

International Panel for Climate Change. (2021). *IPCC sixth assessment report*. https://www.ipcc.ch/report/ar6/wg1/. Accessed 15 Jan 2024.

IPCC. (1994). https://archive.ipcc.ch/ipccreports/tar/wg1/247.html

IPCC. (2021). *Emission factor database, EFDB*. https://www.ipcc-nggip.iges.or.jp/EFDB/main.php. Accessed 2 Feb 2024.

ISO. (2018a). *ISO 14067:2018. Greenhouse gases – Carbon footprint of products – Requirements and guidelines for quantification*. https://www.iso.org/standard/71206.html. Accessed 15 Jan 2024.

ISO. (2018b). *ISO 14064-1:2018. Greenhouse Gases – Part 1: Specification with guidance at the organization level for quantification and reporting of greenhouse gas emissions and removal*. https://www.iso.org/standard/66453.html. Accessed 15 Jan 2024.

ISO. (2024). *ISO 14000 family. Environmental management*. https://www.iso.org/standards/popular/iso-14000-family. Accessed 2 Feb 2024.

Istrate, I.-R., Juan, R., Martin-Gamboa, M., Domínguez, C., García-Muñoz, R. A., & Dufour, J. (2021). Environmental life cycle assessment of the incorporation of recycled high-density polyethylene to polyethylene pipe grade resins. *Journal of Cleaner Production, 319*, 128580. https://doi.org/10.1016/j.jclepro.2021.128580

Jenkin, M. (2015). Millennials want to work for employers committed to values and ethics. *The Guardian*. https://www.theguardian.com/sustainable-business/2015/may/05/millennials-employment-employers-values-ethics-jobs. Accessed 1 Feb 2024.

Martin-Gorriz, B., Gallego-Elvira, B., Martinez-Alvarez, V., & Maestre-Valero, J. F. (2020). Life cycle assessment of fruit and vegetable production in the Region of Murcia (south-east Spain) and evaluation of impact mitigation practices. *Journal of Cleaner Production, 265*, 121656. https://doi.org/10.1016/j.jclepro.2020.121656

Mazzi, A. (2020). Chapter 1: Introduction. Life cycle thinking. In J. Ren & S. Toniolo (Eds.), *Life cycle sustainability assessment for decision-making* (pp. 1–19). Elsevier. https://doi.org/10.1016/B978-0-12-818355-7.00001-4

Molthan-Hill, P., Robinson, Z. P., Hope, A., Dharmasasmita, A., & McManus, E. (2020). Reducing carbon emissions in business through Responsible Management Education: Influence at the micro, meso- and macro-levels. *The International Journal of Management Education, 18*(1), 100328. https://doi.org/10.1016/j.ijme.2019.100328

Muthu, S. S. (2020). Chapter 6: Estimating the overall environmental impact of textile processing: Life cycle assessment of textile products. In S. S. Muthu (Ed.), *The textile institute book series, assessing the environmental impact of textiles and the clothing supply chain* (2nd ed., pp. 105–129). Woodhead Publishing. https://doi.org/10.1016/B978-0-12-819783-7.00006-5

O'Neill, D. W. (2023). Herman E. Daly (1938–2022). *Nature Sustainability, 6*, 118–119. https://doi.org/10.1038/s41893-022-01041-0

OJEU (Official Journal of European Union). (2003). Directive 2003/87/EC of the European Parliament and of the Council of 13 October 2003 establishing a scheme for greenhouse gas emission allowance trading within the Community and amending Council Directive 96/61/EC. http://data.europa.eu/eli/dir/2003/87/oj

OJEU (Official Journal of European Union). (2023a). Regulation (EU) 2023/956 of the European Parliament and of the Council of 10 May 2023 establishing a carbon border adjustment mechanism. http://data.europa.eu/eli/reg/2023/956/oj

OJEU (Official Journal of European Union). (2023b). Commission Implementing Regulation (EU) 2023/1773 of 17 August 2023 laying down the rules for the application of Regulation (EU) 2023/956 of the European Parliament and of the Council as regards reporting obligations for the purposes of the carbon border adjustment mechanism during the transitional period. http://data.europa.eu/eli/reg_impl/2023/1773/oj

OpenLCA Nexus. (2024). *Databases*. https://nexus.openlca.org/databases. Accessed 2 Feb 2024.

Ormazabal, M., Jaca, C., & Puga-Leal, R. (2014). Analysis and comparison of life cycle assessment and carbon footprint software. In J. Xu, V. Cruz-Machado, B. Lev, & S. Nickel (Eds.), *Proceedings of the Eighth international conference on management science and engineering management* (Advances in intelligent systems and computing, 281). Springer. https://doi.org/10.1007/978-3-642-55122-2_131

Ouikhalfan, M., Lakbita, O., Delhali, A., Assen, A. H., & Belmabkhout, Y. (2022). Toward net-zero emission fertilizers industry: Greenhouse gas emission analyses and decarbonization solutions. *Energy & Fuels, 36*, 4198–4223.

Pandey, D., Agrawal, M., & Pandey, J. S. (2011). Carbon footprint: Current methods of estimation. *Environmental Monitoring and Assessment, 178*, 135–160. https://doi.org/10.1007/s10661-010-1678-y

Pearse, P. H. (1991). Scarcity of natural resources and the implications for sustainable development. *Natural Resources Forum, 15*, 74–79. https://doi.org/10.1111/j.1477-8947.1991.tb00112.x

Polonsky, M. J., Carlson, L., Grove, S., & Kangun, N. (1997). International environmental marketing claims: Real changes or simple posturing? *International Marketing Review, 14*(4), 218–232. https://doi.org/10.1108/02651339710173426

Rahman, S. U., & Nguyen-Viet, B. (2023). Towards sustainable development: Coupling green marketing strategies and consumer perceptions in addressing greenwashing. *Business Strategy and the Environment, 32*(4), 2420–2433. https://doi.org/10.1002/bse.3256

Rana, R. L. (2020). La carbon footprint nella filiera agroalimentare. In F. Contò & M. Fiore (Eds.), *Ragionando di Sviluppo Locale: Una Lettera "Nuova" di Tematiche "Antiche"* (pp. 115–139). FrancoAngeli.

Rana, R. L., Andriano, A. M., Giungato, P., & Tricase, C. (2019). Carbon footprint of processed sweet cherries (*Prunus avium* L.): From nursery to market. *Journal of Cleaner Production, 227*, 900–910. https://doi.org/10.1016/j.jclepro.2019.04.162

Rana, R. L., Bux, C., & Lombardi, M. (2023). Carbon footprint of the globe artichoke supply chain in Italy: From agricultural production to industrial processing. *Journal of Cleaner Production, 391*, 136240. https://doi.org/10.1016/j.jclepro.2023.136240

Rauscher, M. (1994). On ecological dumping. *Oxford Economic Papers, 46*, 822–840.

Ritchie, H., & Roser, M. (2021). Environmental impacts of food production. 2021. https://ourworldindata.org/environmental-impacts-of-food#citation. Accessed 30 Jan 2024.

Ritchie, H., Rodés-Guirao, L., Mathieu, E., Gerber, M., Ortiz-Ospina, E., Hasell, J., & Roser, M. (2023). *Population growth*. OurWorldInData.org. https://ourworldindata.org/population-growth. Accessed 16 Jan 2024.

Roche Diagnostics. (2024). *How to gain a competitive advantage through environmental sustainability*. https://diagnostics.roche.com/global/en/article-listing/how-to-gain-a-competitive-advantage-through-sustainability.html. Accessed 1 Feb 2024.

Rubinstein, R. Y., & Kroese, D. P. (2016). *Simulation and the Monte Carlo Method*. Wiley.

Schneider Electric. (2023). *Data center lifecycle CO2e calculator* (Version 6, 27 November 2023).

Selicato, G. (2022). EU proposal for a Carbon Border Adjustment Mechanism: an advanced tool to combat 'carbon leakage', a new EU own resource of 'moral suasion' for third Countries? *Review of European and Comparative Law, 50*, 25–37.

Sevigné Itoiz, E., Gasol, C. M., Farreny, R., Rieradevall, J., & Gabarrell, X. (2013). CO2ZW: Carbon footprint tool for municipal solid waste management for policy options in Europe. *Inventory of Mediterranean countries. Energy Policy, 56*, 623–632. https://doi.org/10.1016/j.enpol.2013.01.027

Shahzad, T., Shah, S. T. H., Rais, S. I., Mansoor, A., & Zaman, K. (2022). People's attitude towards willingness-to-pay for environmental protection in Pakistan. *Environmental Science and Pollutution Research, 29*, 52635–52654. https://doi.org/10.1007/s11356-021-18177-9

Society of Environmental Toxicology and Chemistry. (2024). *Life cycle assessment*. https://www.setac.org/explore-science/methods-and-approaches/life-cycle-assessment.html. Accessed 15 Jan 2024.

Tarnovskaya, V. (2023). Sustainability as the source of competitive advantage. How sustainable is it? In P. N. Ghauri, U. Elg, & S. M. Hånell (Eds.), *Creating a sustainable competitive position: Ethical challenges for international firms* (International business and management, 37) (pp. 75–89). Emerald Publishing Limited. https://doi.org/10.1108/S1876-066X20230000037005

UNEP. (2024). https://www.epa.gov/ghgemissions/understanding-global-warming-potentials

United Nations. (2015). Resolution adopted by the general assembly on 25 September 2015. Transforming our World: The 2030 agenda for sustainable development. https://documents-dds-ny.un.org/doc/UNDOC/GEN/N15/291/89/PDF/N1529189.pdf?OpenElement. Accessed 15 Jan 2024.

United Nations Environmental Protection Agency. (2023). *Scope 1 and scope 2 inventory guidance*. https://www.epa.gov/climateleadership/scope-1-and-scope-2-inventory-guidance. Accessed 15 Jan 2024.

van Kooten, G. C., & de Vries, F. P. (2013). Carbon offsets. In J. F. Shogren (Ed.), *Encyclopedia of energy, natural resource, and environmental economics* (pp. 6–8). Elsevier. https://doi.org/10.1016/B978-0-12-375067-9.00165-0

Wackernagel, M., & Rees, W. (1996). Our ecological footprint: reducing human impact on the earth. New Society Publishers, Philadelphia: United States.

Williams, I., Kemp, S., Coello, J., Turner, D. A., & Wright, L. A. (2012). A beginner's guide to carbon footprinting. *Carbon Management, 3*(1), 55–67. https://doi.org/10.4155/cmt.11.80

World Business Council for Sustainable Development. (2011). *CO2 and energy accounting and reporting standard for the cement industry. The cement CO2 and energy protocol*. https://docs.wbcsd.org/2011/05/CSI-CO2-Protocol.pdf. Accessed 12 Feb 2024.

World Conservation Union. (1991). *Caring for the earth. A strategy for sustainable living*. World Conservation Union.

Zhao, R., Wu, D., & Patti, S. (2020). A bibliometric analysis of carbon labeling schemes in the period 2007–2019. *Energies, 13*, 4233. https://doi.org/10.3390/en13164233

Carbon Footprint Assessment of Agricultural Production Systems Under the Variable System Boundary Scenarios

Ru Chen

1 Introduction

The carbon footprint (CF) is a widely used tool for estimating or measuring the overall emissions of greenhouse gases (GHG), typically in terms of CO_2 equivalents (Wiedmann & Minx, 2008). While according to the Global Footprint Network (GFN), CF can be equated to the demand on "CO_2 land," or the fixation of biomass used to offset GHG emissions from fossil fuel combustion through photosynthesis (Galli et al., 2014), it is essential to note that the definition of "Ecological Footprint," which is closely related to CF, has not altered the public understanding of CF as a measure of carbon dioxide equivalents. However, what exactly is a "carbon footprint?" Wiedmann and Minx (2008) suggested a scientific definition: "a measure of the exclusive total amount of carbon dioxide emissions that are directly and indirectly caused by an activity or are accumulated over the life stages of a product," while other GHGs (especially methane and nitrous oxide) are usually included in the measurement range in subsequent research practices (Chen et al., 2020a). CF is heavily driven by the estimated amount of GHG emissions that can be attributed to human activities, a metric that has direct correlations with the level of climate responsibility each individual or organization must shoulder. Accurate CF accounting is essential for advancing research on climate change mitigation and adaptation strategies, and there are two widely used methods to account for the CF—the "bottom-up" process analysis (PA) and the "top-down" Environmental Input-Output (EIO) analysis. Both methods aim to capture the full life cycle impacts of an activity, but they require clear system boundaries to avoid under-counting or double-counting

R. Chen (✉)
Bay Area International Business School, Beijing Normal University, Zhuhai, China

S. S. Muthu (ed.), *Carbon Footprint Assessments*, Environmental Footprints and Eco-design of Products and Processes,
https://doi.org/10.1007/978-3-031-70262-4_9

of the CF. The PA method is prone to truncation error due to the difficulty in identifying appropriate system boundaries, while the EIO method sets the whole economic system as the boundary to address this issue, but at the expense of detail. Despite these limitations, the popularity of the CF method has led to its application in various research areas and the availability of data and resources has enabled its use in practice.

In relation to climate change and its economic or policy implications, the application of CF accounting presents a mysterious yet important aspect, that is, CF extends its role beyond the simple quantification of greenhouse gas emissions. For example, in the successful implementation of emissions reduction policies, CF plays a crucial role in identifying emission hot spots within the product supply chain. A study in the food system showed that meat and dairy products are the most GHG-intensive food types (Garnett, 2011). The main sources of these emissions were found to be from the production process, specifically from intestinal fermentation and fecal management. Notably, beef production is generally recognized as having the highest global warming potential (GWP) (de Vries & de Boer, 2010). Methane emissions from rice cultivation play a major role in crop production systems, contributing on average 0.69% of China's total CF from flooded paddy rice (Yan et al., 2014). Furthermore, the utilization of fertilizers and energy in the agricultural production stage is also a significant source of emissions. In essence, the indicator CF serves as a tool for GHG emission inventory compilers to determine the total quantity of GHG emitted, expressed in terms of CO_2-eq, released into the atmosphere by an individual, organization, process, or product, within a specific system boundary. In an effort to create consistency in the CF accounting process, various GHG accounting standards have been developed, including the GHG Protocol, PAS 2050, IPCC-2006, and various ISO standards such as ISO14067. Moreover, among researchers in agricultural products CF accounting, the LCA theory-based accounting approach holds a high level of esteem.

A representative framework by CF with regard to LCA accounting is suggested as follows: selecting appropriate GHG emitters, defining boundaries, and gathering data related to the emission of these GHG (Pandey et al., 2011). While these steps do offer a path for accounting, they come with inherent uncertainties that are difficult to eliminate entirely. These uncertainties are classified as epistemic and variability uncertainties, which can lead to inconsistent CF results across different studies (Adewale et al., 2018). One of the key uncertainties that can impact the accuracy of results is the use of inconsistent boundaries (Chen et al., 2020a), which are hypothetical lines that encapsulate the activities and materials that are crucial for the core CF categories, and they are influenced by the entity's objectives and characteristics. Agricultural greenhouse gases are primarily emitted from the production of three gases: CO_2, CH_4, and N_2O. By carefully tracking the key steps in the production of these gases within the confines of a defined system boundary, it is possible to approximate the level of GHG emissions from agricultural activities. However, different CF specifications for calculation of GHG emissions and removals may not always align in terms of key source categories. For instance, to encourage more uniform reporting and estimation, the IPCC-2006 merged the separate guidelines for agriculture and land use, land-use change, and forestry into a single document while preserving the IPCC-1996 framework. Compared to alternative methods, the GHG accounting method relying on LCA recognizes and accounts for both managed cropland and the overall emissions generated across all stages of a product's life cycle, including acquisition, production, transportation, storage, operation, utilization, and disposal. Moreover, the PAS 2050 protocol, which excludes food and forest biomass with biological carbon from the scope of GHG accounting, states that products containing this type of carbon will decompose and release CO_2 once more into the atmosphere. However, the inclusion of soil carbon storage in this accounting process will be further considered in future revisions of this specification (The British Standards Institution, 2011).

Therefore, uncertainties may arise in various accounting steps involved in the calculation of the CF of agricultural products. This underscores the significance of exercising total control over these uncertainties throughout the entire calculation process. By meticulously gathering data from the farmers and leveraging the factor coefficient of localization, the uncertainty associated with the CF of agricultural products can be greatly reduced. In this chapter, we strive to underscore the significant influence of system boundary uncertainty by presenting the results of various CF calculations according to different technical specifications, followed by the design of enhanced models aimed at enhancing the precision of the agricultural carbon footprint.

To be specific, a systematic approach is utilized to establish a comprehensive CF accounting model at both the product and organizational levels in Sect. 2, which process includes rigorous formulation of functional units, system boundaries, accounting system formulas, factor coefficients, and activity data, among other aspects. Section 3 provides an overall perspective of the CF of agricultural production systems, and delves into the CF specificities of various agricultural products, system boundaries, and key categories. In Sects. 4 and 5, two crucial characteristics that need to be emphasized in the CF accounting of agricultural production systems are highlighted. Finally, Sect. 6 explores some potential limitations and suggests several promising directions for future research.

2 Modeling Process for Carbon Footprint Assessment

2.1 Functional Unit

Inconsistent units can hinder comparisons between CF calculated at different levels. At the practical operation level of global scale, the CF functional unit is typically defined as the carbon dioxide equivalent (CO_2-eq), encompassing other GHG, such as methane and nitrous oxide, which are respectively converted to carbon dioxide based on their individual global warming potential. We focus on the CF present in the various agricultural production systems used in the cultivation of agricultural products. This involves CF among individual farmers, when they are regarded as a production organization, and also CF of the same agricultural products that may be grown under the management of various farmer's organizations. In order to accurately assess the total CF of the study area, three individual functional units were utilized. The first one was referred to as the agricultural activities' carbon footprint (ACF), which estimated the overall net carbon budget of the project's agricultural activities in terms of the ton of carbon dioxide equivalent (t C-eq). The second unit was named the farm's carbon footprint (FCF), reflecting the total carbon footprint of the farm's cropland area in t C-eq per hectare (t C-eq ha^{-1}). The last and final unit refers to the carbon footprint of the agricultural product (PCF), calculated by converting crop yield into t C-eq per ton (t C-eq t^{-1}). The greenhouse effect of 1 kg CH_4 is equivalent to 25 kg CO_2 (6.8182 kg C), 1 kg N_2O is equivalent to 298 kg CO_2 (81.2727 kg C), and 1 kg CO_2 is equivalent to 0.2727 kg C.

2.2 System Boundary Scenarios

Different accounting specifications for GHG emissions will inevitably amplify the uncertainties associated with CF calculations. Despite the potential benefits of applying LCA theory to enhance the consistency of these calculations, existing literature presents various perspectives and approaches to defining the system boundaries of agriculture-related activities, including the concept of a "cradle to gate" or "cradle to grave" approach (Brandão et al., 2013). However, owing to various epistemic and variability limitations (e.g., lack of knowledge, insufficient data, and diverse variability), the boundaries of these agricultural activity systems significantly change over time (Table 1), thereby posing a challenge in defining a uniform system boundary to be applied across the entire LCA system. Considering these uncertainties, we implemented the commonly employed "farm gate" approach of LCA for agricultural CF, defining the system boundary from the production of raw materials to the harvesting of crops. Furthermore, we developed four scenarios based on different GHG accounting standards and fundamental theories to calculate the CF of agricultural production systems.

Specifically, Scenario 1 (S1) establishes a system boundary restricted to the IPCC-2006 guidelines, concentrating solely on emissions related to cropland managed by farmers. Scenario 2 (S2) expands the system boundary to align with the PAS 2050 protocol, encompassing direct and indirect emissions that arise during agricultural production from the point of cradle to the farm gate. Scenario 3 (S3) is a system boundary based on the C-N cycle of agro-ecosystem, focusing on the carbon and nitrogen exchanges between the cropland and the atmosphere, which significantly impact the generation of GHG emissions and carbon pool. Lastly, Scenario 4 (S4) is an integrated system boundary that combines scenarios 1, 2, and 3 (Fig. 1).

Table 1 Selection of sources or sinks in the system boundary of some literature on agricultural carbon footprint in China

Subsystem of life cycle assessment	Sources or sinks	DOI: 10.1016/ j.jclepro. 2017.02. 172	DOI: 10.1016/ j.jclepro. 2019.06. 122	DOI: 10.1016/ j.jclepro. 2018.05. 115	DOI: 10.1016/j. jclepro.2019. 04.044	DOI: 10.1016/ j.jclepro. 2016. 09.120	DOI: 10.1016/ j.jclepro. 2015.05. 058	DOI: 10.1016/j. jclepro.2017. 05.108	DOI: 10.1016/ j.jclepro. 2016.11. 181	DOI: 10.1016/ j. scitotenv. 2018.07. 104	DOI: 10.1038/ s41598- 017- 04182-x
Raw material production system (RMPS)	Chemical fertilizer GE	Y[1]	Y[a2]	Y[a]		Y[a]	Y[a]	Y[a]	Y[a]	Y	Y[a]
	Pesticide GE		Y[a]	Y		Y[a]	Y[a]	Y[a]	Y[a]		Y[a]
	Plastic film GE					Y[a]	Y[a]	Y[a]	Y[a]		Y[a]
	Diesel/gasoline/coal GE					Y[a]		Y[a]			
	Electricity GE					Y[a]		Y[a]			
	Organic fertilizer GE[3]										Y[a]
Raw material transportation system (RMTS)[4]	Chemical fertilizer GE		Y[a]	Y[a]		Y[a]		Y[a]		Y	Y[a]
	Pesticide GE		Y[a]			Y[a]		Y[a]			Y[a]
	Plastic film GE					Y[a]		Y[a]			Y[a]
	Diesel/gasoline/coal GE					Y[a]		Y[a]			
	Electricity GE					Y[a]		Y[a]			
	Organic fertilizer GE										Y[a]
	Loss of raw materials GE[5]										

(continued)

Table 1 (continued)

Subsystem of life cycle assessment	Sources or sinks	DOI: 10.1016/j.jclepro.2017.02.172	DOI: 10.1016/j.jclepro.2019.06.122	DOI: 10.1016/j.jclepro.2018.05.115	DOI: 10.1016/j.jclepro.2019.04.044	DOI: 10.1016/j.jclepro.2016.09.120	DOI: 10.1016/j.jclepro.2015.05.058	DOI: 10.1016/j.jclepro.2017.05.108	DOI: 10.1016/j.jclepro.2016.11.181	DOI: 10.1016/j.scitotenv.2013.07.104	DOI: 10.1038/s41598-017-04182-x
Agricultural production system (APS) or agro-ecosystem (AE)	liming and urea application CE										
	Pesticide GE	Y[c2]	Y[a]	Y	Y[c]	Y[a]	Y[a]	Y[a]	Y[a]		Y[a]
	Plastic film GE	Y[c]	Y[c]			Y[a]	Y[a]	Y[a]	Y[a]		Y[a]
	Diesel/gasoline/coal GE	Y[c]	Y[c]		Y[c]	Y[a]	Y[c]	Y[a]	Y[c]	Y	Y
	Electricity GE	Y[c]	Y[c]	Y[c]	Y[c]	Y[a]	Y[c]	Y[a]	Y[c]		Y
	Agricultural machinery and equipment GE[6]										
	Agricultural infrastructure application GE[6]										
	Labor GE[7]										
	Chemical fertilizer N_2O-NE	Y	Y[a]	Y	Y	Y	Y	Y	Y[a]	Y	Y

	Organic fertilizer N_2O-NE		Y[c]								Y
	Straw and root stubble N_2O-NE									Y	Y
	AND $N_2O_{deposition}E$			Y[b2]	Y	Y					Y
	Leaching or runoff loss $N_2O_{lrl}E$			Y[b]	Y	Y					Y
	Non-irrigated farmland ME		Y[c]								
	Rice cultivation ME	Y		Y[b]	Y		Y		Y	Y	Y
	AWDD GE[8]			Y_I[8]						Y_I	Y_I
	Crop biomass CS							Y[b]			
	DOM CS										
	SOC CS[9]	Y_{EE}[9]		Y[b]	Y[b]			Y[b]	Y	Y	Y
	Fertilizer application CS			Y[b]	Y[b]					Y	Y
	Seeds, straw, and root stubble CS			Y[b]	Y[b]			Y[b]		Y	Y
	SRH GE							Y[b]			
	Land use change										

(continued)

Table 1 (continued)

Subsystem of life cycle assessment	Sources or sinks	DOI: 10.1016/j.jclepro.2017.02.172	DOI: 10.1016/j.jclepro.2019.06.122	DOI: 10.1016/j.jclepro.2018.05.115	DOI: 10.1016/j.jclepro.2019.04.044	DOI: 10.1016/j.jclepro.2016.09.120	DOI: 10.1016/j.jclepro.2015.05.058	DOI: 10.1016/j.jclepro.2017.05.108	DOI: 10.1016/j.jclepro.2016.11.181	DOI: 10.1016/j.scitotenv.2018.07.104	DOI: 10.1038/s41598-017-04182-x
Agricultural products transportation system (APTS)	Diesel/gasoline/coal GE		Y		Y[c]				Y		
	Electricity GE										
	Transport machinery GE[6]										
	Infrastructure application GE[6]										
	Labor GE[7]										
	Loss of agricultural products GE[5]										
Agricultural products processing system (APPS)	APPS LCA GE/CS		Y						Y		
Agricultural products consumption system (APCS)	Diesel/gasoline/coal GE										
	Electricity GE										
	Labor GE[7]										

	Machinery and equipment GE[6]										
	Infrastructure application GE[6]										
Agricultural waste disposal system (AWDS)	AWDS LCA GE/CS										

Notes: *GE* Greenhouse gas emissions, *CE* CO_2 emissions, *ME* CH_4 emissions, *N_2O-NE N* is converted into N_2O emissions, *AND* Atmospheric nitrogen deposition, $N_2O_{deposition}E$ N_2O emissions from AND, $N_2O_{lrl}E$ N_2O emissions from leaching or runoff loss, *AWDD* Direct disposal of agricultural waste, *DOM* Dead organic matter, *SOC* Soil organic carbon, *SRH* Soil heterotrophic respiration, *LCA* Life cycle assessment, *CS* Carbon sequestration

[1]Y (yes) means to take the corresponding key sources and sinks into account

[2]Indicates that the emission factor (EF) of the whole life cycle is used but not allocated to the subsystem; [b] indicates that the EF adopted is from the experimental measurement; [c] indicates whether it is a whole life cycle EF is unclear due to the lack of description

[3]Only commercial organic fertilizer is included in the RMPS and RMTS, and the manure directly returned to the cropland is added in APS or AE

[4]Please refer to APTS for the details of key emission sources of each raw material in RMTS

[5]Allocation is based on the proportion of local loss quantity on site and disposal quantity entering system AWDS

[6]Machinery, equipment, and infrastructure apportion the CF calculated by LCA to each year according to their service life

[7]It refers to the input of biological energy of labor force such as human or livestock

[8]Allocation is based on the proportion of disposal quantity on site and entering system AWDS. AWDD emissions include straw burning GE (I), agricultural waste burning GE (II), and other disposal methods GE (III)

[9]The change of SOC needs to take the disturbance of deep tillage, rotary tillage, irrigation, and other cultivation systems into account (EE is emission effect, SE is sink effect), but not include the residual carbon of fertilizer application, sowing seeds, and straw returning and root stubble after harvest, which are listed separately

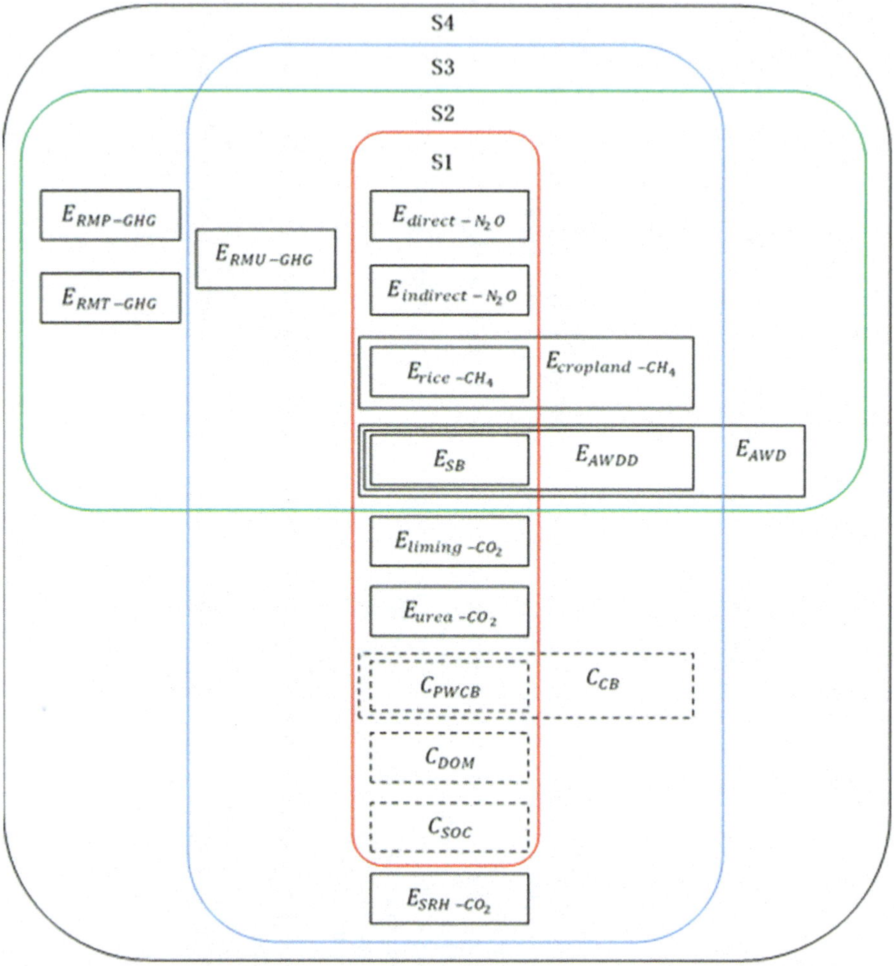

Fig. 1 CF accounting module in multi-system boundary scenarios. (Chen & Kong, 2022)

2.3 Carbon Footprint Accounting System

2.3.1 GHG Emissions from RMPS

GE from RMPS includes the GE from manufacturing of chemical fertilizer, organic fertilizer, pesticide, plastic film, diesel, gasoline, coal and electricity, which can be calculated with Eq. 1:

$$E_{\mathrm{RMP-GHG}} = \sum \Big[\left(\mathrm{CFA}_{ij} + \mathrm{OFA}_{ij} + \mathrm{DA}_j + \mathrm{GA}_j + \mathrm{CA}_j + \mathrm{EA}_j\right) * \mathrm{EFM}_i + \left(\mathrm{PA}_j * \mathrm{CEC}_{\mathrm{up}} + \mathrm{PFA}_j * \mathrm{CEC}_{\mathrm{upf}}\right) * \mathrm{CV}_{\mathrm{ce}} * \mathrm{EFC}_{\mathrm{CI}} \Big] * \mathrm{GWP} \quad (1)$$

where $E_{\mathrm{RMP-GHG}}$ is the total amount GE of RMPS (t C); *i* is indicated the different types; CFA_{ij} is the applied amount of chemical fertilizer i (nitrogen fertilizer, phosphatic fertilizer, and potash fertilizer) for j-th farmer (t); OFA_{ij} is the applied amount of commercial organic fertilizer i (cow manure, pig manure and chicken manure) for j-th farmer (t); DA_j, GA_j, CA_j, and EA_j are the applied amount of diesel, gasoline, coal, and electricity for j-th farmer (t), respectively; EFM_i is EF from manufacturing of different agricultural inputs; $\mathrm{CEC}_{\mathrm{up}}$ (1.272 tce t^{-1}, ECCPCIA and CECEPG, 2014) and $\mathrm{CEC}_{\mathrm{upf}}$ (1.27 tce t^{-1}, ECCPCIA and CECEPG, 2014) are the comprehensive energy consumption of unit pesticide and unit plastic film (ethylene manufacturing and plastic synthesis), respectively; CV_{ce} (29,307 KJ kg^{-1}, CECERG, 2009) is the calorific value of standard coal (ce); EFC_{CI} is the EF of coal application in the chemical industry (23.752 tC TJ^{-1}, NDRC, 2014); PA_j and PFA_j are the applied amount of pesticide and plastic film for j-th farmer (t), respectively; and GWP is the warming potential of greenhouse gases.

2.3.2 GHG Emissions from RMTS

GE from RMTS can be calculated with Eq. 2:

$$E_{\mathrm{RMT-GHG}} = \sum \mathrm{RM}_{\mathrm{ij}} * \mathrm{EFT}_{\mathrm{i}} * \mathrm{GWP} \quad (2)$$

where $E_{\mathrm{RMT-GHG}}$ is the total amount GE of RMTS (t C); RM_{ij} is the applied amount of i-th raw material for j-th farmer (t); EFT_i is EF from transportation of different agricultural inputs (more calculation details and values are shown in Table 2).

2.3.3 GHG Emissions from APS or AE

All GE from APS can be calculated with Eq. 3:

$$E_{\mathrm{APS-GHG}} = E_{\mathrm{RMU-GHG}} + E_{\mathrm{direct-N_2O}} + E_{\mathrm{indirect-N_2O}} + E_{\mathrm{cropland-CH_4}} + E_{\mathrm{AWD}} + E_{\mathrm{liming-CO_2}} + E_{\mathrm{urea-CO_2}} + E_{\mathrm{SRH-CO_2}} \quad (3)$$

where $E_{\mathrm{RMU-GHG}}$ is the GE emissions from raw material utilization process that occurs on cropland (t C), including utilization process of pesticide, plastic film, diesel, gasoline, coal, electricity, labor, and agricultural machinery and equipment. Equation of calculation is as follows:

Table 2 Emission or sequestration factors and other coefficients for carbon footprint accounting

Subsystem of life cycle assessment	Key sources and sinks	Types	Emission or sequestration factors and other coefficients	References
RMPS	Chemical fertilizer GE	N	8.21 t CO_2-eq t^{-1}	DOI: 201710.1038/s41598-017-04182-x
		P_2O_5	0.636 t CO_2-eq t^{-1}	
		K_2O	0.18 t CO_2-eq t^{-1}	
	Pesticide GE[a]		0.885 t C-eq t^{-1}	Self-calculating
	Plastic film GE[a]		0.884 t C-eq t^{-1}	
	Diesel/gasoline/coal GE[b]	Diesel	5.963 t CO_2-eq tce^{-1}	
		Gasoline		
		Coal	1.123 t CO_2-eq tce^{-1}	
	Electricity GE		0.975 kg CO_2-eq kWh^{-1}	Literature collation by this study
	Organic fertilizer GE[c]	Cattle manure	0.067 t CO_2-eq t^{-1}	
		Pig manure	0.045 t CO_2-eq t^{-1}	Self-calculating
		Chicken manure	0.02 t CO_2-eq t^{-1}	
RMTS	Chemical fertilizer GE	N	0.1 t CO_2-eq t^{-1}	DOI: 201710.1038/s41598-017-04182-x
		P_2O_5	0.06 t CO_2-eq t^{-1}	
		K_2O	0.05 t CO_2-eq t^{-1}	
	Pesticide GE[d]		1.624 t C-eq t^{-1}	Self-calculating
	Plastic film GE[e]		0.529 t C-eq t^{-1}	
	Diesel/gasoline/coal GE	Diesel	0.600 t CO_2-eq t^{-1}	DOI: 10.1016/j.agsy.2016.09.004
		Gasoline		
		Coal	0.037 t CO_2-eq t^{-1}	Literature collation by this study
	Electricity GE[f]		0.004 kg CO_2-eq kWh^{-1}	Self-calculating
	Organic fertilizer GE[g]		0.178 t CO_2-eq t^{-1}	

APS or AE	urea application CE[h]		$0.2\ t\ C\text{-}eq\ t^{-1}$		IPCC (2006)
	liming application CE[h]		$0.12\ t\ C\text{-}eq\ t^{-1}$		
	Pesticide GE[i]		$0.0195\ t\ CO_2\text{-}eq\ ha^{-1}$		Self-calculating
	Diesel/gasoline/coal GE[j]	Diesel	$0.845\ t\ C\text{-}eq\ t^{-1}$		
		Gasoline	$0.798\ t\ C\text{-}eq\ t^{-1}$		
		Coal	$0.45\ t\ C\text{-}eq\ t^{-1}$		
	Agricultural machinery and equipment GE[k]	Combine harvesting	$0.153\ t\ CO_2\text{-}eq\ ha^{-1}$		Ecoinvent V2
		Solids manure loading and spreading	$0.003\ t\ CO_2\text{-}eq\ t^{-1}$		
		Sowing by machine	$0.022\ t\ CO_2\text{-}eq\ ha^{-1}$		
		Tillage, rotary cultivator	$0.075\ t\ CO_2\text{-}eq\ ha^{-1}$		
		Tillage, rolling	$0.023\ t\ CO_2\text{-}eq\ ha^{-1}$		
		Planting by machine	$0.095\ t\ CO_2\text{-}eq\ ha^{-1}$		
	$EFNC_i$	Rice	$0.003\ t\ NE\ t^{-1}\ N^{-1}$		NDRC (2014)
		Wheat	$0.0074\ t\ NE\ t^{-1}\ N^{-1}$		
		Maize	$0.0074\ t\ NE\ t^{-1}\ N^{-1}$		
		Apple	$0.0075\ t\ NE\ t^{-1}\ N^{-1}$		
	Organic fertilizer $N_2O\text{-}N$ E[l]		N content	SOC content	Literature collation by this study
		Cattle manure	$0.018\ t\ N\ t^{-1}$	$0.189\ t\ C\ t^{-1}$	
		Pig manure	$0.026\ t\ N\ t^{-1}$	$0.232\ t\ C\ t^{-1}$	
		Chicken manure	$0.017\ t\ N\ t^{-1}$	$0.121\ t\ C\ t^{-1}$	

(continued)

Table 2 (continued)

Subsystem of life cycle assessment	Key sources and sinks	Types		Emission or sequestration factors and other coefficients			References
	Seeds, straw, and root stubble N_2O-N E			Rice	Wheat	Maize	NDRC (2014)
		DWR_k		0.855	0.87	0.86	
		SNC_k		0.0075	0.0052	0.0058	
		HI_k		0.489	0.434	0.438	
		RSR_k		0.125	0.166	0.17	
		CNC_k		0.01	0.014	0.017	
		HLR_k		0.015	0.018	0.02	
	AND $N_2O_{deposition}$E[m]			$QFAND_i$	$EFAND_i$		Self-calculating
		NF[n]	Inside	0.021 t N ha^{-1}	0.0074 NE N^{-1}		
			Outside	0.04 t N ha^{-1}	0.01 NE N^{-1}		
		PF[n]	Inside	0.015 t N ha^{-1}	0.003 NE N^{-1}		
			Outside	0.04 t N ha^{-1}	0.01 NE N^{-1}		
	Leaching or runoff loss N_2O_{lrl}E[m]	$QFLRL_i$		0.024 t N ha^{-1}			
		$EFLRL_i$		0.0075 NE N^{-1}			
	Cropland ME[o]	RO		NF/kg C_eq ha^{-1}	Wheat	−1.45	Literature collation by this study[15]
					Maize	−3.6	
					Apple	−0.75	
				PF	0.176 t C_eq ha^{-1}		
		RO+OF		NF	+16.33%		
				PF	0.0485 t C_eq t^{-1}		
		RO+NT		NF	0%		
				PF	−30%		
		RO+SR		NF	+33.11%		
				PF	+22.95%		

	SRH GE[o]	RO	PWCB	5.92 t C_eq ha^{-1}		
			AHCB	$1.0598e^{0.0012NPPa}$		
		RO+OF	NF	+23%		
			PF	−8%		
		RO+ NT	NF	−39.1%		
			PF	−7%		
	AWDD GE	Straw burning NE	0.011 t C_eq t^{-1}			Self-calculating
		Straw burning ME	0.035 t C_eq t^{-1}			
		Firewood burning NE	0.006 t C_eq t^{-1}			
		Firewood burning ME	0.019 t C_eq t^{-1}			
		Plastic burning GE	0.228 t C_eq t^{-1}			
	Crop biomass CS	CCF_k	Rice	Wheat	Maize	Literature collation by this study
			0.45	0.45	0.471	
	SOC CS	RO	NF	+0.30 t C_eq ha^{-1}		Literature collation by this study[15]
			PF	+0.31 t C_eq ha^{-1}		
		RO+OF		Self-calculating		
		RO+SR				
		RO+BF	NF	+2.55%		
			PF	+1.89%		
		RO+ NT	NF	+0.71 t C_eq ha^{-1}		
			PF	+0.68 t C_eq ha^{-1}		
AWDS	AWDS LCA GE/CS[p]		Reference AWDD GE			

Notes: Due to the inadequate estimation of EFs of localization in the existing literature:
[a]We indirectly estimated the EFs for pesticide and plastic film according to current literature
[b]Our calculation of the EFs for the mining, processing, and manufacturing of diesel, gasoline, and coal was based on NDRC (2014)

(continued)

[c]EF of cattle manure production is calculated based on LCA, but the other two are based on experimental measurements (direct emissions) (using pig manure to represent organic fertilizer in this subsystem)
[d]Calculation by subtracting the EF of pesticide processing and manufacturing in RMPS according to the research results (Audsley et al., 2014)
[e]Calculation by subtracting the EF of plastic film manufacturing in RMPS based on IREEA
[f]Our calculation was based on NDRC (2014), GF from produced and used by power transmission and distribution equipment are allocated to total electricity consumption in China in 2005
[g]Calculation by subtracting the EF of organic fertilizer manufacturing in RMPS (using pig manure to represent organic fertilizer in this subsystem) (Zhang et al., 2017)
[h]This emission source is considered because the CO_2 removed from the atmosphere in the urea manufacturing process is estimated in the industrial process, but it is not included in the RMPS fertilizer factor, so the emission source should be removed according to the scenario of PLCA
[i]Our calculation was based on the research results (Audsley et al., 2014)
[j]Our calculation was based on NDRC (2014)
[k]Including machinery production, storage infrastructure, fuel combustion emissions, and tire aberration emission (Adewale et al., 2016)
[l]N content and SOC content are input to the Eq.5 and Eq.15, respectively
[m]Our calculation was based on NDRC (2014) in Shaanxi Province
[n]NF is non-irrigated farmland (vegetables are not included), and PF is paddy field
[o]RO, Routine operation; OF, Organic fertilizer; BF, Balanced fertilization; NT, No-tillage; SR, Straw returning; NPPa, Net primary productivity of per unit area
[p]Agricultural waste is assumed to be disposed of as an initial fuel for energy production

$$E_{\text{RMU-GHG}} = \sum \Big[\left(Pa_j + \text{DA}_j + \text{GA}_j + \text{CA}_j \right) * \text{EFU}_i + \left(\text{Ha}_j + \text{OFA}_j + \text{Sa}_j + \text{Ta}_j + \text{Ca}_j \right) * \text{EFAME}_i \Big] * \text{GWP} \quad (4)$$

where Pa_j is the applied area of pesticide on cropland for j-th farmer (ha) and EFU_i is EF from utilization of different agricultural inputs (more calculation details and values are shown in Table 2). Ha_j, Sa_j, Ta_j, Ca_j are the area of cropland for agricultural machinery operation in harvesting, sowing, tillage, and chopping for j-th farmer (ha), respectively; EFAME_i is EF from the applied of agricultural machinery and equipment (Table 2). The use of electricity drives mechanical operation, and almost all emissions are covered in RMPS and RMTS, so it is not calculated in the APS. The burying, recycling, and burning of plastic film after utilization were calculated in AWDD and AWDS. N_2O emissions from managed soils (t C) can be calculated with Eqs. 5 and 8:

$$E_{\text{direct-N}_2\text{O}} = \sum \left[\left(\text{CFNA}_j + \text{OFNA}_j + \text{SRSNA}_j + \text{CLN}_j \right) * \text{EFNC}_i \right] * \text{GWP} \quad (5)$$

where CFNA_j is the application of N from chemical fertilizer for j-th farmer (t); OFNA_j is the application of N from organic fertilizer for j-th farmer (t) (nitrogen content is shown in Table 2); SRSNA_j is the application of N from the returning straw and root stubble residue for j-th farmer (t); CLN_j is the application of N from the crop loss for j-th farmer (t); and EFNC_i is the EF from different nitrogen sources (Table 2). SRSNA_j can be obtained by Eq. 6:

$$\text{SRSNA}_j = \sum \Big[\text{CY}_{kj} * (1/\text{HI}_k - 1) * \text{DWR}_k * \text{SNC}_k * \text{SGR}_{kj} + \text{CY}_{kj}/\text{HI}_k * \text{DWR}_k * \text{SNC}_k * \text{RSR}_k \Big] \quad (6)$$

where CY_{kj} is the yield of crop k for j-th farmer (t); HI_k is the harvest index of crop k; DWR_k is dry weight ratio of crop k; SNC_k is nitrogen content of crop k; SGR_{kj} is the straw-grain ratio of crop k for j-th farmer (%); and RSR_k is the root-shoot ratio of crop k. The value of the coefficients in formula 6 can be obtained from Table 2 and our survey data. CLN_j can be obtained by Eq. 7:

$$\text{CLN}_j = \sum \left[\text{CY}_{kj} * \text{HLR}_k/(1 - \text{HLR}_k) * \text{CNC}_k \right] \quad (7)$$

where HLR_k is harvest loss ratio of crop k; CNC_k is nitrogen content of crop k. $E_{\text{indirect-N}_2\text{O}}$ is the indirect N_2O emissions from managed soils (t C), including N_2O emissions from atmospheric nitrogen deposition and N_2O emissions from leaching or runoff loss, which can be calculated with Eq. 8:

$$E_{\text{indirect}-N_2O} = \sum\left[(\text{QFAND}_i * \text{EFAND}_i + \text{QFLRL}_i * \text{EFLRL}_i) * \text{Ca}_{kj}\right] * \text{GWP} \quad (8)$$

where QFAND_i is the quantity factor (QF) of atmospheric nitrogen deposition (inside or outside of the cropland); EFAND_i is the EF of atmospheric nitrogen deposition; QFLRL_i is the QF of leaching or runoff loss; EFLRL_i is the EF of leaching or runoff loss; and Ca_{kj} is the cultivated area of crop k for j-th farmer (ha). The CH_4 emissions from cropland are divided into two land types (t C): Non-irrigated farmland and paddy field, which can be calculated with Eq. 9:

$$E_{\text{cropland}-CH_4} = \sum\left[(\text{MEFRO}_\text{i} + \text{EFCM}_\text{i}) * \text{Ca}_\text{kj}\right] * \text{GWP} \quad (9)$$

where MEFRO_i is the CH_4 EF of routine operation on the cropland and EFCM_i is the EF of cropland management. All the initial values of these factor data can be obtained from Table 2, which were from the literature we collected based on the data integration method. E_{AWD} includes GE from AWDD (E_{AWDD}) and AWDS (E_{AWDS}) (t C), which depends on the proportion of agricultural waste allocated by farmers between the two subsystems. It can be calculated with Eq. 10:

$$E_{\text{AWD}} = E_{\text{AWDD}} + E_{\text{AWDS}} = \sum\left[(\text{AWDDR}_j + \text{AWDSR}_j) * \text{AWA}_j * \text{EFAW}_i\right] * \text{GWP} \quad (10)$$

where AWDDR_j is the ratio of direct disposal of agricultural waste for j-th farmer (%); AWDSR_j is the ratio of agricultural waste disposal in the AWDS for j-th farmer (%); AWA_j is the amount of agricultural waste for j-th farmer (t); and EFAW_i is EF of different agricultural wastes (Table 2). The distribution proportion coefficients are obtained from our survey data. The calculation method of $E_{\text{liming}-CO_2}$ and $E_{\text{urea}-CO_2}$ are consistent with formula 2, and heterotrophic soil respiration from AE ($E_{\text{SRH}-CO_2}$) is consistent with formula 9 (EF and highlights can be obtained from Table 2).

2.3.4 Carbon sequestration from APS or AE

Carbon sinks of AE are from the carbon sequestration of crops, dead organic matter, and mineral soils, which can be calculated by the method of carbon stock changes over a period of time. C_{CB} can be calculated with Eq. 11:

$$C_{\text{CB}} = C_{\text{PWCB}} + C_{\text{AHCB}} \quad (11)$$

where C_{PWCB} and C_{AHCB} are the carbon stock changes of perennial woody crops biomass (PWCB, apple trees in this study) and annual herbaceous crops biomass (AHCB, rice, wheat, and maize in this study) in one growth season (t C), respectively. The calculation method of carbon sequestration from apple trees that we adopted the research of Wu et al. (2012), who estimated the dynamics of the net carbon sink in apple orchards in China. Due to the long-term retention of apple tree branch, stem, and root in the AE, we estimate their carbon stock changes compared with the previous period, and the new growth biomass in the current period includes the transferred fruits and leaves as DOM; the allometric equation is set as follows:

$$
\begin{aligned}
C_{\text{PWCB}-\text{apple}} = \sum \Big[\text{AA}_{aj} * 0.46 * \Big(& \text{DMB}_a - \text{DMB}_{a-1} + \text{DMS}_a - \text{DMS}_{a-1} \\
& + \text{DMR}_a - \text{DMR}_{a-1} \Big) + 0.46 * \text{DMF}_\text{a} - \text{AA}_\text{aj} * \text{CPB}_\text{a} \Big] / 1000 \\
\text{DMB}_\text{a} &= 0.124 * {\text{SD}_\text{a}}^{2.468} \\
\text{DMS}_\text{a} &= 0.178 * {\text{SD}_\text{a}}^{2.202} \\
\text{DMR}_\text{a} &= 0.159 * {\text{SD}_\text{a}}^{1.998} \\
\text{DMF}_\text{a} &= 0.127 * \text{AY}_\text{j} \\
\text{SD}_\text{a} &= 0.888\text{a} - 1.414
\end{aligned}
\tag{12}
$$

where AA_{aj} is the amount of apple trees for *j*-th farmer; a is the average age of the trees (yr); DMB_a, DMS_a, DMR_a, and DMF_a are the dry matter of branch, stem, root, and fruit (kg), respectively; CPB_a is the carbon content in pruned branches of an apple tree with different ages (Wu et al., 2012) (kg C tree^{-1}); and SD_a is the stem diameter of the trees (cm). We set its fitting equation according to Wu et al. (2012), where R^2 is 0.927 and $P < 0.01$ and AY_j is the yield of the apple for j-th farmer (kg). C_AHCB can be calculated with Eq. 13 (t C):

$$
C_\text{AHCB} = \sum \text{CCF}_k * \text{CY}_{kj} * \text{DWR}_k \tag{13}
$$

where CCF_k is the conversion fraction of dry matter to carbon content.

C_DOM are from the carbon sequestration of fallen leaves from apple trees (t C), which can be calculated with Eq. 14:

$$
\begin{aligned}
C_\text{DOM} &= \sum \text{AA}_{aj} * 0.46 * \text{DML}_a / 1000 \\
\text{DML}_a &= 0.16 * {\text{SD}_a}^{1.312}
\end{aligned}
\tag{14}
$$

where DML_a is the dry matter of leaves of apple trees (kg).

C_SOC are from the carbon sequestration of straw and root stubble returning from annual herbaceous crops, pruned branches returning from apple trees, and fertilizer application (t C), which can be affected by farmland management measures. The equation of calculation is as follows:

$$C_{SOC} = \sum \Big[CY_{kj} * (1/HI_k - 1) * DWR_k * CCF_k * SGR_{kj} + CY_{kj}/HI_k * DWR_k * CCF_k * RSR_k + CY_{kj} * HLR_k/(1 - HLR_k) * DWR_k * CCF_k + AA_{aj} * CPB_a * SGR_{kj} + OFA_{ij} * COF_i + (CSFRO_i + CSFCM_i) * Ca_{kj} \Big] \quad (15)$$

where COF_i is the SOC content of organic fertilizer; $CSFRO_i$ is the carbon sequestration factors (CSF) of routine operation (chemical fertilizer application, tillage, irrigating, etc.) on the cropland; and $CSFCM_i$ is the CSF of cropland management. It is worth noting that the loss of C caused by organic matter decomposition is considered in E_{SRH-CO_2} and $E_{cropland-CH_4}$. The value of the coefficients in formula 11, 12, 13, 14 and 15 can be obtained from Table 2 and our survey data.

2.3.5 CF Accounting System Under the Variable System Boundary Scenarios

To calculate the net carbon budget, that is ACF, under the four system boundary scenarios outlined above, the following equations were set:

$$ACF_{S1-j} = E_{direct-N_2O} + E_{indirect-N_2O} + E_{rice-CH_4} + E_{liming-CO_2} + E_{urea-CO_2} + E_{SB} - C_{PWCB} - C_{DOM} - C_{SOC} \quad (16)$$

$$ACF_{S2-j} = E_{RMP-GHG} + E_{RMT-GHG} + E_{RMU-GHG} + E_{direct-N_2O} + E_{indirect-N_2O} + E_{cropland-CH_4} + +E_{AWD} \quad (17)$$

$$ACF_{S3-j} = E_{RMU-GHG} + E_{direct-N_2O} + E_{indirect-N_2O} + E_{cropland-CH_4} + E_{AWDD} + E_{liming-CO_2} + E_{urea-CO_2} + E_{SRH-CO_2} - C_{CB} - C_{DOM} - C_{SOC} \quad (18)$$

$$ACF_{S4-j} = E_{RMP-GHG} + E_{RMT-GHG} + E_{RMU-GHG} + E_{direct-N_2O} + E_{indirect-N_2O} + E_{cropland-CH_4} + E_{AWD} + E_{SRH-CO_2} - C_{CB} - C_{DOM} - C_{SOC} \quad (19)$$

where ACF_{Si-j} is the CF of the j-th farmer's agricultural activities at each of the four designated system boundary scenarios (i = 1,2,3,4); $E_{direct-N_2O}$ is the direct N_2O emissions from managed soils, including the conversion of N into N_2O, that is, N_2O-NE during chemical fertilizer and organic fertilizer application, and N_2O-NE in the returning straw, root stubble residue, and crop loss; $E_{\text{indirect}-N_2O}$ is the indirect N_2O emissions from managed soils, encompasses emissions from both atmospheric nitrogen deposition and leaching or runoff losses; $E_{\text{rice–}CH_4}$ is a subset of $E_{cropland–CH_4}$, which is the CH_4 emissions from rice cultivation; $E_{\text{liming–}CO_2}$ and $E_{\text{urea–}CO_2}$ are the CO_2 emissions associated with liming and urea application to managed soils; E_{SB} represents the quantity of CH_4 and N_2O emitted from the on-site burning of straw; C_{PWCB} and C_{CB} denote the change in C stocks within perennial woody crops and the quantity of crop biomass (the former being a subset of the latter); C_{DOM} and C_{SOC} concern the specific changes in C stock status of dead organic matter and mineral soil; $E_{\text{RMP}-\text{GHG}}$, $E_{\text{RMT}-GHG}$, and $E_{\text{RMU}-\text{GHG}}$ are the GHG emissions released throughout the entire process of raw material production, transportation, and utilization that should be considered, including both direct and indirect emissions of carbon dioxide, methane, and nitrous oxide, which are released as a result of all fertilizers, pesticides, plastic film, diesel, gasoline, coal, and electricity throughout their life cycles. Additionally, the labor, machinery, equipment, and infrastructure each year distribute the LCA-calculated CF based on their respective service life; E_{AWD} is the GHG emissions from disposal of agricultural waste (AWD) (E_{SB} and E_{AWDD} is a subset of it, respectively); and $E_{\text{SRH–}CO_2}$ is CO_2 emissions from the heterotrophic soil respiration. Therefore, the following equations were used to calculate FCF and PCF for the four system boundary scenarios in terms of cropland area and crop yield, respectively:

$$\text{FCF}_{Si-j} = \text{ACF}_{Si-j}/\text{CA}_j \tag{20}$$

$$\text{PCF}_{\text{Si}-\text{j}} = \text{ACF}_{Si-j}/\text{CY}_j \tag{21}$$

where Ca_j is the cropland area of the j-th farmer and CY_j is the crop yield of the j-th farmer.

2.4 *Case Description and Data Collection*

The Shaanxi Province is situated in the heartland of China and covers an expansive area of 205,624.3 square kilometers (Fig. 2). It incorporates the two major river systems, the Yellow River and the Yangtze River, and encompasses three distinct climatic regions, namely the northern subtropical, warm temperate, and middle temperate zones. Known as a significant cradle of China‘s agricultural culture, Shaanxi boasts of rich agricultural resources, diverse climate conditions, and rich genetic resources. According to the National Climate Change Plan (2014–2020) in China, Shaanxi Province has issued the Implementation Plan of Low-Carbon Pilot, since it was included in the first low-carbon pilot areas in 2010. The plan aims to

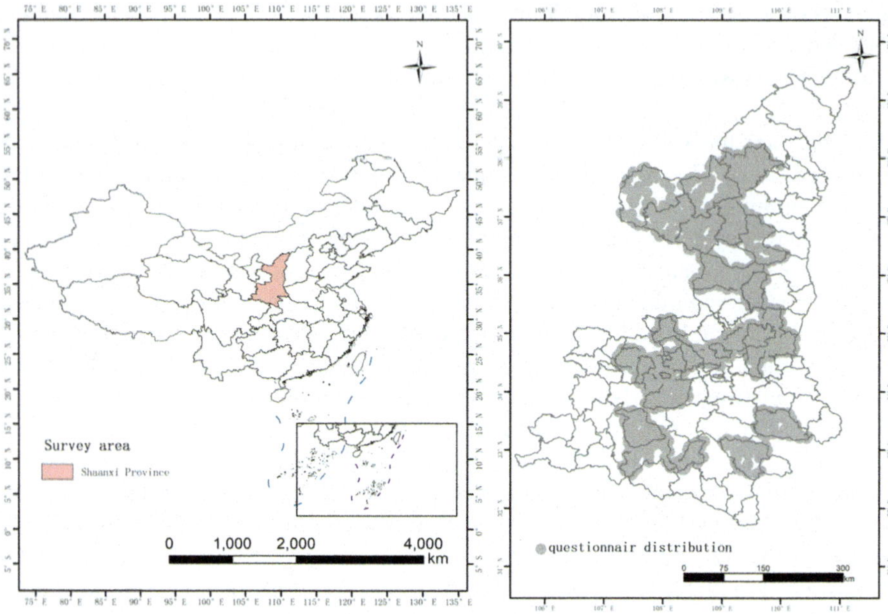

Fig. 2 Geographical location and sampling distribution of the study area

accelerate the promotion and application of low-carbon technologies and promote the emission reduction target of carbon dioxide, methane, nitrous oxide, hydrofluorocarbons, perfluorocarbons, sulfur hexafluoride, and other major greenhouse gases.

When it comes to gathering information for the determination of agricultural activities, as a general rule, it is recommended to prioritize the use of official data provided by established institutions or periodicals. This can be done to guarantee the credibility and accuracy of the data acquired (IPCC, 2006). Additionally, in instances where there is a necessity to enhance the precision of the carbon footprint assessment, a survey questionnaire can be designed and distributed to gather raw data that is in accordance with the research goals (Zhang et al., 2017). Therefore, we carried out a questionnaire survey related to the agricultural output of wheat, maize, rice, and apple cultivators in China's Shaanxi Province in 2019, which meticulously examined their input of agricultural raw materials, adopted cultivation methods, planting scope, and crop yield. The total planting scale for these activities was 367.75 hectares, from the dataset which was obtained including 225 wheat, 273 maize, 96 rice, as well as 211 apple production activities.

3 Carbon Footprint Status of Agricultural Production Systems

3.1 *ACF, FCF, and PCF of Agricultural Products*

Results for the ACF, FCF, and PCF values of wheat, maize, rice, and apple differed among four different scenarios, suggesting significant variations, despite the fact that all four crops had similar characteristics as per the findings of Fig. 3. During the S1 and S3 scenarios, wheat crops experienced a carbon surplus, with the former (S1) exceeding the latter (S3) by 260.426 tons of carbon (which is equivalent to 2.628 tons of carbon per hectare of FCF, and 0.477 tons of carbon per ton of PCF). Comparing wheat's CF in S2 and S4, a significant disparity was observed that the former emitted 153.009 tons of carbon more than the latter, as well as 1.554 tons of carbon per hectare of FCF, and 0.28 tons of carbon per ton of PCF. Similarly, for CF of maize, rice, and apple, there were carbon surpluses in S1, S3, and S4. The largest ACF in S1 was 668.346 t, 57.866 t, and 1140.15 t more carbon than the smallest in S4, and 2.582 tons of carbon per hectare of FCF (0.299 tons of carbon per ton of PCF), 2.943 tons of carbon per hectare of FCF (0.369 tons of carbon per ton of PCF), and 10.719 tons of carbon per hectare of FCF (0.734 tons of carbon per ton of PCF), respectively. Regarding carbon emissions in S2 for CF of maize, rice, and apple, the overall impact on farmers' agricultural production systems is seen as a carbon sink when adopting the system boundaries of scenarios 1 and 3. It is worth noting, however, that the order of carbon sink effects of crops across the ACF, FCF, and PCF was not consistent, with perennial woody crops being identified as the most favorable for carbon sequestration. By implementing the system boundaries of S2, the effect of carbon source is more prominent, with CF contributing significantly to apple production. Moreover, the carbon source or sink effect of the crops produced by farmers within scenario 4 boundaries varies according to the specific crop types. In this scenario, the carbon sink effect is realized for maize, rice, and apple, while it is non-existent for wheat.

3.2 *CF Inventory of S1-4 in Agricultural Production Systems*

The distribution structures of the ACF in RMPS, RMTS, APS or AE, and ADDS are graphically illustrated in Fig. 4, which provides evidence of the inherent consistency with the system boundaries stipulated by various GHG accounting specifications. In S1 and S3, the ACF of wheat, maize, rice, and apple to function as a carbon sink was enhanced to 100% due to the specific system boundary constraints imposed by the IPCC-2006 and C-N cycle theory, which exclusively considered agricultural land under managed cultivation as the system boundaries. Similarly, within S2 and S4, the ACF was spread among several distinct subsystems, guided by the principles of LCA theory. The highest percentage of ACF was observed in RMPS (59.03%, rice

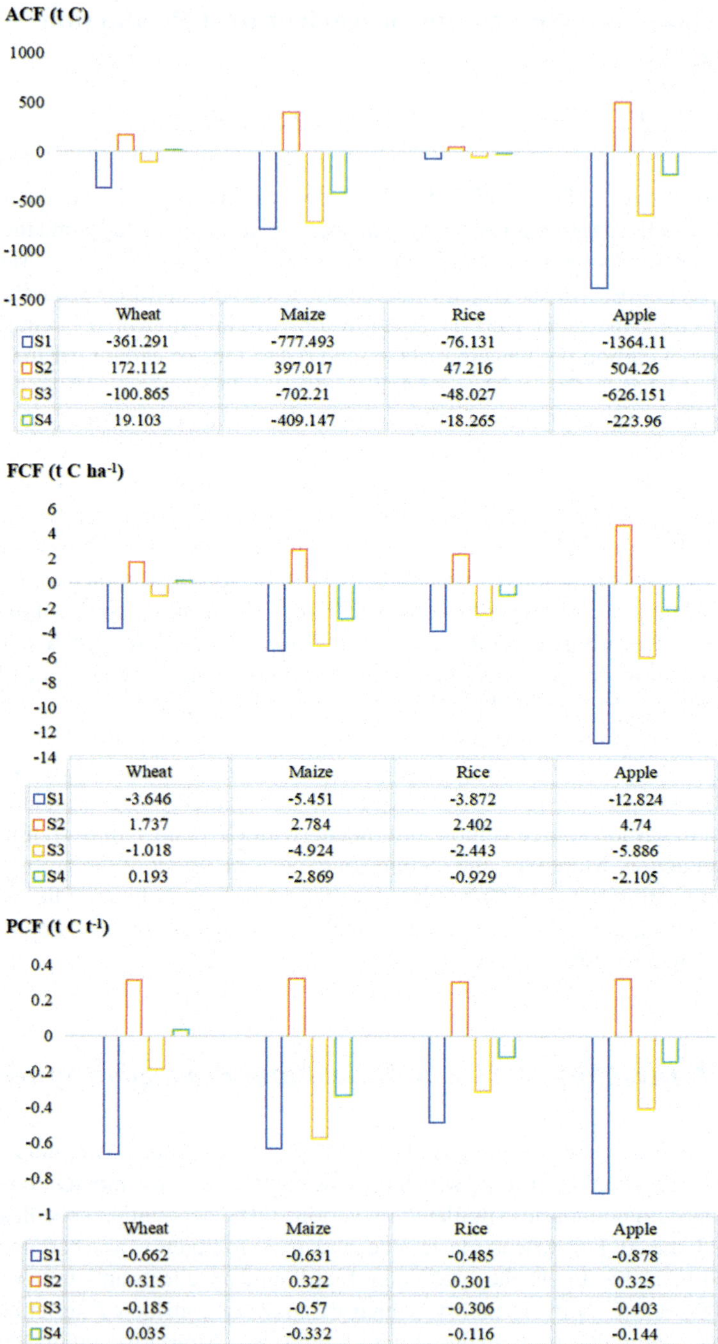

ACF (t C)

	Wheat	Maize	Rice	Apple
S1	-361.291	-777.493	-76.131	-1364.11
S2	172.112	397.017	47.216	504.26
S3	-100.865	-702.21	-48.027	-626.151
S4	19.103	-409.147	-18.265	-223.96

FCF (t C ha^{-1})

	Wheat	Maize	Rice	Apple
S1	-3.646	-5.451	-3.872	-12.824
S2	1.737	2.784	2.402	4.74
S3	-1.018	-4.924	-2.443	-5.886
S4	0.193	-2.869	-0.929	-2.105

PCF (t C t^{-1})

	Wheat	Maize	Rice	Apple
S1	-0.662	-0.631	-0.485	-0.878
S2	0.315	0.322	0.301	0.325
S3	-0.185	-0.57	-0.306	-0.403
S4	0.035	-0.332	-0.116	-0.144

Fig. 3 ACF, FCF, and PCF in the production process of agricultural products

Fig. 4 Distribution of ACF among subsystems in S1-4 (t C)

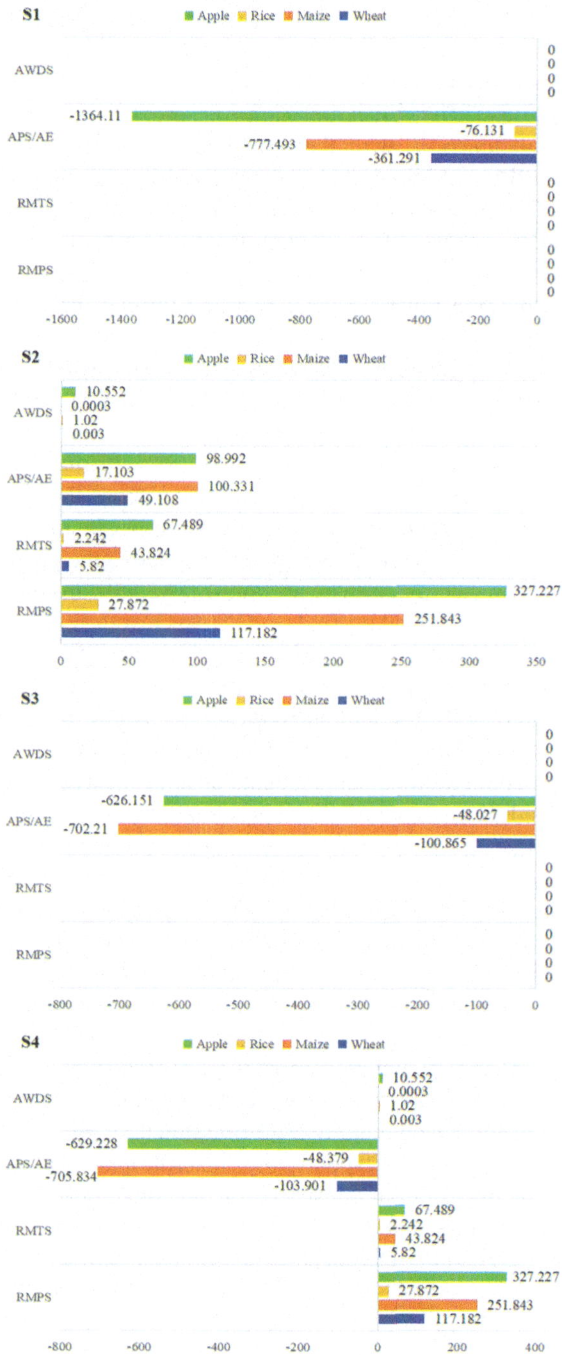

~68.08%, wheat) in S2, followed by APS or AE (19.63%, apple ~36.22%, rice), and the distribution pattern of these distinct crops was consistent, exhibiting a clear sequence of RMPS, APS or AE, RMTS, and AWDS. Furthermore, yielding an earth-positive effect on carbon turnover, crops in APS or AE in S4 demonstrated a carbon sink function, with the remainder exhibiting a carbon source function. Notably, RMPS exhibited the greatest contribution to GHG emissions (80.74%, apple ~95.27%, wheat), followed by RMTS (4.73%, wheat ~16.65%, apple). Moreover, APS or AE offsets a significant amount of GE, effectively reducing the GHG emissions of the production processes of maize, rice, and apple by 2.379, 1.607, and 1.553 times, respectively, and turning these processes from carbon sources to carbon sinks.

In S1, the direct N_2O emissions from wheat, maize, and apple represented the majority of GE, contributing 60.46%, 68.37%, and 87.41% of the total GE, respectively. In contrast, CH_4 emissions from rice production represented the greatest proportion of GE, contributing 63.89% of the total GE. Applying the IPCC-2006 guidelines, soil carbon sequestration from the agricultural fields emerged as the predominant offset mechanism from wheat, maize, and rice in S1. Moreover, the biomass of perennial woody crops yielded the greatest proportion of carbon sequestration (the apple trees account for 82.55% in this study). In S2, the generation of primary material utilized in APS was responsible for the majority of GE, particularly for wheat (68.04%), maize (63.35%), rice (59.03%), and apple (64.88%). Furthermore, nitrogenous fertilizer (45.52% for wheat, 44.61% for maize, 43.26% for rice, and 62.17% for apple) and diesel or gasoline (42.37% for wheat, 34.06% for maize, 53.38% for rice, and 20.5% for apple) production contributed the most to the GE in RMPS. Additionally, methane absorption was observed on the non-irrigated farmland. Heterotrophic soil respiration in the cultivation of crops in S3 (S4) substantially influences the overall GE during the production of wheat, maize, rice, and apples, which accounts for 89.6% (72.32%), 83.63% (57.32%), 82.4% (63.38%), and 87.44% (58.5%) of the total GE, respectively. Moreover, the majority of the carbon sequestration in apple (82.55%) and wheat (64.54%), maize (62.74%) and rice (58.99%) cultivation, respectively, in S3 and S4, was largely attributed to the substantial changes in carbon storage within the cropland, specifically the crops biomass and soil.

Therefore, the predominant order of the effects of the four primary carbon sources was as follows: heterotrophic soil respiration, raw material production, raw material utilization, and directly N_2O emission (CH_4 emissions from cropland) in wheat (rice) production. Correspondingly, heterotrophic soil respiration, raw material production, raw material transportation, and directly N_2O emission were the primary factors during maize and apple production. The detailed contents of the CF inventory of S1-S4 can be referenced from Tables 3, 4, 5 and 6.

Table 3 CF inventory of S1 in FAP (t C)

Type	$E_{direct-N_2O}$	$E_{indirect-N_2O}$	$E_{rice-CH_4}$	$E_{urea-CO_2}$	E_{SB}	C_{PWCB}	C_{DOM}	C_{SOC}
Wheat	16.838	5.923	0.000	3.037	2.050	0.000	0.000	389.138
Maize	43.286	8.524	0.000	3.624	7.879	0.000	0.000	840.807
Rice	1.657	0.999	6.844	0.352	0.861	0.000	0.000	86.843
Apple	65.479	6.358	0.000	3.077	0.000	1187.877	36.909	214.236

Table 4 CF inventory of S2 in FAP (t C)

Type	$E_{RMP-GHG}$	$E_{RMT-GHG}$	$E_{RMU-GHG}$	$E_{direct-N_2O}$	$E_{indirect-N_2O}$	$E_{cropland-CH_4}$	E_{AWD}
Wheat	117.182	5.820	24.398	16.838	5.923	−0.101	2.053
Maize	251.843	43.824	40.115	43.286	8.524	−0.498	9.923
Rice	27.872	2.242	6.742	1.657	0.999	6.844	0.861
Apple	327.227	67.489	26.795	65.479	6.358	−0.068	10.981

3.3 Comparison and Key Categories of Agricultural Carbon Footprint

The FCF and PCF values estimated for S2 within this research are found to be higher than those previously reported by Yan et al. (2015) and Yue et al. (2017). These differences are due to our broader system boundary and expanded consideration of various emission sources active within China. Moreover, the FCF value calculated for apple in S3 falls within the threshold range reported by Wu et al. (2012). It has been increasingly recognized that the utilization of carbon sinks is a critical component in conducting LCA for assessing the CF of agricultural products. However, significant inconsistencies exist in the calculated outcomes of these assessments. For instance, Xu and Lan (2017), Zhang et al. (2017), and Liu et al. (2018) explored the effects of carbon sources on crop production, whereas She et al. (2017) presented their findings on the carbon sink effects in crop production from a carbon balance perspective. The system boundary scenario detailed in She et al. (2017) has been found to be most comparable to our own. Despite employing a similar system boundary to S4, the comprehensive system presented in this study significantly surpasses that of She et al. (2017), including a broader range of key sources and sinks vital to CF accounting.

The term "key category" refers to a type of CF that is prioritized within the CF inventory due to its significant impact on the estimation of both GHG emissions and removals. The process of identifying key categories within the CF inventory not only assists inventory compilers in efficiently directing their efforts but also significantly enhances their overall estimation capabilities. The key categories are those that, when considered collectively and sorted in descending order of magnitude, represent 95 percent of the total inventory level. This constitutes our approach for identifying key categories within the S4 CF inventory (Chen et al., 2020b), and the primary source categories for wheat, maize, rice, and apple production, as identified by SRH GE in AE, chemical fertilizer GE in RMPS, diesel or gasoline GE in RMPS, diesel or gasoline GE in APS, and directly NE in AE, were found to be prevalent. Similarly, the primary sink categories for these crops, as identified by crop biomass CS and cropland management CS in AE, were also found to be common.

Table 5 CF inventory of S3 in FAP (t C)

Type	$E_{RMU-GHG}$	$E_{direct-N_2O}$	$E_{indirect-N_2O}$	$E_{cropland\text{-}CH_4}$	E_{AWDD}	$E_{urea\text{-}CO_2}$	$E_{SRH\text{-}CO_2}$	C_{CB}	C_{DOM}	C_{SOC}
Wheat	24.398	16.838	5.923	−0.101	2.050	3.037	449.892	213.763	0.000	389.138
Maize	40.115	43.286	8.524	−0.498	8.903	3.624	533.909	499.267	0.000	840.807
Rice	6.742	1.657	0.999	6.844	0.861	0.352	81.727	60.365	0.000	86.843
Apple	26.795	65.479	6.358	−0.068	0.429	3.077	710.803	1187.877	36.909	214.236

Table 6 CF inventory of S4 in FAP (t C)

Type	$E_{RMP-GHG}$	$E_{RMT-GHG}$	$E_{RMU-GHG}$	$E_{direct-N_2O}$	$E_{indirect-N_2O}$	$E_{cropland-CH_4}$	E_{AWD}	E_{SRH-CO_2}	C_{CB}	C_{DOM}	C_{SOC}
Wheat	117.182	5.820	24.398	16.838	5.923	−0.101	2.053	449.892	213.763	0.000	389.138
Maize	251.843	43.824	40.115	43.286	8.524	−0.498	9.923	533.909	499.267	0.000	840.807
Rice	27.872	2.242	6.742	1.657	0.999	6.844	0.861	81.727	60.365	0.000	86.843
Apple	327.227	67.489	26.795	65.479	6.358	−0.068	10.981	710.803	1187.877	36.909	214.236

4 Duality of Agricultural Carbon Effect

The economic conduct of farmers in the procurement of raw materials for crop cultivation is often motivated by the goal of maximizing yield, although it inevitably results in an escalation in the overall GES of the agricultural ecosystem. Concurring with previously conducted research examining the carbon footprint of agriculture in China, the utilization of chemical fertilizers and fossil fuels represents a substantial proportion of the total GE associated with the acquisition of essential raw materials by farmers. It represented the respective percentages of 39.96% to 56.89%, 69.72% to 85.1%, and 72.57% to 89.27% of the GE found in the analyzed crops in $E_{\mathrm{RMP-GHG}}$ and $E_{\mathrm{RMT-GHG}}$, $E_{\mathrm{direct-N_2O}}$ and $E_{\mathrm{RMU-GHG}}$, respectively. In terms of the total GE in S2, they contributed to a considerable proportion, exceeding 68%. A significant body of research has highlighted the fact that nitrogenous fertilizer applications do indeed result in significant greenhouse gas emissions; however, they also lead to an increase in soil carbon sequestration by stimulating crop productivity and augmenting the quantity of crop residue input. It was identified by Dou et al. (2016) that the application of chemical fertilizers may not significantly augment the soil organic carbon storage, but it does enhance the carbon content in the mineral-associated organic matter of aggregates. Furthermore, Liu et al. (2018) conducted research and determined that the soil carbon sequestration from nitrogen fertilization was 12.2 Tg CE year^{-1}, which offset 18.1% of the fertilization-induced GHG emissions.

In the context of CF accounting, it is acknowledged that the significance of agriculture‘s contribution to climate change mitigation cannot be overstated, particularly due to the direct and indirect carbon source effects of farming practices. However, there exists a pervasive misconception that the carbon source effect of agriculture is the sole measure of its contribution to climate change mitigation. Therefore, it is necessary to elucidate the significance of the accompanying carbon sink effect of agriculture, particularly the changes in carbon stocks of vegetation and soil carbon pools in cropland (Grace & Basso, 2012), which can counterbalance the emissions of greenhouse gases and contribute significantly to climate change mitigation. It is, thus, prudent not to solely consider the carbon source effect of agriculture, but also the associated carbon sink effect, in order to fully appreciate the climate change mitigation potential of this sector. Beyond this, the net primary biomass formed through photosynthesis represents a significant contributor to the vegetation carbon pool. Moreover, the economic orchard industry serves as a pivotal contributor to the long-term accumulation of crop biomass and carbon sequestration. In the present study, it was observed that the biomass of apple trees growing in a unit area demonstrated a carbon storage capacity that was, on average, 5.177, 3.19, and 3.637 times higher than that of wheat, maize, and rice, respectively. The aforementioned result is a logical conclusion, as the majority of the biomass of wheat and maize, which are staples of the food chain, and vegetable-based herbal cash crops will enter the consumption chain in a relatively short period of time, thus releasing a portion of the carbon effect. Conversely, the trunk biomass of apple, peach, and plough, which are predominant woody cash crops, will persist or continue to grow over an extended period of time.

Consequently, the system boundary established in compliance with PAS 2050 protocol (S2) presents a degree of rationality. Nevertheless, when contrasted with S1, S3, and S4, which take into account the carbon sink effect of agriculture, the carbon footprint estimation of S2 appears to place disproportionate emphasis on the threat posed by agricultural production to climate change, potentially overlooking the climate positive externalities associated with agricultural production.

5 Coupling of Biogeochemical Cycle Theory and Life Cycle Assessment

The intrinsic correlation between greenhouse gases (e.g., CO_2, CH_4, and N_2O) and carbon storage in agricultural ecosystems (e.g., crop biomass, dead organic matter, and mineral soils) is embedded within the biogeochemical cycling of carbon and nitrogen elements. This observation suggests that tracking the complex trajectory of the biogeochemical cycle of these elements in vegetation, soil, and the atmosphere can further enhance the comprehensive assessment of CF. Previous research has highlighted the significance of heterotrophic soil respiration as an output pathway for the carbon cycle, serving as a major source of GHG emissions (Bond-Lamberty et al., 2018), which in S3 accounted for 89.6% (72.32% for S4), 83.64% (57.32% for S4), 82.4% (63.38% for S4), and 87.44% (58.5% for S4) of the total GE of wheat, maize, rice, and apple in AE, respectively.

On the other hand, the breadth of cross-sectional space of agricultural carbon footprint accounting is broadened by the establishment of a system boundary predicated on the principles of the life cycle assessment (LCA) theory applied to agricultural products. The system boundary encompasses the entire supply chain of agricultural products, encompassing both pre-natal and post-natal production stages. In contrast to the traditional approach, which focuses solely on the direct human-induced effects on greenhouse gas emissions and removals from managed lands, this expanded approach also includes indirect impacts resulting from the use of agricultural inputs, transportation, packaging, and consumption. In addition, through the comprehensive analysis of the entire supply chain, the scope for repetitive calculations is eliminated. For instance, in accordance with the principles of LCA, the emission source of urea is removed since the quantity of carbon dioxide that is captured from the atmosphere during the urea manufacturing process is quantified in the industrial process. And likewise, when conducting an LCA analysis, using the "cradle to grave" system boundary, carbon sequestration associated with commercialized agricultural products (typically referring to the fruits of plants transitioning to the subsequent subsystem) should be purposely excluded from the calculation process. Consequently, the incorporation of the carbon-nitrogen (C-N) cycle into agricultural carbon footprint accounting would mitigate uncertainties in the calculations. The outcomes of the carbon footprint calculations in the S4 scenario demonstrated a significant reduction in the risk of overestimating the carbon source or carbon sink effects.

6 Limitations and Implications

Due to the inherent challenges associated with data collection at the level of farmers, although the greenhouse gas emissions associated with agricultural production in downstream links were indeed considered within the system boundary scenario, the agricultural products processing and consumption systems were excluded from the carbon footprint accounting process. The life cycle assessment design principle was applied, which extends from cradle to farmgate; thus, the carbon footprint of agricultural products from farmgate to grave remains an enigma and resides in a so-called black box. Simultaneously, despite our meticulous consideration and selection of appropriate factor coefficients for this research, the lack of comprehensive research literature, particularly comparative analysis results of factor coefficients from the perspective of big data, presents significant challenges in the application of these results to a broader audience in the field of non-experimental science. In light of the above, the development of more precise fitting equations or simulation models, aimed at estimating the coefficients of the factors, will play an instrumental role in mitigating the uncertainty associated with carbon footprint accounting in the forthcoming future. In addition, from the implications of this case study, we also strongly recommend the following framework (Fig. 5) to reduce the uncertainty generated by the carbon footprint accounting process.

We expanded on the system boundary set in scenario 4 of the study, incorporating principles of life cycle assessment and carbon-nitrogen cycles, focusing on the duality of agriculture in terms of carbon effect. This dual role is essential for a well-rounded analysis of the climate impact of agricultural production. It is questionable whether the upstream agricultural production system‘s greenhouse gas emissions should be distributed to the consumers, namely the farmers, due to the study’s findings that the allocation of climate change responsibility significantly affects the carbon effect of farmers’ agricultural production. Nevertheless, the role of farmers’ agricultural production in acting as a carbon sink remains considerable, despite the fact that they shoulder the responsibility for greenhouse gas emissions during the upstream stages of the agricultural production system, such as raw material production and transportation. Therefore, it is anticipated that future research on the incorporation of carbon sink effects in the accounting process of agricultural product carbon footprints will be enriched with increased analysis and discussion pertaining to the impact of climate policies.

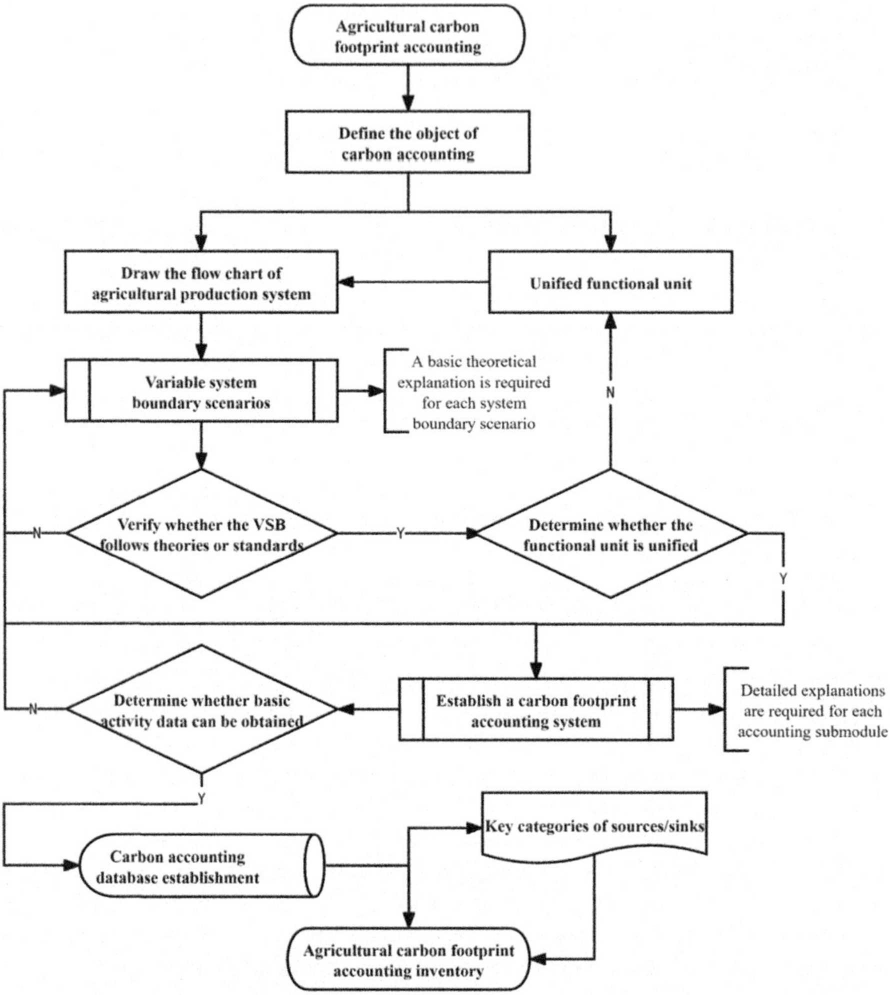

Fig. 5 Self-checking process of agricultural carbon footprint accounting

Acknowledgments This work was supported by the project funded by the National Natural Science Foundation of China [grant number 72203124], and we express our gratitude to the team led by Professor Jiang Zhide of Northwest A & F University in China for their contribution to data acquisition, and express our appreciation to Professor Han Hongyun of Zhejiang University in China for the valuable suggestions for enhancing the manuscript.

References

[CECERG] China Energy and Carbon Emission Research Group. (2009). *The report on China's energy and carbon emissions*. Science Press.

[ECCPCIA] Chinese Industrial Association of Energy Conservation and Clean Production, [CECEPG] China Energy Conservation and Environmental Protection Group. (2014). *China energy conservation and emission reduction development report*. Chinese Economic Press.

[NDRC] The National Development and Reform Commission of the People's Republic of China. (2014). *National greenhouse gas inventory*. China Environmental Science Press.

Adewale, C., Higgins, S., Granatstein, D., Stöckle, C. O., Carlson, B. R., Zaher, U. E., & Carpenter-Boggs, L. (2016). Identifying hotspots in the carbon footprint of a small scale organic vegetable farm. *Agricultural Systems, 149*, 112. https://doi.org/10.1016/j.agsy.2016.09.004

Adewale, C., Reganold, J. P., Higgins, S., Evans, R. D., & Carpenter-Boggs, L. (2018). Improving carbon footprinting of agricultural systems: Boundaries, tiers, and organic farming. *Environmental Impact Assessment Review, 71*, 41. https://doi.org/10.1016/j.eiar.2018.04.004

Audsley, E., Stacey, K., Parsons, D. J., & Williams, A. G. (2014). Estimation of the greenhouse gas emissions from agricultural pesticide manufacture and use. *Uma Ética Para Quantos?, XXXIII*(2).

Bond-Lamberty, B., Bailey, V. L., Chen, M., Gough, C. M., & Vargas, R. (2018). Globally rising soil heterotrophic respiration over recent decades. *Nature, 560*(7716), 80. https://doi.org/10.1038/s41586-018-0358-x

Brandão, M., Levasseur, A., Kirschbaum, M. U. F., Weidema, B. P., Cowie, A. L., Jørgensen, S. V., Hauschild, M. Z., Pennington, D. W., & Chomkhamsri, K. (2013). Key issues and options in accounting for carbon sequestration and temporary storage in life cycle assessment and carbon footprinting. *International Journal of Life Cycle Assessment, 18*(1), 230. https://doi.org/10.1007/s11367-012-0451-6

Chen, R., & Kong, Y. (2022). Redefine carbon footprint performance with the total factor productivity: A case from a regional food production system. *Ecological Indicators, 144*, 109540. https://doi.org/10.1016/j.ecolind.2022.109540

Chen, R., Zhang, R., & Han, H. (2020a). Where has carbon footprint research gone? *Ecological Indicators, 120*, 106882. https://doi.org/10.1016/j.ecolind.2020.106882

Chen, R., Zhang, R., Han, H., & Jiang, Z. (2020b). Is farmers' agricultural production a carbon sink or source?—Variable system boundary and household survey data. *Journal of Cleaner Production, 266*, 122108. https://doi.org/10.1016/j.jclepro.2020.122108

de Vries, M., & de Boer, I. J. M. (2010). Comparing environmental impacts for livestock products: A review of life cycle assessments. *Livestock Science, 128*(1–3). https://doi.org/10.1016/j.livsci.2009.11.007

Dou, X., He, P., Zhu, P., & Zhou, W. (2016). Soil organic carbon dynamics under long-term fertilization in a black soil of China: Evidence from stable C isotopes. *Scientific Reports, 6*. https://doi.org/10.1038/srep21488

Galli, A., Wackernagel, M., Iha, K., & Lazarus, E. (2014). Ecological footprint: Implications for biodiversity. *Biological Conservation, 173*, 121. https://doi.org/10.1016/j.biocon.2013.10.019

Garnett, T. (2011). Where are the best opportunities for reducing greenhouse gas emissions in the food system (including the food chain)? *Food Policy, 36*(Suppl. 1). https://doi.org/10.1016/j.foodpol.2010.10.010

Grace, P. R., & Basso, B. (2012). Offsetting greenhouse gas emissions through biological carbon sequestration in North Eastern Australia. *Agricultural Systems, 105*(1), 1. https://doi.org/10.1016/j.agsy.2011.08.006

IPCC. (2006). In H. S. Eggleston, L. Buendia, K. Miwa, T. Ngara, & K. Tanabe (Eds.), *2006 IPCC guidelines for national greenhouse gas inventories*. Prepared by the National Greenhouse Gas Inventories Programme. Published: IGES, Japan

Liu, W., Zhang, G., Wang, X., Lu, F., & Ouyang, Z. (2018). Science of the total environment carbon footprint of main crop production in China: Magnitude, spatial-temporal pattern and attribution. *Science of the Total Environment, 645*, 1296–1308. https://doi.org/10.1016/j.scitotenv.2018.07.104

Pandey, D., Agrawal, M., & Pandey, J. S. (2011). Carbon footprint: Current methods of estimation. *Environmental Monitoring and Assessment, 178*(1–4), 135. https://doi.org/10.1007/s10661-010-1678-y

She, W., Wu, Y., Huang, H., Chen, Z., Cui, G., Zheng, H., Guan, C., & Chen, F. (2017). Integrative analysis of carbon structure and carbon sink function for major crop production in China's typical agriculture regions. *Journal of Cleaner Production, 162*, 702. https://doi.org/10.1016/j.jclepro.2017.05.108

The British Standards Institution. (2011). *Specification for the assessment of the life cycle greenhouse gas emissions of goods and services*. Publicly Available Specification PAS 2050: 2011.

Wiedmann, T., & Minx, J. (2008). A definition of 'Carbon footprint ISA research report 07-01'. In C. C. Pertsova (Ed.), *Ecological economics research trends* (pp. 1–11). Nova Science Publishers. http://www.censa.org.uk/docs/ISA-UK_Report_07-01_carbon_footprint.pdf

Wu, T., Wang, Y., Yu, C., Chiarawipa, R., Zhang, X., Han, Z., & Wu, L. (2012). Carbon sequestration by fruit trees – Chinese apple orchards as an example. *PLoS One, 7*(6). https://doi.org/10.1371/journal.pone.0038883

Xu, X., & Lan, Y. (2017). Spatial and temporal patterns of carbon footprints of grain crops in China. *Journal of Cleaner Production, 146*, 218–227. https://doi.org/10.1016/j.jclepro.2016.11.181

Yan, M., Cheng, K., Luo, T., & Pan, G. (2014). Carbon footprint of crop production and the significance for greenhouse gas reduction in the agriculture sector of China. In *Assessment of carbon footprint in different industrial*. https://doi.org/10.1007/978-981-4560-41-2_10

Yan, M., Cheng, K., Luo, T., Yan, Y., Pan, G., & Rees, R. M. (2015). Carbon footprint of grain crop production in China – Based on farm survey data. *Journal of Cleaner Production, 104*, 130. https://doi.org/10.1016/j.jclepro.2015.05.058

Yue, Q., Xu, X., Hillier, J., Cheng, K., & Pan, G. (2017). Mitigating greenhouse gas emissions in agriculture: From farm production to food consumption. *Journal of Cleaner Production, 149*, 1011. https://doi.org/10.1016/j.jclepro.2017.02.172

Zhang, D., Shen, J., Zhang, F., Li, Y., & Zhang, W. (2017). Carbon footprint of grain production in China. *Scientific Reports, 7*(1), 4126. https://doi.org/10.1038/s41598-017-04182-x

Offsetting Carbon Footprints: Exploring CCUS Challenges and Solutions Across Diverse Industries in India

Baishali Chakrabarti, Sadichha Jagadale, Hemant Bherwani, and Rajesh Biniwale

1 Introduction

The demand for energy worldwide, predominantly satisfied by burning fossil fuels, is significantly high. In 2020, British Petroleum (BP) noted that 84% of this demand was met by fossil fuels. The trend is shifting toward renewable energy, as indicated by 28 countries announcing climate change countermeasures by the end of 2019, primarily focusing on adopting renewable energy sources (REN21 2020). Nonetheless, transitioning to renewable energy is hindered by various challenges including social, economic, technological, and regulatory factors (Seetharaman et al., 2019). This transition is vital for ecological and environmental reasons, particularly because the combustion of coal significantly contributes to the emission of CO_2. The historical role of CO_2 emissions, such as during the Snowball Earth event around 800–600 million years ago (Crowley et al., 2001), is notable. The release of CO_2 then is thought to have warmed the climate after the Snowball Earth phase, possibly aiding in the development of multi-cellular life (Hyde et al., 2000). While CO_2 is essential for photosynthesis and the sustenance of the Earth's food chain, its current atmospheric concentration, exceeding 350 ppm, is detrimental (Hansen et al., 2008). Carbon emissions are rising globally, accounting for 30% of atmospheric pollutants.

B. Chakrabarti · S. Jagadale
CSIR-National Environmental Engineering Research Institute (CSIR-NEERI), Nagpur, Maharashtra, India

H. Bherwani (✉) · R. Biniwale
CSIR-National Environmental Engineering Research Institute (CSIR-NEERI), Nagpur, Maharashtra, India

Academy of Scientific and Industrial Research (AcSIR), Ghaziabad, Uttar Pradesh, India
e-mail: h.bherwani@neeri.res.in

S. S. Muthu (ed.), *Carbon Footprint Assessments*, Environmental Footprints and Eco-design of Products and Processes,
https://doi.org/10.1007/978-3-031-70262-4_10

India has a population exceeding 1.3 billion, according to a 2019 report by the United Nations (UN). Most greenhouse gas emissions in India come from coal and petroleum used in electricity generation and transportation, highlighting the need for emission reduction measures. CCUS technologies are becoming more prevalent in numerous countries as a cost-effective means to capture and repurpose carbon dioxide across various sectors. Countries like the USA, Russia, and Germany are working to lower CO_2 emissions, including through CCUS strategies. The circular economy, focusing on the restorative and regenerative use of materials, components, and products, seeks to extend their lifespan through remanufacturing, reuse, and recycling, thus maintaining value in consumption and production systems. CO_2 has a range of applications, as detailed in this summary. After use, CO_2 can be recaptured and reintegrated into the circular economy, completing a cycle that promotes sustainability.

The 2021 Status Report by the Global Carbon Capture and Storage (CCS) Institute reveals that 149 million metric tons of CO_2 are being captured and stored globally through various methods. There are 135 CCS facilities worldwide, with 27 in full operation (Global Status of CCS, 2021). A survey by IQAir, a Swiss climate-related organization in partnership with the United Nations Environmental Programme (UNEP), found that three major Indian cities rank among the top ten most polluted globally (The mint article, 2021). In India, the energy sector accounts for 68.7% of greenhouse gas emissions, with agriculture contributing 19.6%, industrial processes 6.0%, changes in land use and forestry 3.8%, and waste 1.9% (Climate Links, 2019). The country's annual CO_2 emissions exceed 2.71 billion tons (Hannah & Max, 2022). However, India has placed relatively little focus on CCUS technologies and research. The Biennial Update Reports (BUR-I and BUR-II) from the MoEFCC in 2015 and 2018 scarcely addressed India's carbon capture activities. The BUR-III, submitted in February 2021, highlighted that CO_2 capture in India is still in its infancy, with only a handful of sectors, such as refineries and steel, executing limited-scale initiatives. The Mission Innovation (MI) initiative, launched at the Conference of Parties (COP-21), marked a significant step for CCS technology research and development. The CEEW identified three critical dependencies for CCUS in India: research and development, finance, and policy (Malyan & Chaturvedi, 2021).

The Long Term-Low GHG Emission Development Strategies (LT-LEDS) report, presented at the COP27 conference in November 2022, emphasized the inevitability of CCUS adoption for climate change mitigation (United Nations Climate Change, 2022). India's potential for CO_2 storage is considerable, estimated between 47 and 572 GT, with deep saline aquifers capable of storing 105–360 Gt and basalt formations around 200 Gt (Dooley et al., 2005; Singh et al., 2006a). This situation underscores the urgent need for concentrated efforts in research and innovation, as well as the development of strategic plans for adopting CCUS technologies in India (Malyan & Chaturvedi, 2021).

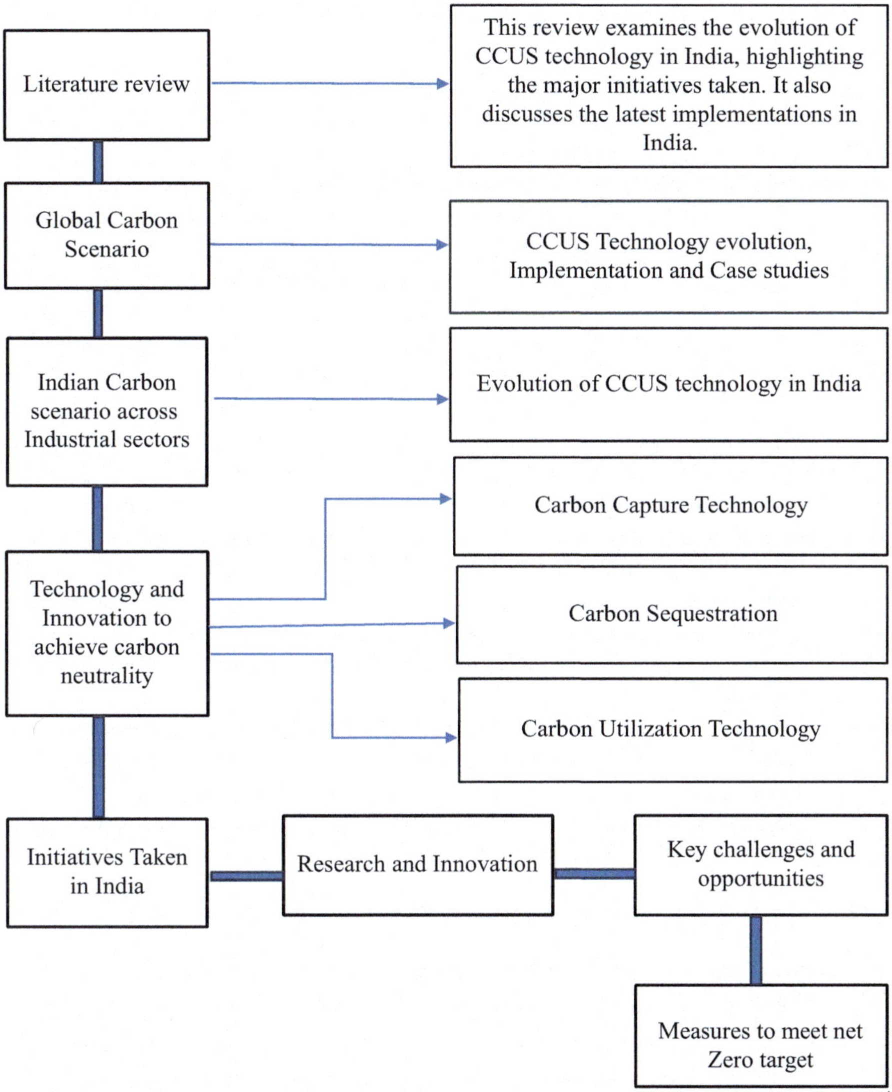

Fig. 1 Structural overview of CCUS development and implementation strategies in India

This chapter undertakes an exhaustive review of the available literature to gauge the scope of research conducted on this topic both internationally and in India as depicted in Fig. 1. The subsequent section presents an analysis of key studies and their findings, serving as a comprehensive resource for gathering all pertinent information on CCUS and related subjects.

2 Literature Review and Methodology

This literature review was conducted using databases, news articles, and search engines, including Global CCS study, DST initiatives, Mission Innovation, BUR, CEEW study, Google Scholar, Submissions made to UNFCC, Tata Steel, TERI report, *The Hindu* article, Google, as well as the IEA reports, New York Springer CCUS papers, and UNECE (United Nations Economic Commission for Europe) reports. Searches used combinations of the following terms: Carbon capture technologies, CO_2 emissions in the world and India, carbon sequestration methods, utilization pathways, barriers in implementation, and India's efforts for CCUS implementation. Over 100 research papers, review papers, articles, and press releases from India and other countries were examined.

This review examines the evolution of CCUS technology in India, highlighting the major initiatives taken. It also discusses the latest implementations in India. We adopted a systematic approach to briefly discuss the overview of CCUS technology, including the initiatives taken, research and innovation, identifying the challenges and opportunities, and the upcoming strategies. CCUS technology and its Contribution to Circular Economy is a topic of discussion for further research. Very few reviews exist on the evolution of CCUS technology in the past years in India. A summary of the major studies conducted in India is given in Table 1. The study of CEEW is one of the reviews that has discussed the scenarios with and without CCS. Another review named "The development of carbon capture and storage (CCS) in India: A Critical Review," by Shaw, R. and Mukherjee, S., describes the CCUS technologies in India and the Research and development taking place. Singh Ajay et al. and Ganguli et al. study India's geological sequestration potential. Goldar et al., Chaturvedi and Malyan, Goel et al., and Devdhas et al. are a few major studies focussing on the implications and perspectives of CCUS technologies in India.

3 Global Industry: The Present Carbon Scenario

Globally, the energy sector accounts for approximately 73.2% of GHG emissions, followed by agriculture, industrial processes, and garbage, accounting for 18.4%, 5.2%, and 3.2%, respectively (refer to Fig. 2) (Ghosh, 2020). Maximum emission from the energy sector is from the industry field, followed by the building and transport sector. The direct GHG emissions in the industrial sector amount to 14.1 $GtCO_2eq$ and rise to 20 $GtCO_2eq$ with the addition of indirect emissions (Bashmakov et al., 2022). According to Bashmakov et al., the demand for materials is a key driver of energy consumption and CO_2 emissions in the industrial sector (Bashmakov et al., 2022).

As of 2021, the Global CCS Institute's 2021 Status Report indicates that 149 million metric tons of CO_2 have been captured and sequestered from various plants. Globally, there are 135 CCS facilities, including sites under construction, advanced

Table 1 Studies on CCUS in India

Sr. no.	Study	Observation
1	Chen et al. (2023)	Implementing CO_2-EOR storage in a mature oil field involves a holistic approach, reservoir static analysis, dynamic modeling, pilot planning, and techno-economic sensitivity evaluations. This comprehensive strategy facilitates the removal of approximately 268,275 metric tons of CO_2 from the environment
2	Shaw and Mukherjee (2022)	Different technologies for capturing and storing potential in India in various pathways are being discussed. The public-private partnership framework is a much-needed incorporation. The coming 15–20 years are the most crucial for the CCUS technology development in India
3	Chaturvedi and Malyan (2022)	CCUS has started gaining momentum in the industrial sectors. The industrial sector is the most crucial sector contributing around 58% share of energy consumption by 2050. Tata Steel, Dalmia cement, ONGC (oil and Natural gas Corporation), Indian oil, NTPC, and many other private and government sectors are leading some of the main projects
4	Goldar and Dasgupta (2022)	A study of the UK and China is discussed, which can be used for policy actions. These countries' staggered approach entails thorough research and development, the implementation of pilot projects, the allocation of funds, and the formulation of regulatory policies. These efforts are all guided by predefined, time-bound objectives. The development and mitigation needs shall be prioritized to evaluate the substantial investments needed for the implementation
5	Devadhas et al. (2022)	This study also underscores the significance of carbon capture and the hydrogen economy. The simultaneous generation of hydrogen and electricity from fossil fuel plants equipped with carbon capture technology emerges as a notably promising technological advancement
6	Vishal et al. (2021a, b)	The study portrays the present possible opportunities for CCUS development. Retrofitting existing industries and power plants shall be the foremost action to reduce emission levels. Lack of risk assessment is another issue identified here. Coordinated policy-making across sectors is necessary
7	Gupta and Paul (2019)	This study addresses the challenges facing CCS implementation in India. Key obstacles identified include insufficient R&D efforts, the necessity for a comprehensive national study on geological storage, the energy penalty, inadequate financing and foreign direct investment (FDI), public opinion, and foreign policies
8	Datta and Krishnamoorti (2019)	The efficiency of using CCUS is evaluated by the levelized cost of electricity (LCOE). Economic comparison of the incorporation of CCUS at various thermal plants and fertilizer plants in Assam are discussed here
9	Ganguli et al. (2016)	This research assessed the potential for CO_2-EOR and storage at the Ankleshwar oil field. As per the research findings, the reservoir exhibits significant potential for the secure and cost-effective geological sequestration of 15.04 million metric tons of CO_2

(continued)

Table 1 (continued)

Sr. no.	Study	Observation
10	Garg and Shukla (2009)	This study analyses the adaptation of CCS in coal and natural gas plants. It is estimated that if 20% of CBM (coalbed methane) potential is captured, it will mitigate around 0.2 Bt CO_2
11	Shackley and Verma (2008)	Research and development activities should be supported for the implementation of low-carbon technologies. A legal and actual framework for CCS in India should be created
12	Singh et al. (2006b)	India's geological structures, encompassing basalt formations, coastal and inland saline aquifers, unextractable coal seams, and depleted petroleum reservoirs, harbor an estimated total carbon dioxide storage potential of 572 billion tons (Bt). Among these, unminable coal deposits alone could serve as a potential CO_2 storage sink, accounting for 150 billion tons

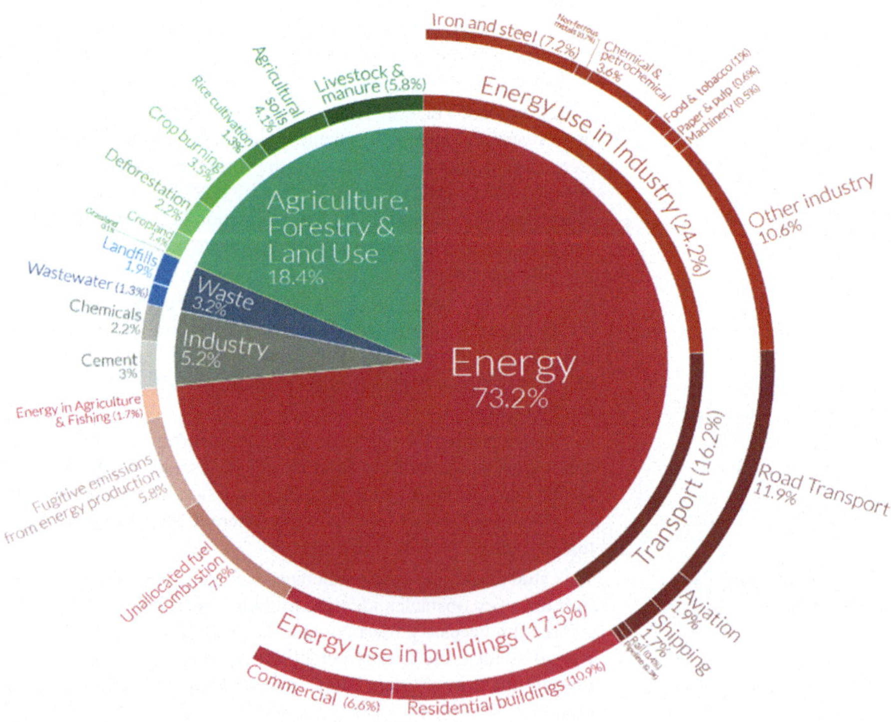

Fig. 2 Global greenhouse gas emissions by sector. (Source: Ghosh, 2020)

development, early development, and operation suspended. This also includes 27 fully operational facilities (Global Status of CCS, 2021). As per the CIAB (Center of Innovative and Applied Bioprocessing) 2019 report, it was found that the cost of decarbonizing the UK power system would be about 50% higher if carbon capture were not available (Pratama & Mac Dowell, 2019). Commercial CCS facilities as of September 2021 by number and total capacity are shown in Table 2.

Table 2 Commercial CCS facility as of September 2021 by number and total capacity (Global CCS Institute, 2022)

	Active	Under construction	Advanced development	Early development	Activity halted	Total
Count of facilities	27	4	58	44	2	135
Capture capacity (Mtpa)	36.6	3.1	46.7	60.9	2.1	149.3

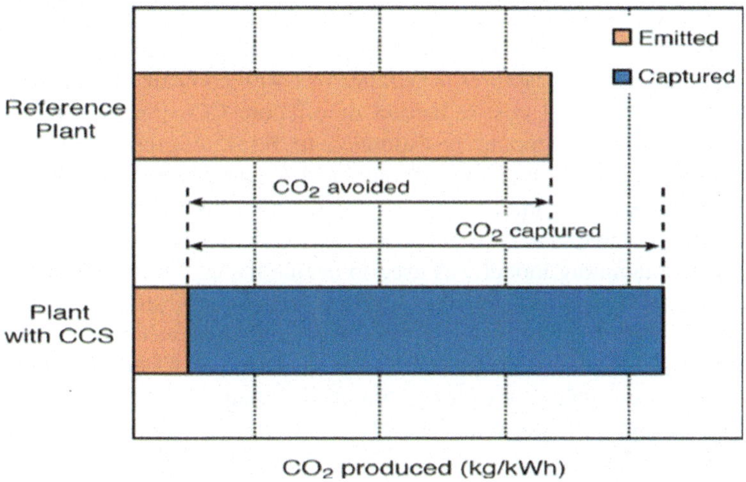

Fig. 3 Carbon dioxide capture in power plants with CCS technology. (Metz et al., 2005)

Many leading countries like the USA, Canada, Germany, Japan, China, and Europe have established and performed innovative Research on applying CCUS in other processes to obtain a zero carbon environment. Some studies identify the efficiency of CCUS technology that can be used to enhance conventional techniques and processes. Eiji Kawai et al. found that CCU could potentially reduce the fossil fuel-based energy required by the petrochemical industry by 32% (Kawai et al., 2022). Additionally, Diaz et al. suggested employing Sequential Supplementary Firing in natural gas Combined Cycle with carbon capture for producing low-carbon electricity and utilizing CO_2 Enhanced Oil Recovery (EOR) in the heat recovery steam generator. This approach resulted in a 50% reduction in flue gas volume and a further 15% reduction in overall capital costs (Diaz et al., 2016). Carbon dioxide capture in power plants with CCS technology is depicted in Fig. 3.

Case Studies

Swedish case study on 28 plants: A Swedish case study conducted on 28 plants estimated the expenses associated with installing and operating amine-based post-combustion CO_2 capture across all manufacturing facilities in Sweden. The Marginal Abatement Cost Curve indicated an estimated range for CO_2 capture from 40 €/t CO_2 to 110 €/t CO_2, depending on the emission source. Additionally, transportation and storage costs may vary from 25 to 40 €/t CO_2. The study also suggests that partial capture of CO_2 could reduce capturing costs and serve as an economical option for CCUS utilization. However, the impact of partial capture on transportation and storage costs requires further investigation (Johnsson et al., 2020).

Brevik Cement plant: The iCCS (Integrated CCS) methodology and tool are employed to evaluate the techno-economic and environmental performances of the Brevik cement plant in Norway. Nine CCS chains were evaluated. The full and alternative chains lead to a cost estimate of around 120€/t CO_2 and 150€/t CO_2, respectively. The cost-reduction potential of advanced amine capture and two membrane-based captures is estimated as 40% compared to the MEA-based capture. The potential cost reduction of offshore CO_2 pipeline transportation compared to CO_2 shipping is estimated to be 45% greater. EOR shows a 28–35% higher cost reduction potential than CCS storage (Jakobsen et al., 2017).

Turkey carbon capture plant: A case study of fossil-fired power plants in Turkey revealed a comparative analysis between CCU and Carbon trading. Mixed-integer programming model was used to decide the capacities of carbon capture units, the transportation network, and the locations of storage sites. General Algebraic Modeling System (GAMS) was used to develop a model for two coal-fired power plants. Distance between the location of emission and location of utilization/storage is a major factor in deciding between CCU and carbon trading. According to the analysis, CCU is preferred to carbon trading in Turkey (Agrali et al., 2018).

Carbon Return on Investment (CROI): Another study on CCUS assessed the CROI for CCUS platforms. Based on the analysis, there will be a decrease in gross positive emissions from 27 Gt-CO_2 in 2040 to approximately 5 Gt-CO_2 in 2100. Meanwhile, gross negative emissions (i.e., sequestration) will rise from 1 Gt-CO_2 to 15 Gt-CO_2 over the same timeframe. The corresponding minimum CROI value will vary from 2040 to 2100 (Singh et al., 2021).

Machine learning in CCUS technology: Y. Yan et al. explored the utilization of machine learning in CCUS technology in their report, highlighting its effectiveness in screening solvents and adsorbents, as well as in process modeling and optimization. This approach has the potential to reduce development time, as well as capital and operating costs for CO_2 capture. The report also recommends focusing on chemical looping combustion for CO_2 capture (Yan et al., 2021).

4 The Status of Carbon in Indian Industrial Sectors

In India, the energy sector is responsible for 68.7% of GHG emissions, with agriculture, industrial processes, land-use change and forestry, and waste contributing 19.6%, 6.0%, 3.8%, and 1.9%, respectively (Anand & Kaushal, 2021). Annually, India emits over 2.71 billion tons of CO_2 (Hannah et al., 2020). In a survey

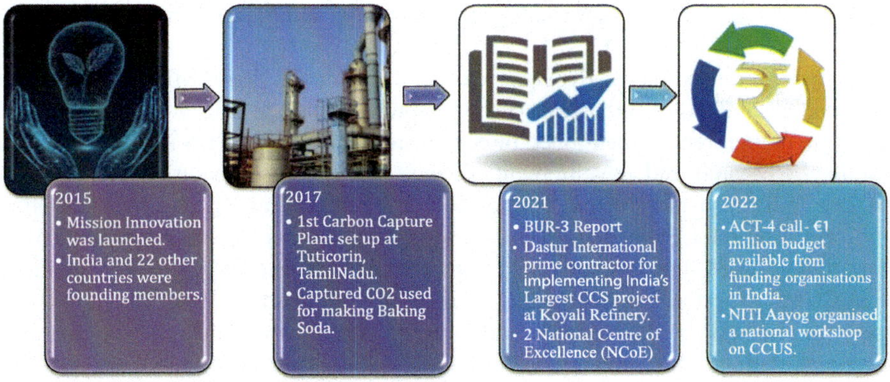

Fig. 4 Summary of evolution of CCUS technology in India

conducted by IQAir (a Switzerland-based climate-related organization and a technology partner of the United Nations Environmental Programme (UNEP)), three Indian megacities were recently classified among the world's top ten polluting cities (Mahato & Ghosh, 2020). Even if India's per capita CO_2 emissions are very low, India still ranks the world's third-largest producer of CO_2 (Sengupta et al., 2022). To achieve the targets under Paris Climate Agreement, mitigation technologies are the need of the hour. These technologies must be adopted at an accelerated pace (Datta & Krishnamoorti, 2019).

The discussion surrounding CCUS in India continues to evolve, with its significance well-established in both energy and climate policy domains. Figure 4 outlines the key milestones in the progression of the CCS/CCUS discourse in India. A notable development was the establishment of the Indian CO_2 Sequestration Applied Research (ICOSAR) Network by the Department of Science and Technology (DST), Government of India, in 2007. This initiative aimed to foster research dialogues on CCS applications among various stakeholders (Viebahn et al., 2011, 2014; Chaturvedi & Malyan, 2022). Furthermore, in 2008, the Government of India introduced the National Action Plan on Climate Change (NAPCC), which acknowledged the role of CCS in mitigating emissions from coal-fired power plants. Nonetheless, it also voiced apprehensions regarding the elevated expenses and stability of CO_2 storage facilities (PMCCC, 2008).

In the Second National submitted to the United Nations Framework Convention on Climate Change, India presented a demonstration project for carbon dioxide capture and storage (MOEFCC & UNFCCC, 2012). There has been much less emphasis on CCUS technologies and Research in India. The Biennial Update Report (BUR-I and BUR-II) did not extensively cover India's carbon capture initiatives (MoEFCC, 2015, 2018). However, BUR-III, submitted on Feb 2021, mentioned that CO_2 capture is still in its early stages of implementation in India. In 2017, India's first unsubsidized industrial scale carbon capture plant made a breakthrough and was set up by Carbon Clean. The firm claims that the plant captures 60,000 CO_2 a year, which CO_2 shall be further used to make baking soda (Harrabin, 2017). In the Conference of Parties (COP-21), Mission Innovation (MI) was launched and is considered a milestone for R&D activities on CCS technology. MI identified eight innovation challenges, including carbon capture (Rimmer, 2021). Chaturvedi and Malyan, in their report, mentioned three key areas on which CCUS in India depends, namely, Research and development, finance, and policy (Chaturvedi & Malyan, 2022). India also participated in the ACT call, which raised funding of approx. 8 crores. The LT-LEDS (Long Term-Low GHG Emission Development Strategies) report, submitted during the COP27 conference on November 14, 2022, mentions that CCUS adoption is inevitable to mitigate climate change (MoEFCC, 2022). As per the analysis by Chaturvedi and Malyan, CCS technology would assist in accommodating approximately 30% of fossil fuel's share in primary energy, compared to only 5% without CCS availability (Chaturvedi & Malyan, 2022). There are also tremendous numbers of sequestration sites available in India.

5 Technologies and Innovations Paving the Way for Carbon Neutrality

Many companies like Carbon Clean, Axens, Aker Carbon Capture, Carbon Engineering, C Capture, CO_2 CAPSOL, Decarbontek LLC, Entropy, etc. have developed their CCUS technologies and are adopted globally. Carbon Clean established India's first commercial CCUS plant (Global CCS Institute, 2022). These technologies available are discussed below.

5.1 Carbon Capture Technologies

CO_2 has been captured for over 80 years, but this captured CO_2 was vented into the atmosphere. The following are available methods used worldwide to capture the CO_2 gas emitted. Each method has advantages and disadvantages (Shaw & Mukherjee, 2022; UNT Digital Library, 2018; Wilcox, 2012). IPCC has identified four basic carbon capturing systems: industrial process streams, post-combustion capture, oxyfuel combustion capture, and pre-combustion capture. The choice of technology to be established depends on factors like the extent of purity required and conditions such as temperature, the pressure of CO_2, and the concentration of impurities (Rubin et al., 2012). Sixty to eighty-five percent reduction in GWP (Global Warming Potential) can be achieved through the use of carbon capture technologies, with the highest reduction achieved by oxyfuel combustion (Petrescu & Cormos, 2017).

5.1.1 Post-combustion

Post-combustion carbon capture involves the separation of CO_2 from the flue gases generated from large-scale fossil fuel combustion processes, such as boilers, cement kilns, and industrial furnaces (Wilberforce et al., 2021). Typically, 3–15% by volume of CO_2 is captured using liquid solvent-based methods (Wang et al., 2011). Some additional gases (sulfur dioxide, nitrogen oxides, and particulate matter) are also formed during combustion. These pollutants must be removed to achieve efficiency and a clean gas stream. Nowadays, chemical reaction with amine-based solvents such as monoethanolamine (MEA), diethanolamine (DEA), and N-methyl diethanolamine (MDEA) is used for capturing (Rubin et al., 2012). The flue gas is cooled at temperatures 40-60 °C and taken to the absorber column, where a chemical solvent is introduced to bond with CO_2 gas. This solvent is then pumped to the stripper column and the CO_2 is stripped off by heating at 100-140 °C (Wilberforce et al., 2021).

The captured CO_2 is subsequently compressed and transported through pipelines to storage or utilization sites. This technology is widely preferred in retrofitting existing power plants. This is the oldest technology in the carbon capture sector. Users can easily retrofit this system in new and old plants, but capturing efficiency is less than other technologies. Moreover, the technology is associated with large parasitic loads and high-energy generation costs. Countries like Norway and Canada have employed post-combustion plants of 0.3 million tons/year and one million tons/year, respectively (Shaw & Mukherjee, 2022; UNT Digital Library, 2018; Wilcox, 2012).

5.1.2 Pre-combustion

The pre-combustion capture process separates CO_2 from the primary fuel before the combustion is completed to produce a mixture of carbon monoxide, carbon dioxide, and hydrogen ("synthesis gas"). After the removal of particulate impurities from the syngas, a 2-stage water–gas shift reactor converts carbon monoxide to CO_2 and H_2, leaving H_2- and CO_2-rich gas mixture (Rubin et al., 2012; Wilberforce et al., 2021). This can also be applied to power plants using natural gas. Although fuel conversion costs are higher than in the post-combustion process, a high quantity of CO_2 is captured (15–60%). This technology is used in Integrated Gasification Combined Cycle (IGCC) power plants. IGCC power plants have enhanced power generation efficiency (Al-Mamoori et al., 2017). Another leading technology is Integrated Re-formed Combined Cycle (IRCC), which contains an auto-thermal reformer (ATR), two water-Gas shift (WGS) reactors, one pre-reformer, and a single CO_2 capture section (Shaw & Mukherjee, 2022).

The adsorption process's cost for capturing CO_2 generated by an IGCC plant is approximately \$60/ton. US-DOE (US Department of Energy) Research is being carried out to reduce this cost to \$30/ton (US-DOE). In an IGCC plant, it is estimated that the cost of CO_2 avoided is 26–62 USD/tCO_2 for the OECD (Organisation for Economic Co-operation and Development) region. The avoided CO_2 cost for pulverized coal plants is higher than IGCC plants (55 USD/tCO_2) (Davidson, 2011).

5.1.3 Oxyfuel

In this process, pure oxygen, obtained using an air separation unit (ASU), is utilized instead of air for combustion, significantly reducing the CO and SO_2 contents. This involves burning fossil fuels in pure oxygen, resulting in nitrogen-free flue gas and condensation containing only CO_2 and H_2O (Wilberforce et al., 2021). This process incorporates an ASU, carbon dioxide purification, and compression unit (CPU), which has been found to be more economical compared to post- and pre-combustion technologies (Rubin et al., 2012; Yadav & Mondal, 2022). It is considered a cost-effective solution for new plants, although the cost may increase for retrofitted old plants.

The primary objective of employing oxyfuel combustion in a coal-fired power plant is to generate flue gas with significantly high concentrations of CO_2 and water vapor, thereby facilitating the separation or capture of CO_2 from the flue gas. The combustion temperature is typically limited to 1300–1400 °C and 1900 °C in oxyfuel coal-fired boilers (Metz et al., 2005). Oxygen with a purity of 97% is utilized for combustion, although trace amounts of nitrogen and argon may still be present (Kanniche et al., 2010). According to a review by Yadav and Mondal (2022), carbon capture is more cost-effective in oxy-combustion capture compared to other methods. Moreover, CO_2 with a purity of 90% can be directly compressed and stored, setting it apart from alternative techniques (Yadav & Mondal, 2022).

5.1.4 Direct Air Capture (DAC)

USA, Switzerland, and China have adopted DAC technology, with the USA alone having a capturing potential of 1MtCO_2/year (McQueen et al., 2021). According to IEA 2022 report, 18 DAC plants are present worldwide (IEA, 2022a, b). This technology uses a solvent-based approach (McQueen et al., 2021) and is classified into solid and liquid DAC. Solid DAC (S-DAC) requires solid adsorbents operating at ambient to low vacuum pressure and at a medium temperature of 80–120 °C (IEA, 2022a, b). Solid sorbents used in S-DAC are metal–organic frameworks (MOFs), zeolites, activated carbon, silica materials, carbon nanotubes, porous organic polymers, and carbon molecular sieves (McQueen et al., 2021).

Liquid DAC (L-DAC) relies on a basic aqueous solution like potassium hydroxide, operating between 300 °C and 900 °C (IEA, 2022a, b). A pellet reactor, calciner, and slaker are used to carry out the process, as shown in Fig. 5 (Erans et al., 2022). There are very few commercially operating plants. Most DAC plants in the USA, Canada, and Europe are largely used for drink carbonation. Enterprises like

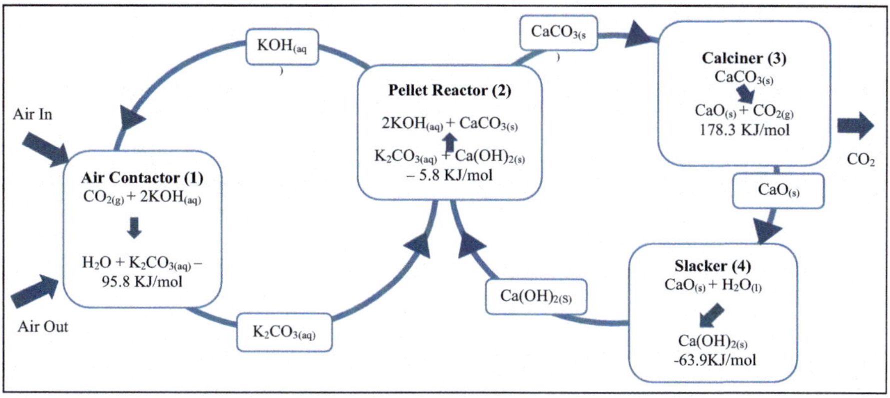

Fig. 5 Carbon engineering DAC process. (Adapted from Erans et al., 2022)

Microsoft, Shopify, and Swiss Re are investing in direct air capture to counterbalance their emissions. This technology works in a chemical loop that repeatedly reuses the same captured chemicals (carbon engineering).

The technology is energy-intensive and more expensive than others. The cost of DAC, assessed by Cameron Heprun et al., ranges between \$600 and \$1000 per ton of CO_2 (Hepburn et al., 2019). The investment expenditure for the liquid and solid systems is estimated at approximately \$80 per ton of CO_2 and \$200 per ton of CO_2, respectively. The operation and maintenance cost ranges between \$40/tCO_2 and \$80/tCO_2 for liquid systems and \$5/tCO_2 to \$50/tCO_2 for solid systems (Ozkan et al., 2022). DAC is compatible with solar, wind, and battery-dominated sustainable energy systems (Breyer et al., 2019). The high cost of capture and high sorbent regeneration energy are the major blockages in the large-scale development of this technology (Sodiq et al., 2022).

5.1.5 Chemical Looping

The Chemical looping combustion (CLC) process occurs within a two-reactor system: the fuel reactor (FR) and the air reactor (AR). Combustion occurs in FR, and oxygen carriers are regenerated in AR. Chemical looping employs metal oxide as an oxygen carrier. Fuel (gaseous form) reacts with metal oxides in the fuel reactor (Mattisson & Lyngfelt, 2001). The metal oxide undergoes reduction to its metallic form, producing CO_2 and H_2O, as shown in Fig. 6. Reduction and oxidation of the metals are continuous in a loop (Czakiert et al., 2022). Nowadays, combined oxides are used to improve oxygen carrier characteristics. This method has the highest capturing efficiency (98–100%) compared to the other technologies (Abuelgasim et al., 2021). The CO_2 captured is in pure form, which is relatively easy to store and doesn't require further processing. A significant concern revolves around the energy

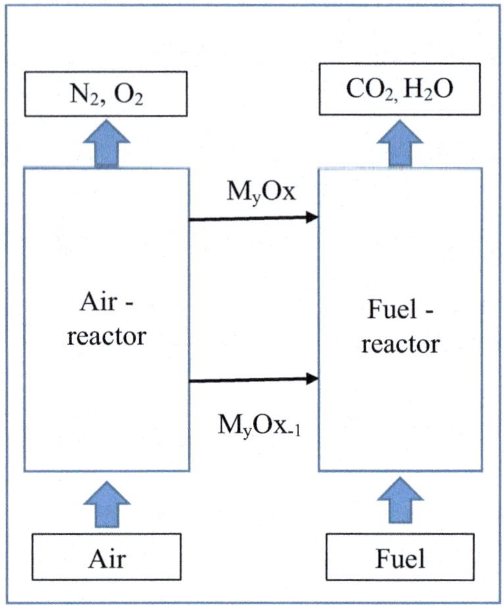

Fig. 6 Chemical looping combustion (CLC). MyOx and MyOx-1 symbolize oxidized and reduced oxygen carriers. (Adapted from: Mattisson & Lyngfelt, 2001)

needed to cycle solid metal between chambers. Earlier conversations implied that smaller nanoparticles of oxygen carriers enhance the reactivity of the carrier particles.

5.1.6 Microalgae-Based Carbon Capture

In recent times, the study of microalgae sequestration has gained considerable attention. While biotic sequestration techniques have been implemented, their widespread adoption is limited due to land availability issues and ecological considerations. These methods largely hinge on afforestation or reforestation strategies (Smith & Torn, 2013). Microalgae sequestration emerges as a sustainable and viable option to these traditional approaches. It is particularly effective for direct application at emission sources and holds promise in the transportation sector. Not only does it capture carbon, but it also contributes to biofuel generation from microalgae, creating a circular economy (Onyeaka et al., 2021).

Additionally, microalgae can play a dual role in wastewater treatment plants by forming a symbiotic relationship with bacteria. In this process, the CO_2 released by bacteria during waste decomposition is absorbed by microalgae for sequestration. Concurrently, the oxygen produced by microalgae aids bacterial survival. This synergy not only facilitates waste breakdown but also generates algal biomass. However, this system still faces challenges, including suboptimal CO_2 dispersion in water and poor algal growth in industrial effluents. Addressing these issues could lead to the development of cost-effective and efficient wastewater treatment and sequestration plants (Viswanaathan et al., 2022).

When conditions are favorable, microalgae can multiply rapidly, demonstrating a CO_2 capture efficiency that exceeds terrestrial plants by a factor of 10–50. Remarkably, 1 kg of microalgal biomass is capable of fixing 1.38 kg of CO_2. This makes it possible to cultivate microalgae using CO_2 derived from industrial emissions or flue gases emitted by large-scale sources. Thomas et al. (2016) conducted a review that suggests that directly utilizing flue gases for the growth of microalgae could be an effective approach. To ensure optimal microalgal growth, it is necessary to remove certain components from flue gases, such as SOx and NOx.

While the expense associated with cultivating microalgae artificially has been a concern, this can be balanced by the diverse applications of microalgae, such as in the creation of renewable fuels, organic fertilizers, and wastewater treatment processes (Osman et al., 2020).

5.2 Carbon Sequestration

A combination of processes can lead to long-term storage benefits (IPCC, 2018). Carbon sequestration can be differentiated into mainly two types: (1) biotic and (2) abiotic. India, too, has a significant amount of storage capacity. ONGC, NGRI (National Geophysical Research Institute), and NALCO (National Aluminium Company Limited) are a few companies that have initiated studies toward sequestration in the oil fields in India (Shaw & Mukherjee, 2022). More such studies are needed in the coming years. There is the risk of increased seismic activity due to the injection of CO_2 underground (Dai et al., 2016). Risk also assists in the transportation of CO_2 through pipelines. Leaks of pipelines can cause harmful human exposure to plants and animals (Price et al., 2007). Figure 7 illustrates the sequestration pathways currently available for research and development.

5.2.1 Biotic Sequestration

Biotic sequestration involves a medium where carbon is stored through the interaction of plants and animals. This process is particularly efficient because it relies on living organisms and minimizes the need for external energy inputs. Photosynthesis plays a crucial role in this form of carbon capture, a process also known as phytoremediation. Plants serve as effective CO_2 sequestration agents, with different species being more suited to specific environments, such as urban or industrial areas (Govindaraju et al., 2021).

Sequestration in the oceans is another significant method, with oceans being rich in plant life and playing a crucial role in global carbon capture. The photosynthesis of phytoplankton in these zones sequesters about 45 Pg of carbon per year (Lal, 2008). This carbon is then deposited on the ocean floor as these phytoplanktons die. Oceans have absorbed 25% of anthropogenic CO_2, helping to mitigate extreme climate events. However, this absorption has led to ocean acidification (Turley

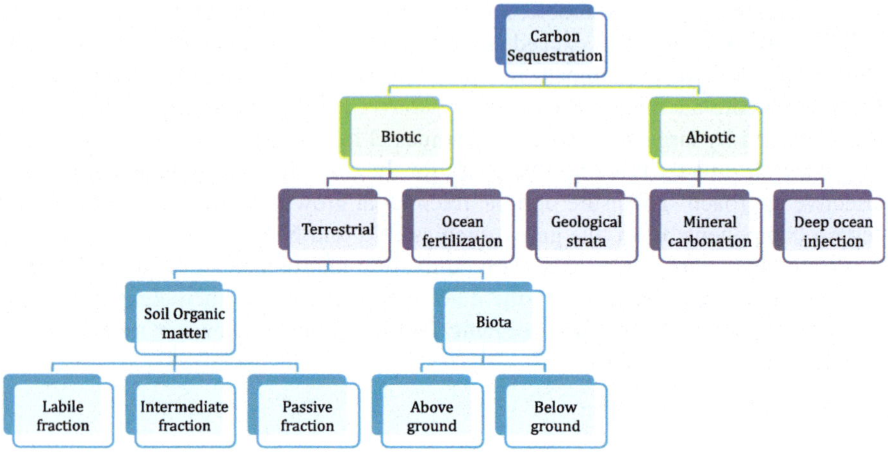

Fig. 7 Sequestration pathways. (Adapted from Kambale & Tripathi, 2010)

et al., 2010). A proposed method to enhance phytoplankton growth involves ocean fertilization with iron (Kambale & Tripathi, 2010), which could boost phytoplankton proliferation, increasing photosynthetic activity and overall biomass, leading to greater CO_2 consumption and sequestration upon the death of these phytoplanktons.

In contrast, terrestrial sequestration is globally the most widespread form of carbon storage, varying in intensity depending on latitude. The intensity peaks near the tropical zone (0–10° N/S) and diminishes toward the poles. The quantity of storage is directly correlated with the accessibility of biomass, which fluctuates with latitude. This sequestration occurs through photosynthesis, wherein plants capture atmospheric CO_2 using sunlight and water to produce food and release oxygen (Lal, 2008). As a component of the global carbon cycle, this natural sequestration has been crucial in maintaining the planet's homeostasis.

5.2.2 Abiotic Sequestration

Abiotic, unlike biotic sequestration, does not involve biological components. Its pace, intensity, and frequency are adjustable, potentially making it a more viable option due to the ability to develop and refine technologies for increased effectiveness. Global studies have estimated the total geological sequestration capacity to be between 10,000 and 30,000 billion tons (BT) of CO_2, although these figures are highly speculative and uncertain (Budinis et al., 2018).

Different studies have provided varied estimates of sequestration potential. Brown et al. (2013) utilized numerical modeling to estimate the sequestration potential of the Florida Panhandle at 4.55 billion tons (BT). Van der Meer and Yavuz (2009) calculated the capacity of the Rotliegend Formation in the Netherlands using an empirical equation based on storage efficiency, aquifer capacity, porosity, and CO_2 density at depth, arriving at a total value of 104.12 million tons (MT). The capacities of four sandstone formations: Mt. Simon (USA), Basal (USA and Canada), Bunter (UK), and Rotliegend (The Netherlands), were determined using 3D flow modelling, with calculated values of 13.3, 16.2, 2.23, and 0.33 billion tons (BT), respectively (Shaw and Mukherjee 2022).

Vangkilde-Pedersen et al. (2009), as part of the "GeoCapacity" Project, provided estimates of the geological sequestration capacity for Europe. They assessed diverse factors including aquifer size, capacity, thickness, and CO_2 density at different levels. The total capacity for potential European sites was estimated at 325 billion tons (BT) in aquifers, 30 BT in hydrocarbon fields, and 1.5 BT in coal fields. Faiz et al. (2007) computed the storage potential of coal seams in the Southern Sydney Basin, Australia, yielding estimates of 350 million tons (MT) (with full capacity-based methane recovery) and 175 MT (with half capacity-based methane recovery).

Additionally, there are regional efforts such as the North American Carbon Storage Atlas and the CO_2 Atlas for the Norwegian Continental Shelf to gauge geological storage potentials (Ringrose, 2020). Abiotic sequestration comprises three primary categories (Kambale & Tripathi, 2010; Lal, 2008): oceanic injection, geological storage, and mineralization.

- India's Storage Potential:

Some industries that have initiated setting up CCS facilities include NALCO, ONGC, BHEL (Bharat Heavy Electricals Limited), and APGENCO (Andhra Pradesh Power Generation Corporation Limited) (Gupta & Paul, 2019). According to research conducted in various parts of India, the country has a potential storage capacity of 47-572 GT (Viebahn et al., 2015). Deep saline aquifers have a storage capacity of 105–360 Gt. Basalt formations have a 200 Gt capacity (Dooley et al., 2005; Singh et al., 2006b).

The results from the analysis of carbon sequestration in India by Vishal V. et al. indicated a great potential. This study consists of four storage pathways: Storage through EOR, Enhance Coalbed Methane recovery (ECBMR), and storage in deep saline aquifers and basalt formations. Storage potential in saline aquifers is at 291 Gt; through basalt, it ranges from 97 to 316 Gt. The EOR and ECBMR have a low storage capacity of 3.4 Gt and 3.7 Gt, respectively (Vishal et al., 2021a, b). Another study by Abhishek Gupta and Akshay Paul listed the potential of commercial CO_2 storage plants. The Aonala Urea plant in UP is estimated to have an absorption capacity of 450 TPD (Tons per day). The Jagdishpur (Orissa) urea plant has a potential of 150 TPD and Phulpur (Uttar Pradesh) urea plant is identified to have a potential of 450 TPD (Gupta & Paul, 2019)

-

5.3 Carbon Utilization Technologies

Utilization is an important parameter to classify CCUS technology as a part of the CE and for the successful implementation of CCUS at scale. Carbon can be utilized to build technology in various paths, including uptake, conversion, and

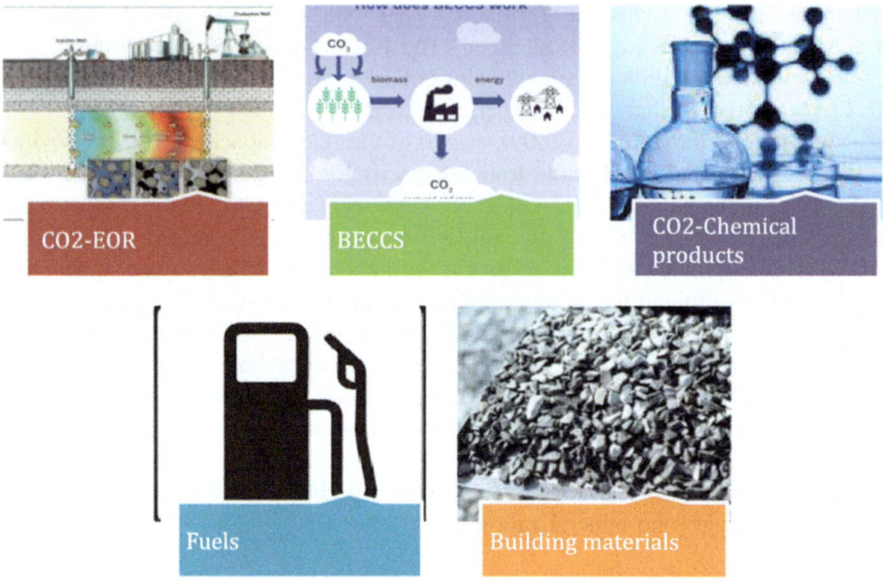

Fig. 8 CO_2 utilization pathways

mineralization. Cameron Heprun et al., in their study on CO_2 utilization, mentioned different pathways like chemicals, EOR, BECCS (bioenergy with carbon capture and storage), fuels, microalgae fuels, and building materials for utilizing carbon (Hepburn et al., 2019). Figure 8 identifies the different CO_2 utilization pathways, which can be explored in future research activities.

The captured CO_2 is supplied for EOR, one of the most extensively performed in many developing countries (Alina & Ekaterina, 2022). An EOR survey for the year 2020 revealed that 142 EOR projects were operational in the USA (Wallace, 2021). The injected CO_2 will reduce the viscosity of oil, increasing production. Currently, EOR might contribute to efficiently removing CO_2 from the atmosphere. However, there exists a risk of a net increase in CO_2 emissions if the EOR is not associated with a direct air capture system (Hepburn et al., 2019). EOR is not completely efficient in trapping utilized CO_2. Carbon8, a British firm, has developed a technology that uses CO_2 and other waste materials to form lightweight aggregates for construction work (Carbon8, products). CO_2 is also applied in making materials such as baking soda, aerated drinks, and chemicals like methanol and urea. The main blockage in utilizing captured carbon is the efficiency of the CO_2, source of CO_2, market potential, cost, and the energy required (Seetharaman et al., 2021).

5.4 *Initiatives Taken in India*

- Various Indian organizations, institutes, and agencies have implemented the following policies and initiatives:

Initially, DST took the initiative in the years 2006–2007, conducting global workshops to tackle research and development hurdles in the CCS domain. It also established an independent network, Indian CO_2 Applied Research (ICOSAR), to analyze CCS applications in India and launched the National Program on Carbon Sequestration in 2008 (Shackley & Verma, 2008; Shaw & Mukherjee, 2022). The ICOSAR network connects researchers, policymakers, and stakeholders to collaborate on innovations and development activities.

In 2015, Mission Innovation was announced in COP-21. This program focuses on clean energy, research, development, and demonstration initiatives to accelerate clean energy innovations (Mission Innovation, 2019). In 2017, Carbon Clean (a US-based firm) began a preliminary installation stage of the first non-subsidized Industrial CCS plant at the Tuticorin coal-powered plant. The captured CO_2 is converted into baking soda, having a capacity of 60,000 tons of CO_2/year (Harrabin, 2017). Later in the year 2018, DST and DBT joint the Funding Opportunities Announcement (FOA) in the Carbon Capture Innovation Challenge (IC#3). The IC#3 is headed by Mexico, Saudi Arabia, and the United Kingdom and was launched in 2018 (Goldar & Dasgupta, 2022). The Carbon Capture Innovation Challenge aims to advance CCUS technology development to reduce costs and improve performance (FOA, 2018). A dialogue has been initiated between the US Department of Energy (US-DoE) and DST on supercritical carbon dioxide (sCO_2) and carbon capture utilization and storage (CCUS) technologies. India has engaged in the multilateral platform for Accelerating CCUS Technologies (ACT-3 call), creating avenues for potential collaboration between the USA and India (DST, 2020). A budget of €1 million was available from funding organizations in the ACT3 call published in June 2020 (ACT-3 call, 2020).

In 2021, DST established India's first two National Centers of Excellence in the areas of CCUS. The National Center of Excellence in Carbon Capture and Utilization (NCoE-CCU) at the IIT Bombay, Mumbai, and the National Center in Carbon Capture and Utilization (NCCCU) at JNCASR, Bengaluru, are among these two NCoEs. These NCoEs aim to streamline the capture and mapping of ongoing R&D and innovation endeavors while fostering networks of professionals, industries, and relevant participants. These centers will serve as multi-disciplinary, long-term hubs for research, design development, collaboration, and capacity-building initiatives in the field of CCU.

In January 2021, Dastur International (Prime contractor), Inc. (Ridgewood, NJ); Air Liquide Global E&C Solutions (Air Liquide); and the Bureau of Economic Geology (BEG) at the University of Texas at Austin have been chosen to conduct the design and feasibility study for India's largest industrial CCUS project, with a capacity of 13.7 million tons per year (m.t./yr.). Koyali Refinery of Indian Oil Corporation Ltd. (IOCL) is part of a project funded by the United States Trade

Fig. 9 Tata Steel Jamshedpur Amine-Based Carbon Capture plant. (Tata Steel, 2021)

and Development Agency (USTDA). NITI Aayog organized a National Workshop on CCUS on March 30, 2022, on CCUS, which accentuated and proposed launching a National Mission on CCUS and added that the CCUS projects would become viable once CO_2 Capture was done at the pithead.

Tata Steel, at its Jamshedpur plant, in collaboration with Carbon Clean, has inaugurated India's inaugural carbon capture plant, which directly captures CO_2 from blast furnace. The captured CO_2 will be reused at the site, promoting a circular economy (Tata Steel, 2021). Tata Steel Jamshedpur Amine-Based Carbon Capture Plant is depicted in Fig. 9. The plant has a capture capacity of 5 tons of CO_2 per day (5 tons/day)). Amine-based CO_2 capture technology is used in this plant (Devadhas et al., 2022).

DST, Ministry of Science and Technology, Government of India has been listed as the funding organization in the ACT-4 call published on May 2022, having an indicative budget of ₹8 Crore (€ 1 million). However, this amount is insufficient to meet the Net Zero targets. In July 2022, the Ministry of Petroleum and Natural Gas (MoPNG) formulated a draft 2030 roadmap for CCUS targeting upstream E&P companies (MoPNG, 2022). The draft proposes appropriate policy measures to mitigate the long-term risks associated with CO_2 storage, which may entail migration or leakage years or even decades after operations cease.

Dalmia Cement has unveiled intentions to manufacture 100% low-carbon cement by 2031. The company has devised a US$ 405 million investment plan for CCU (IBEF, 2022). In September 2022, NTPC Energy Technology Research Alliance (NETRA) chose Carbon Clean and Green Power Int. Pvt. Ltd. to establish a carbon

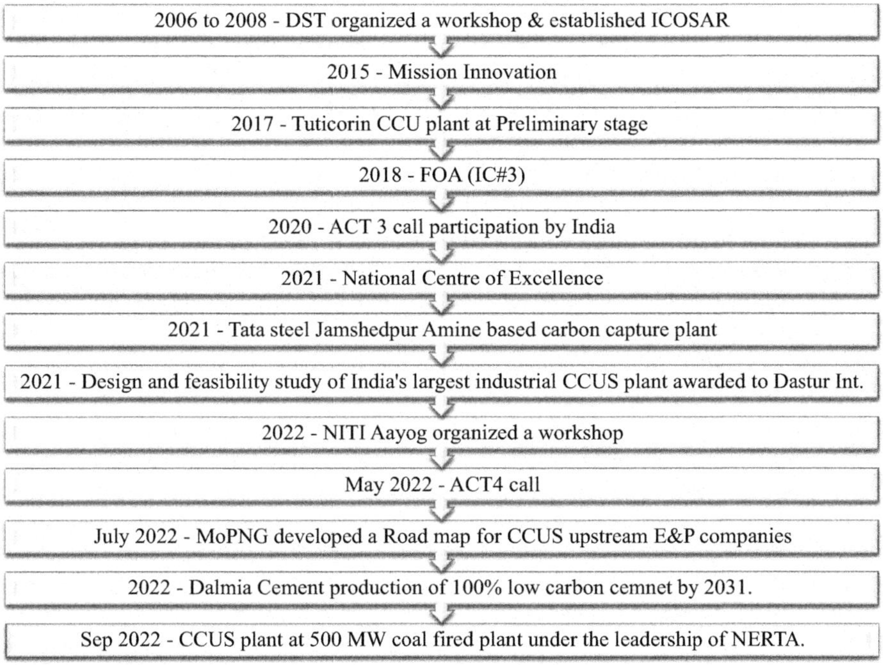

Fig. 10 Steps taken by government, institutes, and private organizations

capture plant at the 500 MW coal-fired Vindhyachal Super Thermal Power station located in Madhya Pradesh (Carbon Clean, 2022). On February 17, 2023, MOEFCC released a notification mentioning the 13 activities finalized for trading carbon credits under the Article 6.2 mechanism. CCUS comes under the removal activities category. These activities will help adopt emerging technologies and even help mobilize international finance in India. For the first 3 years, these activities are used and can be revised or updated by NADAIPA (National Designated Authority for the Implementation of the Paris Agreement) (MOEFCC, 2023). Steps taken by government, institutes, and private organizations are explained in Fig. 10.

Individual industries, universities, and organizations are carrying out numerous initiatives within the country to develop and upgradation of CCUS technologies, many of which are government-aided. Key initiatives are highlighted here:

- JNCASR, Bengaluru, spearheaded the project aimed at developing integrated technologies for converting CO_2 into value-added products. A pilot-scale plant with a capacity to convert 300 kg CO_2/day into methanol has been successfully operated. Additionally, novel catalysts have been engineered to produce hydrocarbon fuels with high energy content and a high octane rating derived from CO_2, achieving a selectivity of 80% (DST, MoU).

- The IISER, Kolkata team, in collaboration with DST, has showcased a method for synthesizing innovative solid adsorbents, particularly for CO_2 capture and utilization. They have uncovered unique nanoparticles or microparticles, porous covalent bonded organic nanotubes, and COF-coated zeolite. These materials effectively capture CO_2 within their micro- and mesoporous voids (Ashok Kumar et al., 2021; Shibani et al., 2022).
- CSIR-IICT (Hyderabad) is working on designing, synthesizing, and evaluating Combined Sorbent Catalyst Material (CSCM) for CO_2 Capture. In January 2022, IICT, under the MI project supported by DST, successfully commissioned the FBR facility to test the performance of dual functional materials for SESMR (Sorption Enhanced Steam Methane Reforming) in the fluidized bed reactor system. SESMR offers the removal of in situ CO_2 (Ministry of Science & Tech, 2022).
- The team at IIT Delhi is striving to develop a novel hierarchical catalyst for the one-pot conversion of CO_2-rich synthesis gas to dimethyl ether, and they are conducting scale-up studies. In this endeavor, the team has successfully formulated an innovative and cost-effective catalyst for achieving higher conversion rates of syngas rich in CO_2 (IIT Delhi Newsletter, 2020). The Government of India has announced the National Policy on Biofuels-2018, which, among other objectives, aims to foster the development of alternative fuels and reduce import dependency (MoEFCC 2021). NTPC and IIT-Guwahati have joined forces to design a highly energy-efficient CO_2 capture system for power plants. This system will be tested using industrial flue gas (Goel et al., 2019).
- Sequestration studies conducted by various government institutions, corporations, and agencies are mentioned below:

The research team at IIT Bombay is spearheading an initiative for a comprehensive large-scale assessment of the CO_2-Enhanced Oil and Natural Gas Recovery potential in key sedimentary basins in India. This initiative has identified four distinct storage pathways with significant potential: CO_2-EOR, ECBMR, storage in deep saline aquifers and basaltic formations. The work has led to the first-ever detailed countrywide assessment of geological potential for storage of carbon dioxide at a value of 400–600 Gt (Atanu & Saurav, 2022).

NALCO is pursuing bio-sequestration at its coal-powered plant in Orissa, collaborating with the high-end biotechnology company Indo-Can Technology Solutions (ICTS) to achieve its objectives (Shaw & Mukherjee, 2022). NGRI has demonstrated its capabilities in leading the CO_2 EOR project with ONGC and permanent sequestration in Deccan basalt. The mission mode project submitted by NGRI focuses on assessing India's vast coal basins and depleted oil and gas reservoirs, which can be used as a sequestration site. A systematic study has been initiated to understand the Deccan basalt of western India for structural and permanent sequestration (DST & UKRI, 2022). A model was developed to evaluate the potential for CO_2 capture via the EOR pathway in the Ankleshwar oil field, with data sets supplied by ONGC initially (Ganguli et al., 2016).

6 Research and Innovation

Improvement and innovation in the previous technologies are needed to enhance efficiency and make the technologies economically viable for adoption. CCUS can play a key role in reducing emissions from iron, steel, cement, chemicals, and other sectors with the research and innovation needed in several areas (Gaps, IEA, 2019). The policy system and technological advancement will be the deciding factor for the success of this technology (Chaturvedi & Malyan, 2022). India, with its large population, faces some of the most significant environmental challenges, making it a major concern that requires urgent attention (Bherwani et al., 2022). The research and innovation activities and their relevance to India are highlighted in Table 3.

7 Key Challenges and Opportunities

- The considerable expense poses a major obstacle to the widespread implementation of CCS technologies. According to OECD (2008), Brazil, Russia, India, and China have significant barriers to trade in carbon abatement technology. This group of countries frequently imposes tariffs exceeding 10% on burdensome pre-shipment inspection and informal "additional payments" (Economics, 2009).
- There is a risk of seismic activity due to the underground storage of CO_2. A study on this risk and many other storage factors is being carried out. According to Energy Technology Perspectives 2020, crucial challenges for acquiring CO_2 storage technologies include storage and associated infrastructure, coordination across various sectors and stakeholders, uncertainties regarding long-term ownership and liability for stored CO_2, unexplored insurance and finance markets, as well as public concerns related to storage, particularly onshore, in certain regions (IEA, 2020a, b; MoPNG, 2022).
- Most novel technology is financially unviable or underused; the government and private sector's involvement in innovation in new technologies and financial aspects can be a milestone for achieving Net zero targets. Achieving global climate objectives requires a rapid transition to lower carbon development paths, particularly in heavy sectors like steel and cement, for effective decarbonization.
- A large amount of energy is needed to compress CO_2 and uphold high pressure along pipelines, and the pipelines themselves entail significant costs. Impurities present in the CO_2 stream, such as water, can result in pipeline damage and pose risks of leaks and explosions as the compressed fluid rapidly transitions into a gas. However, most green financing institutions in India are smaller and more interested in investing in renewable energy and energy efficiency projects. Lack of awareness and access to technical knowledge, less participation of building owners and stakeholders, and false perception that energy efficiency is costly are a few points that need to be focused on. Besides, some technologies require

Table 3 Research and innovation activities and their relevance to India

Research and innovation	Research and innovation gap	Relevance to India
Cement and steel industry: Numerous CCS technologies exist for implementation in the cement sector, with the most advanced options being chemical absorption post-combustion, and oxyfuel capture (IEA, 2019)	CO_2 capture has the potential to achieve capture yields of up to 95%, and its widespread adoption could decrease emissions from the clinker production process, where reduction options are limited (IEA, 2019)	The cement and steel sectors are the most vital to focus on for emerging countries like India. Alternative methods of CCUS technology, keeping in mind the country's development, must be researched. Dalmia Cement, a prominent cement manufacturer in India, secured the top position within the worldwide cement industry, according to the Climate Disclosure Project (CDP) for its commendable efforts in promoting environmental sustainability (Chaturvedi & Malyan, 2022)
Industrial hubs with shared CO_2 storage and infrastructure are emerging as fresh investment prospects. Availing industrial hub, the financial cost and commercial risk are reduced by separating the capture, storage, and utilization components providing continuous operation. This can also boost employment (IEA, 2020a, b)	Only 12 CCUS Hubs developed in countries like Australia, Europe, and the USA (IEA, 2020a, b). More such hubs needed to be constructed to aid developing countries financially. For fostering hub development, governments must initially pinpoint opportunities in collaboration with potential investors and then formulate business models that incentivize their establishment	A major cause of concern is India's financial setback in CCUS technology. The government might establish an industrial hub through public-private partnerships to manage finances effectively
Emerging technologies and alternative methods for utilizing or recycling CO_2 beyond EOR, such as the production of synthetic fuels or building materials, are on the rise, potentially increasing the demand for CO_2. In June 2019, Japan unveiled a Carbon Recycling Roadmap outlining opportunities to commercialize CO_2 utilization technologies in the coming decade (METI, 2021). Various prize initiatives have been conducted to encourage the advancement of CO_2 conversion technologies, recognizing the most innovative applications of CO_2 utilization		The private sector is critical to fostering CCUS growth in any country. Financial issues may be effectively handled by encouraging private sector participation

(continued)

Table 3 (continued)

Research and innovation	Research and innovation gap	Relevance to India
The Faster Innovation Case, a variant of the Sustainable Development Scenario, aims to determine the additional clean energy technology innovation required beyond the Sustainable Development Scenario level. Its goal is to accelerate the attainment of net-zero emissions by 2050 (IEA, 2020a, b)	Approximately two-thirds of the emissions reductions from CCUS by 2070 in the Sustainable Development Scenario, compared to the Stated Policies Scenario, come from technologies currently in the prototype or demonstration stage. Accelerating innovation is crucial to ensure that key applications become commercially feasible within the coming decade (IEA, 2020a, b)	The time lags also apply to India, and a systematic method is necessary to prevent this time lag

special infrastructure. This infrastructure may necessitate significant additional investment, posing challenges for developing countries in finding sufficient capital. Figure 11 highlights major barriers and opportunities.

7.1 Financial Aspect

- As per CEEW analysis, the estimated cumulative investments of 10 trillion USD are needed to attain net zero by 2070. Governments and private sector partnerships are much needed (Nagar, 2022). For 100 km transportation, the estimated cost for the year 2011 is $ 2/ton. As per Indian conditions, the distance is 350 km, and the estimated cost is $ 7.5/ton (Viebahn et al., 2015).
- DST has engaged in the ACT, which aids in project finance. ACT aims to expedite the development of CCUS by advancing and refining CCUS technology through targeted funding of innovation and research initiatives (ACT-3 call, 2020). ACT provides funding for the shortlisted proposals. Funding agencies from Canada, Denmark, France, Germany, Greece, India, Italy, the Netherlands, Norway, the Nordic Region, Romania, Spain, Switzerland, Turkey, the UK, and the USA constitute ACT members. Indian funding organizations were allocated ₹8 crores in funding for each of the ACT 3 and 4 calls.
- Blended finance involves leveraging funding to draw private capital into emerging market sectors, like the green industry or climate technology. Typically, it is spearheaded by development finance and public funds. Blended finance encompasses a range of financial sources and mechanisms, including risk reduction or guarantees, preferential debt and equity, performance-based incentives, and co-financing with public funds (Tamiksha et al., 2020).

Fig. 11 Major barriers and opportunities

8 Pathways to a Carbon-Neutral Future

The UNECE has initiated measures to assist countries in adopting CCUS technologies and achieving carbon neutrality. These efforts center around three primary objectives (UNECE, 2021). They include:

- To raise awareness: CCUS should be recognized as an essential mitigation option to mitigate climate change.
- To accept technology: This technology shall be encouraged in the industrial sector where energy is dependent on renewable sources.
- To finance project: Create a funding mechanism for CCUS.

Considering the scenarios in India, the following are the measures that can be adapted to meet the net zero targets:

More Centre of Excellence can be formed to examine domestic capabilities and study climatic development data. They will also oversee research and development and coordinate efforts across industries. On the establishment of the Industrial Hubs, employment opportunities will be developed. Training centers and skill improvement camps can be conducted for the workforce hiring required in the industrial hubs.

Considering the Indian scenarios, systematic and planned approach needed to be adopted, and the timeline for each section should be fixed. Demonstration projects play a crucial role in strengthening stakeholder confidence in the technology and acquiring a better understanding of its uncertainties. Moreover, awareness among individuals, researchers, and policymakers should be raised regarding CCUS technology.

Promote alternative funding sources for innovative low-carbon technologies in industries such as cement, steel, chemicals, transportation, and others, utilizing export credit agencies and multilateral development banks.

A cost-benefit analysis based on Indian environmental conditions can be performed, which may be useful in selecting appropriate technologies to achieve maximum efficiency in implementing CCUS projects. The risk of each technique as per the Indian scenario and the technology readiness level (TRL) of each technique should be analyzed, which will decide the technology to be adopted. The need for accounting of carbon removal efficiencies for the carbon market shall be a trading option for private owners, PSUs, etc.

According to global studies, the geological capacity for CO_2 sequestration ranges from 10,000 to 30,000 billion tons (BT). Biotic and abiotic sequestration can be compared regarding efficiency, risk, and long-term reliability. Abiotic sequestration is considered more feasible than biotic sequestration due to its potential for development and correction. Addressing uncertainties, environmental variations, and extreme events requires promoting research and innovation endeavors.

Mapping suitable geological formations for CO_2 storage using surface geophysical and borehole data is critical for determining the quantity of CO_2 that may be stored. Moreover, site characterization and de-risking of CO_2 sequestration for safe storage are an important study area where attention may be directed, and risk assessments can be carried out. Identifying geological formations for permanent sequestration and its risk assessment must be researched and studied.

Launching a field observatory to test the real injection scenarios in different geological formations. A few representative blocks can be tested initially based on the ease of operation/accessibility of the source and sink for the pilot project. Initial testing and observations should be carried out, and further actions should be planned accordingly. Strategies for the future as depicted in Fig. 12.

Collaboration: Norway, the USA, the UK, and Australia are leading the geological sequestration research and could be India's potential partners. The USA and Australia have expertise in unconventional formations like Shale and Coal beds, while Norway has established its leadership in saline aquifers. The UK has also worked on the North Sea geological sequestration projects. India should participate in such collaboration, which will benefit CCUS technology implementation.

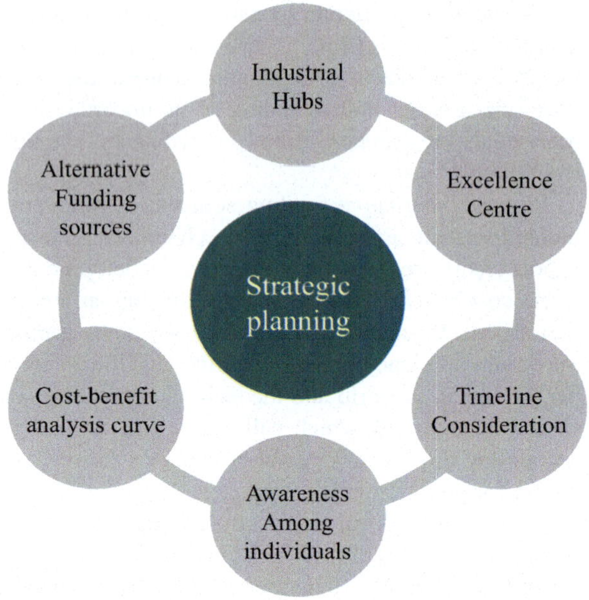

Fig. 12 Strategies for the future

9 Conclusion

The climate change issue has become a serious global concern. Especially the GHG emissions dreadfully affect the environment. It is estimated that by 2030, carbon emissions will spike to over 5.3 BT in India. Despite India's comparatively low per-person CO_2 emissions, the country maintains its position as the third-largest global emitter of CO_2. CCUS is a technology implemented across the globe to counteract climate change effects by mitigating carbon emissions. Developed countries like the USA, Germany, Canada, Norway, Europe, etc. are a few leading countries in adopting this technology. CCUS has been a topic of discussion in India for the past two decades, and there are some key initiatives taken to implement this technology. However, these are insufficient to achieve targets under the Paris climate agreement. To endorse CCUS in the global circular economy, numerous renewable energy sources will be required. CCUS technology and its contribution to circular economy is a topic of discussion for further research.

Our study focuses on the evolution of CCUS within the context of India (Sect. 5.4). we examine the growth of CCUS technology alongside current research and development efforts in India. Despite global advancements, the reception of CCS technology among Indian stakeholders has been notably cautious. Within this chapter, we navigate through the significant technical and logistical barriers hindering the widespread implementation of CCUS technology within the Indian landscape, detailed in Sect. 7. Furthermore, we outline major steps taken to adopt CCUS technologies in India:

1. The crucial step to carbon neutrality was when India announced joining Mission Innovation in 2015 (COP21).
2. India's LT-LEDS report claims that India will rely heavily on CCS and negative emissions technologies to achieve the Net Zero target. Government should extend its support to R&D in carbon capture and utilization.
3. CCUS has also been mentioned as one of the 13 activities finalized for trading carbon credits by MOEFCC.
4. ₹8 Crore (€ 1 million) budget sanctioned to India in ACT 3 and 4 call.
5. Tata Steel and Carbon Clean initiated the implementation of CCUS at different plants.
6. Dalmia Cement has revealed its intention to manufacture 100% low-carbon cement by 2031.

Research and development should be encouraged at different administrative levels to overcome challenges such as technological advancement, sequestration liability, and transportation and storage risks. Different technologies adopted for carbon capture are briefly discussed (Sect. 5). Implementing CCUS technology and focusing on recapturing and reutilization shall help achieve the circular economy strategies. Studies are carried out to evaluate the sequestration potential and EOR activities. MI-DST/DBT, ONGC, NTPC, NALCO, NITI Aayog, Government universities, and Individual industries have implemented numerous initiatives to develop and upgrade the CCUS technology. The research and innovations and their relevance to Indian industries have been discussed in this chapter (Sect. 6). The way forward and the strategies to advance the development of this technology are also suggested (Sect. 8).

Although India has initiated steps to adopt CCUS technologies, it is still in the nascent stages of development and deployment. While there are challenges such as cost, infrastructure, and public acceptance, the long-term advantages cannot be ignored. There exists a lot of scope for the growth of CCUS technology influencing India's carbon market. CCUS may be the fastest and most effective option to address the high carbon emissions, but large-scale adoption of innovative technologies is the need of the hour.

References

Abuelgasim, S., Wang, W., & Abdalazeez, A. (2021). A brief review for chemical looping combustion as a promising CO_2 capture technology: Fundamentals and progress. *Science of the Total Environment, 764*, 142892.

ACT-3 call, 20th July 2020, Version 3. Available online: https://dst.gov.in/sites/default/files/DST%20Jointly%20Launches%20Accelerating%20CCUS%20Technologies%20%28ACT%29%20Call.pdf. Accessed 12 Dec 2022.

Agrali, S., Uctug, F. G., & Turkmen, B. A. (2018). An optimization model for carbon capture & storage/utilization vs. carbon trading: A case study of fossil-fired power plants in Turkey. *Journal of Environmental Management, 215*, 305–315.

Alina, I., & Ekaterina, K. (2022). CC(U)S initiatives: Prospects and economic efficiency in a circular economy. *Energy Reports, 8*(Supplement 1), 1295–1301. https://doi.org/10.1016/j.egyr.2021.11.243. ISSN 2352-4847.
Al-Mamoori, A., Krishnamurthy, A., Rownaghi, A. A., & Rezaei, F. (2017). Carbon capture and utilization update. *Energy Technology, 5*(6), 834–849.
Anand, A., & Kaushal, P. (2021). The need for life cycle assessment in India for achieving the sustainable development goals (SDGs). *Academia Letters, 2*.
Malyan, A., & Chaturvedi, V. Aug'2021, Carbon Capture, Utilization, and Storage (CCUS) in India. https://www.ceew.in/sites/default/files/ceew-study-on-the-role-of-carbon-capture-utilization-and-storage-in-india.pdf
Ashok Kumar, M., Saikat, B., Himadri, S. S., Kaushik, D., Indrajit, G., Mercedes, L.-M., Carlos, C., Paolo, F., Gowd, E. B., Vijayaraghavan, R. K., & Rahul, B. (2021). Crystallizing sub 10 nm covalent organic framework thin films via interfacial–residual concomitance. *Journal of the American Chemical Society, 143*(49), 20916–20926. https://doi.org/10.1021/jacs.1c09740
Atanu, M., & Saurav, C. (2022). *CCUS, policy framework and its deployment mechanism in India*. NITI Aayog, M. N. Dastur & Co. (P) Ltd.
Bashmakov, I. A., Nilsson, L. J., Acquaye, A., Bataille, C., Cullen, J. M., Fischedick, M., & Tanaka, K. (2022). Chapter 11: Climate change 2022: Mitigation of climate change. In *Contribution of working group III to the sixth assessment report of the intergovernmental panel on climate change*.
Bherwani, H., Nair, M., Niwalkar, A., Balachandran, D., & Kumar, R. (2022). Application of circular economy framework for reducing the impacts of climate change: A case study from India on the evaluation of carbon and materials footprint nexus. *Energy Nexus, 5*, 100047.
Breyer, C., Fasihi, M., Bajamundi, C., & Creutzig, F. (2019). Direct air capture of CO_2: A key technology for ambitious climate change mitigation. *Joule, 3*(9), 2053–2057.
Brown, C. J., Poiencot, B. K., Hudyma, N., Albright, B., & Esposito, R. A. (2013). An assessment of geologic sequestration potential in the panhandle of Florida USA. *Environmental Earth Sciences, 71*(2), 793–806. https://doi.org/10.1007/s12665-013-2481-1
Budinis, S., Krevor, S., Dowell, N. M., Brandon, N., & Hawkes, A. (2018). An assessment of CCS costs, barriers, and potential. *Energy Strategy Reviews, 22*, 61–81. https://doi.org/10.1016/j.esr.2018.08.003
Carbon Clean, Carbon capture begins at NTPC's power plant in India, 29 Sept'2022. Available online: https://www.carbonclean.com/news/ntpc-power-plant-india (Press release). Accessed 12 Nov 2022.
Carbon Engineering, Direct Air Capture Article, Available online: https://carbonengineering.com/our-technology/. Accessed 20 June 2022.
Carbon8, products article. Available online: https://www.carbon8.co.uk/products. Accessed 20 June 2022.
Chaturvedi, V., & Malyan, A. (2022). Implications of a net-zero target for India's sectoral energy transitions and climate policy. *Oxford Open Climate Change, 2*(1), kgac001.
Chen, P., Bose, S., Selveindran, A., & Thakur, G. (2023). Application of CCUS in India: Designing a CO_2 EOR and storage pilot in a mature field. *International Journal of Greenhouse Gas Control, 124*, 103858.
Climate Links, Jan'2019, GHG emissions factsheet: India. https://www.climatelinks.org/resources/greenhouse-gas-emissions-factsheet-india
Crowley, T. J., Hyde, W. T., & Peltier, R. (2001). CO_2 levels required for deglaciation of a "Near-Snowball" Earth. *Geophysical Research Letters, 28*(2), 283–286. https://doi.org/10.1029/2000GL011836
Czakiert, T., Krzywanski, J., Zylka, A., & Nowak, W. (2022). Chemical looping combustion: A brief overview. *Energies, 15*(4), 1563.
Dai, Z., Viswanathan, H., Middleton, R., Pan, F., Ampomah, W., Yang, C., & White, M. (2016). CO_2 accounting and risk analysis for CO_2 sequestration at enhanced oil recovery sites. *Environmental Science & Technology, 50*(14), 7546–7554.

Datta, A., & Krishnamoorti, R. (2019). Opportunities for a low carbon transition-deploying carbon capture, utilization, and storage in Northeast India. *Frontiers in Energy Research, 7*, 12.

Davidson, R. (2011). Pre-combustion capture of CO_2 in IGCC plants. IEA Clean Coal Centre, 571(98), 572.

Devadhas, G. G., Prabha, D. M. S. R., Gayathri, A. R., Shinu, M. M., & Dhanoj, M. (2022). Carbon capture, utilization and storage (CCUS)-the energy source of the future. In *2022 third international conference on intelligent computing instrumentation and control technologies (ICICICT)* (pp. 1748–1752). IEEE.

Diaz, A. G., Fernández, E. S., Gibbins, J., & Lucquiaud, M. (2016). Sequential supplementary firing in natural gas combined cycle with carbon capture: A technology option for Mexico for low-carbon electricity generation and CO_2 enhanced oil recovery. *International Journal of Greenhouse Gas Control, 51*, 330–345.

Dooley, J. J., Kim, S. H., Edmonds, J. A., Friedman, S. J., & Wise, M. A. (2005). A first-order global geological CO_2-storage potential supply curve and its application in a global integrated assessment model. In *Proceedings of the 7th international conference on greenhouse gas control technologies* (Vol. 1, pp. 573–581). Elsevier Science Ltd. https://doi.org/10.1016/B978-008044704-9/50058-6. ISBN 9780080447049.

DST & UKRI. (2022). *Carbon capture utilization & storage in India and the U.K.*

DST, Ministry of Science & Technology, July'2020, Ministerial meeting of Indo-US Strategic Energy Partnership highlight major accomplishments, prioritizes new cooperation areas (Press Release). https://pib.gov.in/PressReleasePage.aspx?PRID=1639577

DST MoU. Available online: https://dst.gov.in/jncasr-signs-mou-incubated-company-scaling-technology-reducing-co2-methanol-other-useful-chemicals. Accessed 12 Oct 2022.

Economics, C. (2009, January). *Are IPRs a barrier to the transfer of climate change technology?* Study of Copenhagen Economics and the Danish IPR Company APS for the European Commission, Copenhagen.

Erans, M., Sanz-Pérez, E. S., Hanak, D. P., Clulow, Z., Reiner, D. M., & Mutch, G. A. (2022). Direct air capture: Process technology, techno-economic and socio-political challenges. *Energy & Environmental Science, 15*(4), 1360–1405.

Faiz, M. M., Saghafi, A., Barclay, S. A., Stalker, L., Sherwood, N. R., & Whitford, D. J. (2007). Evaluating geological sequestration of CO_2 in bituminous coals: The southern Sydney Basin, Australia as a natural analogue. *International Journal of Greenhouse Gas Control, 1*(2), 223–235. https://doi.org/10.1016/s1750-5836(07)00026-6

Funding Opportunity Announcement (FOA) in Carbon Capture Innovation Challenge (I.C. #3), Ministry of science & technology, Govt. of India, 2018. Available online: http://dst.gov.in/sites/default/files/FOA-IC3.pdf. Accessed 31 Oct 2022.

Ganguli, S. S., Dimri, V. P., & Vedanti, N. (2016). Integrated flow simulation, rock physics & geomechanics identifies CO_2-EOR and storage potential at Ankleshwar, India. In *78th EAGE conference and exhibition 2016* (Vol. 2016, No. 1). EAGE Publications B.V.

Gaps, IEA Innovation. (2019). *Key long-term technology challenges for research, development and demonstration.* Technology Report—May.

Garg, A., & Shukla, P. R. (2009). Coal and energy security for India: Role of carbon dioxide (CO_2) capture and storage (CCS). *Energy, 34*(8), 1032–1041.

Ghosh, I. (2020). *A global breakdown of greenhouse gas emissions by sector.*

Global CCS institute, State Of The Art: CCS Technologies. (2022). https://www.globalccsinstitute.com/wp-content/uploads/2022/05/State-of-the-Art-CCS-Technologies-2022.pdf

Global Status of CCS. (2021). *CCS accelerating to net zero.* Global CCS Institute. https://www.globalccsinstitute.com/wp-content/uploads/2021/11/Global-Status-of-CCS-2021-Global-CCS-Institute-1121.pdf

Goel, M., Sudhakar, M., & Shahi, R. V. (Eds.). (2019). *Carbon capture, storage and utilization: A possible climate change solution for energy industry.* CRC Press.

Goldar, A., & Dasgupta, D. (2022). Exploring carbon capture, utilisation and storage in the Indian context: ICRIER G20 policy series-run up to 2023.

Govindaraju, M., Banu, J. F., & Goel, M. (2021). CO 2 sequestration through phytoremediation techniques with special emphasis on urban forestry to mitigate climate change impact. In M. Goel, T. Satyanarayana, M. Sudhakar, & D. P. Agrawal (Eds.), *Climate change and green chemistry of CO_2 sequestration* (pp. 263–271). Springer. ISBN: 978-981-16-0028-9.

Gupta, A., & Paul, A. (2019). Carbon capture and sequestration potential in India: A comprehensive review. *Energy Procedia, 160*, 848–855. https://doi.org/10.1016/j.egypro.2019.02.148

Hannah Ritchie and Max Roser, India CO_2 country profile. https://ourworldindata.org/co2/country/india. Aniruddha Bhattacharjee, Shreeshan V, July'2022. Carbon capture and removal: not all it is cracked up to be. https://carboncopy.info/carbon-capture-and-removal-not-all-it-is-cracked-up-to-be/#:~:text=India%2C%20which%20emits%20about%202.65,%25)%20and%20USA%20(15%25)

Hannah, R., Max, R., & Pablo, R. (2020). *CO_2 and greenhouse gas emissions*. Published online at OurWorldInData.org. Retrieved from: https://ourworldindata.org/co2-and-greenhouse-gas-emissions (Online Resource)

Hansen, J., Sato, M., Kharecha, P., Beerling, D., Berner, R., Masson-Delmotte, V., & Zachos, J. C. (2008). Target atmospheric CO_2: Where should humanity aim? *arXiv preprint arXiv*, 0804.1126. https://doi.org/10.2174/1874282300802010217

Harrabin, R. (2017). Indian firm makes carbon capture breakthrough. *The Guardian*. Retrieved from: https://www.theguardian.com/environment/2017/jan/03/indian-firm-carbon-capture-breakthrough-carbonclean

Hepburn, C., Adlen, E., Beddington, J., Carter, E. A., Fuss, S., Mac Dowell, N., & Williams, C. K. (2019). The technological and economic prospects for CO_2 utilization and removal. *Nature, 575*(7781), 87–97.

Hyde, W. T., Crowley, T. J., Baum, S. K., & Peltier, W. R. (2000). Neoproterozoic 'snowball Earth' simulations with a coupled climate/ice-sheet model. *Nature, 405*(6785), 425–429.

IBEF (India Brand Equity Foundation), Feb'2022, Cement. Available online: https://www.ibef.org/download/1650431302_cement-february-2022.pdf. Accessed 17 Aug 2022.

IEA. (2019). *Global energy review 2019*. IEA, Paris. https://www.iea.org/reports/global-energy-review-2019. Accessed 29 Aug 2024.

IEA. (2020a). *CCUS in Clean energy transitions*. IEA. https://www.iea.org/reports/ccus-in-clean-energy-transitions. License: CC BY 4.0.

IEA. (2020b). *Energy technology perspectives 2020*. IEA. https://www.iea.org/reports/energy-technology-perspectives-2020. License: CC BY 4.0.

IEA. (2022a). *Direct air capture*. IEA. https://www.iea.org/reports/direct-air-capture, License: CC BY 4.0.

IEA. (2022b). *Global energy review: CO_2 emissions in 2021-global emissions rebound sharply to highest ever level*. IEA

IIT Delhi, Jan 2020, Newsletter, Available online: https://home.iitd.ac.in/uploads/newsletters/Newsletter%20January%202020.pdf. Accessed 20 Apr 2023.

Intergovernmental Panel on Climate Change (IPCC). (2018). *Global Warming of 1.5°C*. Available online: https://www.ipcc.ch/sr15/chapter/chapter-5/. Accessed 14 May 2023.

Jakobsen, J., Roussanaly, S., & Anantharaman, R. (2017). A techno-economic case study of CO_2 capture, transport and storage chain from a cement plant in Norway. *Journal of Cleaner Production, 144*, 523–539.

Johnsson, F., Normann, F., & Svensson, E. (2020). Marginal abatement cost curve of industrial CO_2 capture and storage–a Swedish case study. *Frontiers in Energy Research, 8*, 175.

Kambale, J. B., & Tripathi, V. K. (2010). Biotic and abiotic processes as a carbon sequestration strategy. *Journal of Environmental Research and Development, 5*(1), 240–251.

Kanniche, M., Gros-Bonnivard, R., Jaud, P., Valle-Marcos, J., Amann, J. M., & Bouallou, C. (2010). Pre-combustion, post-combustion and oxy-combustion in thermal power plant for CO_2 capture. *Applied Thermal Engineering, 30*(1), 53–62.

Kawai, E., Ozawa, A., & Leibowicz, B. D. (2022). Role of carbon capture and utilization (CCU) for decarbonization of industrial sector: A case study of Japan. *Applied Energy, 328*, 120183.

Lal, R. (2008). Carbon sequestration. *Philosophical Transactions of the Royal Society B: Biological Sciences, 363*, 815–830.

Mahato, S., & Ghosh, K. G. (2020). Short-term exposure to ambient air quality of the most polluted Indian cities due to lockdown amid SARS-CoV-2. *Environmental Research, 188*, 109835.

Mattisson, T., & Lyngfelt, A. (2001). *Capture of CO_2 using chemical-looping combustion* (pp. 163–168). Scandinavian-Nordic Section of Combustion Institute.

McQueen, N., Gomes, K. V., McCormick, C., Blumanthal, K., Pisciotta, M., & Wilcox, J. (2021). A review of direct air capture (DAC): Scaling up commercial technologies and innovating for the future. *Progress in Energy, 3*(3), 032001.

METI (Ministry of Economy, Trade and Industry), July'2021, Roadmap for carbon recycling technologies revised. Available online: https://www.meti.go.jp/english/press/2021/0726_003.html. Accessed 20 Apr 2023.

Metz, B., Davidson, O., De Coninck, H. C., Loos, M., & Meyer, L. (2005). *IPCC special report on carbon dioxide capture and storage*. Cambridge University Press.

Ministry of Science & technology, 11th April'2022, New materials & processes for carbon capture and Utilization could show new light for global warming challenge (Press Release). Available online: https://pib.gov.in/PressReleasePage.aspx?PRID=1815616. Accessed 20 Apr 2023.

Mission Innovation. (2019). *Report of the mission innovation carbon capture, utilization and storage experts' workshop*. Mission Innovation.

MoEFCC. (2015). *India: First biennial update report to the United Nations framework convention on climate change*. Ministry of Environment, Forest and Climate Change, Government of India.

MoEFCC. (2018). *India: Second biennial update report to the United Nations framework convention on climate change*. Ministry of Environment, Forest and Climate Change, Government of India.

MoEFCC. (2021). *India: Third biennial Uudate report to the United Nations framework convention on climate change*. Ministry of Environment, Forest and Climate Change, Government of India.

MoEFCC. (2022). *India's long-term low-carbon development strategy*. Ministry of Environment, Forest and Climate Change, Government of India.

MoEFCC. (2023). *Activities finalized to be considered for trading of carbon credits under Article 6.2 mechanism to facilitate transfer of emerging technologies and mobilise international finance in India*. https://pib.gov.in/PressReleaseIframePage.aspx?PRID=1900216

MOEFCC & UNFCCC. (2012). Ministry of Environment, & United Nations Framework Convention on Climate Change (Organization). (2012). India, Second National Communication to the United Nations Framework Convention on Climate Change. Ministry of Environment and Forests, Government of India.

MoPNG. (2022). (The Ministry of Petroleum and Natural Gas), Government of India, 28th July'2022, Draft 2030 Roadmap for Carbon Capture Utilization and Storage (CCUS) for Upstream E&P Companies Available online: https://mopng.gov.in/files/article/articlefiles/Draft_UFCC_Roadmap_2030_v3.pdf. Accessed 20 Apr 2023.

Nagar, P. (2022). *Critical analysis of green finance in India* (p. 72). Gurugram University Business Review (GUBR).

OECD. (2008). OECD Annual Report 2008. OECD Publishing, Paris. https://doi.org/10.1787/annrep-2008-en. Accessed 29 Aug 2024.

Onyeaka, H., Miri, T., Obileke, K., Hart, A., Anumudu, C., & Al-Sharify, Z. T. (2021). Minimizing carbon footprint via microalgae as a biological capture. *Carbon Capture Science & Technology, 1*, 100007. https://doi.org/10.1016/j.ccst.2021.100007

Osman, A. I., Hefny, M., Abdel Maksoud, M. I. A., Elgarahy, A. M., & Rooney, D. W. (2020). Recent advances in carbon capture storage and utilisation technologies: A review. *Environmental Chemistry Letters, 19*(2), 797–849. https://doi.org/10.1007/s10311-020-01133-3

Ozkan, M., Nayak, S. P., Ruiz, A. D., & Jiang, W. (2022). Current status and pillars of direct air capture technologies. *Iscience, 25*, 103990.

Petrescu, L., & Cormos, C. C. (2017). Environmental assessment of IGCC power plants with pre-combustion CO_2 capture by chemical & calcium looping methods. *Journal of Cleaner Production, 158*, 233–244.

PMCCC, P. (2008). National Action Plan on climate change.

Pratama, Y. W., & Mac Dowell, N. (2019). *The role and value of CCS in different national contexts.* Report for the Coal Industry Advisory Board (CIAB).

Price, P. N., McKone, T. E., & Sohn, M. D. (2007). *Carbon sequestration risks and risk management.*

REN21. (2020). Renewables 2020 Global Status Report (Paris: REN21 Secretariat).

Rimmer, M. (2021). Carbon capture and storage: Intellectual property, innovation policy, and climate change. In *Carbon capture and storage in international energy policy and law* (pp. 181–203). Elsevier.

Ringrose, P. (2020). *How to store CO_2 underground: Insights from early-mover CCS projects.* Springer. ISBN 978-3-030-33113-9.

Rubin, E. S., Mantripragada, H., Marks, A., Versteeg, P., & Kitchin, J. (2012). The outlook for improved carbon capture technology. *Progress in Energy and Combustion Science, 38*(5), 630–671.

Seetharaman, K. M., Patwa, N., & Saravanan, G. Y. (2021). Breaking barriers in deployment of renewable energy. *Heliyon, 5*, e01166. https://doi.org/10.1016/j.heliyon.2019e01166

Sengupta, S., Adams, P. J., Deetjen, T. A., Kamboj, P., D'Souza, S., Tongia, R., & Azevedo, I. M. (2022). Subnational implications from climate and air pollution policies in India's electricity sector. *Science, 378*(6620), eabh1484.

Shackley, S., & Verma, P. (2008). Tackling CO_2 reduction in India through use of CO_2 capture and storage (CCS): Prospects and challenges. *Energy Policy, 36*(9), 3554–3561.

Shaw, R., & Mukherjee, S. (2022). The development of carbon capture and storage (CCS) in India: A critical review. *Carbon Capture Science & Technology, 2*, 100036.

Shibani, M., Kaushik, D., Surojit, B., Neethu, T., Gowd, E. B., Ajithkumar, T. G., Reddy, C. M., & Rahul, B. (2022). Dual nanomechanics in anisotropic porous covalent organic framework janus-type thin films. *Journal of the American Chemical Society, 144*(1), 400–409. https://doi.org/10.1021/jacs.1c10263

Singh, A. K., Mendhe, V. A., & Garg, A. (2006a). CO_2 sequestration potential of geologic formations in India. In *Proceedings, Trondheim.*

Singh, A. K., Mendhe, V. A., & Garg, A. (2006b). CO_2 storage potential of geologic formations in India. In *8th greenhouse gas technology conference, Trondheim, Norway.*

Smith, L. J., & Torn, M. S. (2013). Ecological limits to terrestrial biological carbon dioxide removal. *Climatic Change, 118*, 89–103. https://doi.org/10.1007/s10584-012-0682-3

Sodiq, A., Abdullatif, Y., Aissa, B., Ostovar, A., Nassar, N., El-Naas, M., & Amhamed, A. (2022). A review on progress made in direct air capture of CO_2. *Environmental Technology & Innovation, 29*, 102991.

Tamiksha, S., Ritu, A., Jalpa, M., Saumya, M., July'2020, Transitioning India's steel and cement industries to low carbon pathways, SNAPFI country study, TEERI.

Tata steel, 2021, Press release, Tata Steel commissions India's first plant for CO_2 capture from Blast Furnace gas at Jamshedpur, 14th Sept'2021. Available online: https://www.tatasteel.com/media/newsroom/press-releases/india/2021/tata-steel-commissions-india-s-first-plant-for-co2-capture-from-blast-furnace-gas-at-jamshedpur/. Accessed 20 Apr 2023.

The mint article, Nov'2021, Most polluted cities in India. https://www.livemint.com/news/india/apart-from-delhi-these-2-indian-metros-also-feature-in-top-10-world-s-most-polluted-cities-11636772613577.html

Thomas, D. M., Mechery, J., & Paulose, S. V. (2016). Carbon dioxide capture strategies from flue gas using microalgae: A review. *Environmental Science and Pollution Research, 23*, 16926–16940. https://doi.org/10.1007/s11356-016-7158-3

Turley, C., Blackford, J., Hardman-Mountford, N., Litt, E., Llewellyn, C., Lowe, D., et al. (2010). Carbon uptake, transport and storage by oceans and the consequences of change. *Issues In Environmental Science and Technology, 29*, 240.

UNECE. (2021). *Carbon capture, use and storage (CCUS)*. Available online: https://unece.org/sites/default/files/2021-03/CCUS%20brochure_EN_final.pdf. Accessed 20 Apr 2023.

United Nations Climate Change, Nov'2022, India's Long-Term Low-Carbon Development Strategy. https://unfccc.int/documents/623511

UNT Digital Library, Aug'2018, *Carbon capture and sequestration (CCS) in the United States*. Available online: https://digital.library.unt.edu/ark:/67531/metadc1248265/m1/. Accessed 20 Apr 2023.

US-DOE, Office of Fossil Energy and Carbon Management, Pre-Combustion Carbon Capture Research, Available online: https://www.energy.gov/fecm/science-innovation/carbon-capture-and-storage-research/carbon-capture-rd/pre-combustion-carbon#:~:text=Pre%2Dcombustion%20capture%20refers%20to,pressure%20to%20form%20synthesis%20gas. Accessed 20 Apr 2023.

van der Meer, L. G. H., & Yavuz, F. (2009). CO_2 storage capacity calculations for the Dutch subsurface. *Energy Procedia, 1*(1), 2615–2622. https://doi.org/10.1016/j.egypro.2009.02.028

Vangkilde-Pedersen, T., et al. (2009). Assessing European capacity for geological storage of carbon dioxide–the EU Geo Capacity project. *Energy Procedia, 1*(1), 2663. https://doi.org/10.1016/j.egypro.2009.02.034

Viebahn, P., Höller, S., Vallentin, D., Liptow, H., & Villar, A. (2011). Future CCS implementation in India: A systemic and long-term analysis. *Energy Procedia, 4*, 2708–2715.

Viebahn, P., Vallentin, D., & Höller, S. (2014). Prospects of carbon capture and storage (CCS) in India's power sector – An integrated assessment. *Applied Energy, 117*, 62–75. https://doi.org/10.1016/j.apenergy.2013.11.054

Viebahn, P., Vallentin, D., & Höller, S. (2015). Prospects of carbon capture and storage (CCS) in China's power sector–An integrated assessment. *Applied Energy, 157*, 229–244.

Vishal, V., Chandra, D., Singh, U., & Verma, Y. (2021a). Understanding initial opportunities and key challenges for CCUS deployment in India at scale. *Resources, Conservation and Recycling, 175*, 105829.

Vishal, V., Verma, Y., Chandra, D., & Ashok, D. (2021b). A systematic capacity assessment and classification of geologic CO_2 storage systems in India. *International Journal of Greenhouse Gas Control, 111*, 103458.

Viswanaathan, S., Perumal, P. K., & Sundaram, S. (2022). Integrated approach for carbon sequestration and wastewater treatment using algal–Bacterial Consortia: Opportunities and challenges. *Sustainability, 14*(3), 1075. https://doi.org/10.3390/su14031075

Wallace, M. (2021). *The US CO_2 enhanced oil recovery survey (EOY 2020)*.

Wang, M., Lawal, A., Stephenson, P., Sidders, J., & Ramshaw, C. (2011). Post-combustion CO_2 capture with chemical absorption: A state-of-the-art review. *Chemical Engineering Research and Design, 89*(9), 1609–1624.

Wilberforce, T., Olabi, A. G., Sayed, E. T., Elsaid, K., & Abdelkareem, M. A. (2021). Progress in carbon capture technologies. *Science of the Total Environment, 761*, 143203.

Wilcox, J. (2012). Introduction to carbon capture. In *Carbon capture*. Springer. https://doi.org/10.1007/978-1-4614-2215-0_1

Yadav, S., & Mondal, S. S. (2022). A review on the progress and prospects of oxy-fuel carbon capture and sequestration (CCS) technology. *Fuel, 308*, 122057.

Yan, Y., Borhani, T. N., Subraveti, S. G., Pai, K. N., Prasad, V., Rajendran, A., et al. (2021). Harnessing the power of machine learning for carbon capture, utilisation, and storage (CCUS)–a state-of-the-art review. *Energy & Environmental Science, 14*(12), 6122–6157.

Climate Response and Nature-Based Solutions: Evaluating the Impact of the Belt and Road Initiative in Latin America

Rafael Ferraz dos Santos, Victor Hugo Souza de Abreu, Sandra Oda, and Andrea Souza Santos

1 Introduction

The Belt and Road Initiative (BRI) was launched by the Chinese government in 2013 and institutionalized in 2015. Over time, it has become increasingly comprehensive, involving more countries from different continents. The primary objective of this initiative is to strengthen political coordination, develop infrastructure in the transportation and energy sectors, eliminate trade barriers, and promote financial cooperation and integration. The BRI has gained significant prominence as a symbol of China's investments abroad, especially in the context of logistical and energy infrastructure. Its central goal is to catalyze Chinese economic activities and promote global connectivity. The BRI aims to address logistical challenges in strategic areas, create new trade opportunities, expand financial reach, and consider environmental and sociocultural spheres (Renwick et al., 2018).

Initially, Latin America was not included in the BRI when the Chinese government published its policy document in 2016. However, in 2017, at the Belt and Road Forum in Beijing, Chinese President Xi Jinping referred to Latin America as a natural extension of the twenty-first Century Maritime Silk Road. The following year, at the China-Community of Latin American and Caribbean States (CELAC) Ministerial Forum, China invited Latin American countries to participate in the BRI (Barbosa & Santos, 2021). As China's presence in Latin America continues to grow, there are still ambiguities and contradictions in this bilateral approach, primarily due to the prevalence of projects characterized by extractive approaches embedded in complex environmental, social, and political contexts.

R. F. dos Santos · V. H. S. de Abreu (✉) · S. Oda · A. S. Santos
Federal University of Rio de Janeiro (Universidade Federal do Rio de Janeiro—UFRJ), Transport Engineering Programme (Programa de Engenharia de Transportes—PET), COPPE-UFRJ, Rio de Janeiro, Brazil
e-mail: victor@pet.coppe.ufrj.br

S. S. Muthu (ed.), *Carbon Footprint Assessments*, Environmental Footprints and Eco-design of Products and Processes,
https://doi.org/10.1007/978-3-031-70262-4_11

Therefore, it is crucial to monitor the development and direction of these initiatives from the outset to either prevent or improve them and ensure sustainable development. For instance, lithium extraction, essential for electric vehicle adoption and predominantly concentrated in South America, illustrates a complex issue. This practice can contradict the fundamental principles of the Sustainable Development Goals (SDGs), specifically Goals 8 (Decent Work and Economic Growth), 9 (Industry, Innovation, and Infrastructure), 11 (Sustainable Cities and Communities), and 13 (Climate Action). The challenge lies in balancing resource extraction with equitable and sustainable practices that align with these global objectives.

In 2021, China implemented the "Opinions on the Establishment and Improvement of the Mechanism for Realizing the Value of Ecological Products," actively promoting nationwide the appreciation of ecological products. The crux of this approach lies in compensating those who safeguard the environment, levying fees on those utilizing it, and imposing fines on those causing harm. Through compensations, usage fees, and penalties, the policy aims to guide regions in adopting sustainable development practices, seeking the convergence of ecological protection and socio-economic progress. Notably, China has recently witnessed several emblematic cases related to the Realization of Ecosystem Value, exemplified by projects like ecological tourism in conservation areas and forest carbon sequestration trading in Zhejiang Province. These projects vividly demonstrate how Nature-Based Solutions (NbS) align with China's vision of valuing ecological products, offering a practical avenue for achieving ecological civilization in China and realizing its "dual carbon" goals (Xu et al., 2023).

Given the evolving dynamics and the increasingly close ties between the Asian giant and Latin American countries, it becomes imperative to scrutinize the purposes, benefits, and potential contradictions arising from this long-standing yet evolving relationship. Over the past 5 years, particularly in the realms of sustainability, environmental preservation, and the mitigation of climate issues, this relationship has taken on new dimensions. Therefore, the objective of this study is to conduct a comprehensive analysis of the BRI and its specific implications for Latin America. The analysis will center on the logistical dimension of the BRI and transportation infrastructure, placing special emphasis on environmental and sustainability aspects such as NbS, the 2030 Agenda, and the Sustainable Development Goals (SDGs). Finally, the study will delve into the geopolitical challenges and tensions stemming from China's escalating involvement in the region, considering the ongoing Sino-American rivalry and the influence of the United States in the area.

The research contributes to the academic discourse, adding depth to the existing literature on Chinese influence in Latin America. It enhances our understanding of the implications of this escalating influence, offering valuable insights that can shape future policies and strategies in Sino-Latin American relations.

2 Methodology

The research methodology employed for this study primarily centered on a comprehensive literature review, as illustrated in Fig. 1. This approach was selected to uphold the research's reliability and validity, as it drew from well-established scientific and academic sources. These sources encompassed a diverse range of materials, including books, articles, theses, and various scholarly works available in the academic literature.

Significantly, a substantial proportion of the selected sources had their publication dates falling within the window of 2019 to 2023. This carefully selected time frame was a deliberate choice aimed at firmly anchoring the research in the latest and most pertinent data available. This becomes particularly essential considering the dynamic nature of the subject matter, which has witnessed remarkable advancements over the preceding 5 years, demanding the most up-to-date information for a comprehensive understanding.

The process of data collection commenced with an initial exploratory reading phase, aimed at aggregating a comprehensive dataset from the identified sources. All selected materials underwent an impartial review to assess their relevance and contribution to the research methodology. In addition to the extensive review, selective reading was employed to complement the overall assessment, ensuring the content's cohesiveness and consistency. This meticulous approach played a pivotal role in confirming the quality and reliability of the materials included in the study. Detailed records were meticulously maintained for each piece of data, starting with the author's name and concluding with the publication date of the source.

In the final phase of the investigation, the amassed information underwent synthesis and analytical reading. This critical step involved a thorough analysis of the data, leading to a deeper comprehension and insight into the subject matter. The

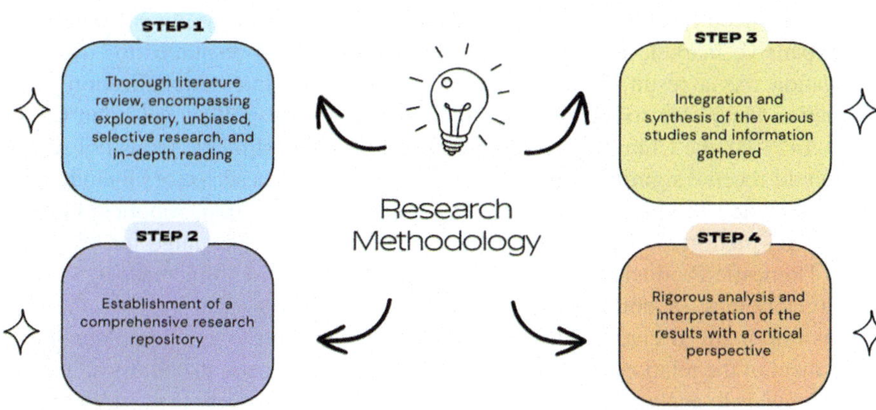

Fig. 1 Methodology

synthesis of this information enabled the development of a well-rounded perspective and a clear understanding of the facts, while also providing insights into potential future developments and implications for all relevant stakeholders.

3 Impact of the Belt and Road Initiative in Latin America: Tensions and Cooperation

The BRI has significant political and economic consequences, but its fundamental principle is to enhance connectivity, both institutional and physical, among countries. Therefore, it would be misguided to analyze it from a narrow perspective of Chinese foreign policy or as a purely commercial endeavor (Abdenur & Levaggi, 2018). In 2017, the first cooperation agreement in the region was inaugurated, with Panama being the pioneer in formalizing its ties with the initiative.

The Central American country, for logistical and transportation reasons, had established diplomatic relations with Beijing only five months prior. In the following years, an additional 18 countries formalized their association, and as seen in Fig. 2, there is a notable concentration in nations geographically oriented toward the Pacific Ocean coasts, reflecting the strategic importance of maritime routes to the west of the continent.

Despite the advancements, gaining the support of the key economies in the region remains a challenge. Notably, Brazil, Colombia, and Paraguay have not formally participated in this cooperative agreement. However, there are indications that this may change, especially in the case of Brazil. The newly elected government of Luiz Inácio Lula da Silva has taken a new direction in foreign policy, resulting in significant progress. Ambassador Celso Amorim, the current Chief of the Special Presidential Advisory Office, stated during his visit to China in March 2023, "There is no reason for Brazil not to join." It is important to note, however, that no formal accession has been finalized (Oliveira, 2023).

The advanced level of technological development in China requires a diversified and expansive market, as well as a wide range of natural resources to sustain its application. As a result, China is moving towards the internationalization of its production factors, introducing geopolitical considerations (Wise & Chonn Ching, 2018). In 2010, the Chinese government identified seven strategic industrial sectors that would receive significant investments and support. These sectors include technologies related to energy efficiency, environmental protection, robotics, biotechnology, 5G platforms, and electric vehicles. These sectors accounted for 8% of the Gross Domestic Product (GDP) in 2015, and it is expected to increase to 15% by 2020 (Gabriele & Jabbour, 2021).

It is important to clarify that this chapter does not aim to provide a qualitative assessment of the relationship between China and Latin American countries. Instead, its objective is to analyze how the progress of the BRI aligns with the historical need to address infrastructure integration gaps in this region. It is worth noting that the

Fig. 2 Chinese partnerships in Latin America linked to the BRI. (Source: Based on Barbosa and Santos (2021))

analysis presented here contradicts theories that view China's growing economic presence in peripheral nations as an example of neocolonialism or neo-imperialism. Such arguments often conflate asymmetric relationships with domination (Pautasso et al., 2020).

As former Chilean Finance Minister Felipe Larraín stated, "For much of Latin America, growing ties with China were driven, above all, by trade and broader economic considerations, not geopolitics" (Larraín & Zhang, 2021). This can largely explain why Chinese presence is perceived with less hostility in Latin America than might be expected when compared to a European or North American perspective (Hogenboom et al., 2022).

4 China, Latin America, and the Perspective of Sustainable Development with Nature-Based Solutions

Currently, all issues related to the Belt and Road Initiative (BRI) are being considered and implemented within the framework of "green development" concepts, fostering significant interdisciplinary efforts to mitigate BRI-related risks (Harlan, 2021). These actions are paramount given that BRI infrastructure often intersects with key conservation priority areas (Narain et al., 2020). Nonetheless, such efforts have yet to translate into substantial action. As of 2020, 65 BRI financiers have been identified, comprising 35 Chinese and 30 international entities, among which only 17 mandate measures to mitigate biodiversity impacts and 12 stipulate achieving a "net gain" (Narain et al., 2020). While China accounts for over 90% of BRI financing in terms of investment value, only one of these investments includes provisions for biodiversity protection (Narain et al., 2020). This financial disproportionality is evident in regions where there is a complete absence of environmentally friendly BRI projects, resulting in a "greenwashing" phenomenon (Harlan, 2021). The urgent commitment of all BRI financiers to conservation and impact mitigation measures is imperative (Narain et al., 2020), ensuring that the BRI's implementation genuinely aligns with green principles (Harlan, 2021).

The incorporation of Nature-based Solutions (NbS) into the environmental cooperation program of China's Belt and Road Initiative (MEE, 2020) reflects China's commitment to the NbS approach at both national and international levels. The concept of NbS first emerged in 2008 to underscore the significant role of biodiversity and ecosystem services in mitigating climate change (World Bank, 2008). Since then, the definition of NbS has been developed and concretized with an ever-deepened understanding of the interdependence between humans and nature, as well as the interconnected challenges of society, by international institutions and organizations (United Nations, 2019). The Chinese Council for International Cooperation on Environment and Development (CCICED), a government-affiliated think tank, recommends the adoption of NbS to achieve domestic climate and biodiversity goals (CCICED, 2019, para. 4.7; CCICED, 2020). Furthermore, China's 2021 Guideline Opinion for climate action and environmental protection now mandates subnational governments to utilize the "general principles" of NbS to protect biodiversity, restore ecosystems, and mitigate and adapt to climate change (MEE, 2021).

For Chinese policymakers, it is important to emphasize that the concept of NbS aligns well with the country's long-standing policies for the management of natural resources and ecosystems. Currently, there are six large-scale reforestation projects across China. These projects aim, in part, to prevent desertification, soil erosion, and dust storms in arid and semi-arid regions (Qiu et al., 2017) and, in part, seek to enhance ecosystem services, improve forest management, protect biodiversity, and strengthen climate resilience. Particularly noteworthy for its scale is the Great Green Wall Program for the North, initiated in 1978 with a scheduled completion date of

2050. This program currently spans 13 provinces and includes at least half a dozen reforestation and land-use projects to sequester carbon dioxide and mitigate climate change (Zhang et al., 2017).

President Xi Jinping's metaphorical expression, highlighting the value of "clear waters and lush mountains as invaluable assets," forms the cornerstone of the Chinese government's strategy in the pursuit of an ecological civilization. Confronted with a rise in domestic greenhouse gas emissions, a burgeoning public demand for a cleaner environment, and local governments' hesitancy to regulate high-emission industries, the central government, led by President Xi, has taken a more proactive stance toward environmental and ecological protection (Liu, 2021).

Cities play a crucial role in accelerating actions to tackle two critical environmental challenges: strengthening the climate resilience of populations and critical infrastructure while preserving and creating habitats to enhance biodiversity. Key documents such as the New Urban Agenda (UN-Habitat, 2020) and the Regional Action Plan for the implementation of the New Urban Agenda in Latin America and the Caribbean (Economic Commission for Latin America and the Caribbean [ECLAC], 2018), along with the Paris Agreement on Climate Change (United Nations Framework Convention on Climate Change [UNFCCC], 2015) and the Post-2020 Global Biodiversity Framework (Convention on Biological Diversity [CBD], 2021), emphasize the need to integrate nature into urban planning and design, recognizing the potential of cities to meet environmental commitments.

This integration can be exemplified by the immense potential of Nature-Based Solutions (NBS), which play a significant role in enhancing natural capital for the benefit of communities. As highlighted by Temmerman et al. (2013), innovation in NBS can be achieved by clustering them in an inclusive environment, empowering and integrating the community. Innovative pilot projects, such as urban farms and the regeneration of urban streams and green areas, incorporate systems like rain gardens, cork material green roofs, and vertical farming. These initiatives promote the removal of water pollutants and connect natural water bodies, providing significant social benefits.

Additionally, as identified by Fan et al. (2017), five types of NBS will be implemented to strengthen the social and economic dynamics of urban communities: (i) creating social ties through a permanent NBS committee, (ii) using NBS to professionally train and provide job opportunities for local citizens, associations, and small and medium enterprises, (iii) raising awareness of urban environmental dynamics through structured educational programs in partnership with local schools and associations, (iv) creating a "Communiversity," and (v) promoting a circular economy among larger social classes. As suggested by Artmann and Sartison (2018), promoting urban agriculture around roadways can also mitigate urban heat, improve air quality, and enhance local food security, while assisting in stormwater management and promoting biodiversity.

These approaches highlight how cities can be platforms for innovation in sustainable and inclusive practices, addressing both climate change and biodiversity loss issues simultaneously, and demonstrating urban adaptability by strategically incorporating nature into urban development.

Although some mitigation projects in Latin America have shown positive outcomes for social adaptation and some adaptation initiatives have contributed to increased carbon stocks, the full exploration of synergies between these strategies has yet to be achieved (Locatelli et al., 2011). Furthermore, climate and forest policies in the region rarely address the interconnections between adaptation and mitigation in the forestry sector, which directly interacts with the urban environment. Forests are critical for climate adaptation and mitigation, providing essential ecosystem services for local adaptation and global carbon sequestration, which is crucial for mitigation strategies. Understanding the interactions between adaptation and mitigation is critical for integrating these approaches into climate and forest policies, a debate reflected in the literature on the synergies and dilemmas between these two approaches, especially at the global level (Locatelli, 2010; McKibbin & Wilcoxen, 2004; Swart & Raes, 2008). Given the implementation of these strategies throughout Latin America, it is vital to investigate how they interact, to understand potential synergies and trade-offs, and how they relate to development plans and institutional structures, aiming to optimize their effectiveness (Klein et al., 2005).

China's presentation of NbS has undeniably generated benefits over the past four decades, showcasing advancements in tree planting in arid and semi-arid regions, improvements in landslide and flood control, and the sequestration of significant amounts of carbon dioxide. Nevertheless, it is imperative to maintain a critical perspective on how China's leadership interprets and implements the concept of NbS. When defined with precision and robustness, NbS holds the potential to drive global sustainability. However, if integrated merely as a broad concept into existing policies, it may end up merely renaming conventional approaches to environmental management (Qi & Dauvergne, 2022).

China's international aspirations and diplomatic engagements have the potential to leverage the benefits proposed by NbS initiatives. Considering Latin America's geographical diversity and its capacity for a cleaner energy matrix, coupled with a geopolitically less tumultuous environment compared to other regions, an opportune scenario emerges for achieving multilateral goals. As emphasized in this chapter, Latin America plays a pivotal logistical and commercial role, suggesting that transportation infrastructure will prominently feature on the agenda in the coming years.

5 Infrastructure in Latin America: The Impact of BRI and Development Outlook

China's increasing economic influence in Latin America is evident through various investment projects, particularly in the infrastructure and transportation sectors (Jenkins, 2017; Liang, 2019). Among these projects are proposals for a canal in Nicaragua that would compete with the Panama Canal, as well as substantial investments in ports such as Balboa in Panama and Santos in Brazil, improving the region's transportation and trade capabilities. These projects, along with others that will be monitored and analyzed throughout this study, highlight China's growing impact on the development of Latin America's infrastructure.

The Water Black Tunnel Project was a plan to connect the Chilean port of Coquimbo to the Argentine province of San Juan by crossing the Andes Mountains. The project serves as an excellent example of how external collaboration could bring about significant benefits. Unfortunately, the planning was suspended in 2018 due to concerns over its cost viability. However, in 2022, following a meeting between the presidents of Chile and Argentina, Gabriel Boric and Alberto Fernández, respectively, the need for an interoceanic corridor between the two countries was emphasized. It is unclear whether they were referring to the Water Black Tunnel Project. Nevertheless, better connectivity between the regions would result in a shorter route to the Pacific Ocean, which is of paramount logistical-strategic importance. This is particularly true for countries such as China, whose transportation of goods on maritime routes is mostly concentrated in areas of constant political and military tension, like the Strait of Hormuz and the Strait of Malacca.

China's prioritization of infrastructure in its foreign agenda is supported by a range of economic, political, and strategic factors. Firstly, infrastructure construction is seen as a key driver of economic growth in China, as exemplified by the popular Chinese saying, "If you want to get rich, build a road." Secondly, infrastructure construction aligns with the global sustainable development agenda, which is supported by various international organizations such as the G20, the New Development Bank (NDB) of BRICS (Brazil, Russia, India, China and South Africa), and the International Monetary Fund (IMF). As a result, China views infrastructure cooperation to promote sustainable development not only in its own economy but also in the relevant region.

Third, China has a significant advantage in terms of experience, financing capacity, and industrial competence in the infrastructure sector, which has prompted it to seek global expansion of its capabilities through initiatives such as the BRI. Finally, Latin America is grappling with a significant infrastructure deficit that is hampering its economies. For example, high transportation costs in countries like Brazil have been identified as critical factors in labor market friction, affecting labor mobility between sectors and the integration of domestic labor markets. As a result, many policymakers in Latin America recognize the value of infrastructure for sustainable development and have expressed interest in receiving Chinese investments to address this gap. As economic ties deepen and China gains a better understanding of the region, there are increasingly strong economic incentives for China to invest in Latin American infrastructure (Niu, 2018).

The Bi-Oceanic Railway project is a long-awaited initiative that aims to connect the Atlantic coast of Brazil with the Pacific coast of Peru. The Brazilian government presented this project to the United States in 2008, hoping for possible cooperation. Recently, representatives from China, Peru, and Brazil have agreed to initiate the feasibility study for the project during Chinese Premier Li Keqiang's visit to Brazil in 2015, which has brought new hope for the project's realization. The project promises significant economic benefits by greatly enhancing regional interconnectivity in Latin America, a region with a history of limited intra-regional trade due to infrastructure deficiencies. The railway will reduce domestic transportation costs by providing access from agricultural production areas to seaports. Additionally, the railway could stimulate trade between Asia and Latin America, making the latter more competitive compared to other regions like Australia by reducing logistical costs (Niu, 2018).

In 2019, Peru solidified its strategic relationship with China by participating in the Chinese Foreign Direct Investment (FDI) initiative. Peru is the second-largest destination for Chinese FDI in the region. The country expects investments in projects such as the Chancay mega-port and others in Ilo and is even considering the possibility of resuming the bi-oceanic railway project. This relationship between Peru and China has been developing throughout the twenty-first century (Liévano, 2019).

Colombia has a unique geographical advantage as it has access to both the Pacific Ocean and the Caribbean Sea, which is considered underutilized by many. The Colombian government's fourth-generation infrastructure program features "Mar 2," the newest highway that connects the Antioquia region to the southwest of the Pacific and Buenaventura, Colombia's largest port. Buenaventura is responsible for 60% of the country's incoming and outgoing goods. The funding for "Mar 2" was secured through a public-private partnership (PPP) with Chinese and Japanese banks. This agreement, signed in 2015, was the first infrastructure PPP in Latin America involving a Chinese financier. The Autopistas Urabá consortium, consisting of the China Harbour Engineering Company (Chec) and five Colombian construction companies, is responsible for constructing the road (Cañón, 2022).

6 Fostering Sustainable Development: China, Latin America and the 2030 Agenda

The United Nations' 2030 Agenda for Sustainable Development was adopted in 2015 by all 193 UN member states to create a global plan for international cooperation. This plan comprises 17 Sustainable Development Goals (SDGs), covering areas such as gender equality, health, poverty eradication, and climate action. It presents vital opportunities for collaboration between China and Latin America. However, achieving these goals requires joint efforts to overcome challenges such as financing, capacity building, and data collection.

In the first 5 years of the implementation of the SDGs, some aspects improved considerably. According to the United Nations Department of Economic and Social Affairs (UNDESA), the number of people living in extreme poverty was decreasing, reaching 8.6% in 2018, compared to 16.0% in 2010 and 36.0% in 1990. In terms of infrastructure, efforts were emphasized in the energy sector. International investments in renewable energy in developing countries also increased significantly, from USD 9.9 billion in 2010 to USD 18.6 billion in 2016, with particularly positive progress for wind, geothermal, and solar energy (Chrzanowski & Zawada, 2023).

The Decade of Action from 2020 to 2030 was supposed to be dedicated to accelerating solutions to meet the goals of the 2030 Agenda and achieve a more sustainable world. Unfortunately, only a few months into this new decade, a global pandemic occurred. Even before the devastating impact of COVID-19, the UN was warning that global efforts were not sufficient to bring about the necessary changes, despite previous reports indicating progress and positive trends in important areas. However, since the onset of the pandemic, progress reports have taken a negative turn. A significant volume of investments had to be redirected to combat the pandemic, thereby disrupting other essential healthcare services.

All countries were compelled to adjust their priorities, resulting in setbacks for environmental and climate-related initiatives. The current status of the implementation of the goals, according to the latest United Nations SDGs Report and the Sustainable Development Report, indicates mostly unsatisfactory results (Sachs et al., 2019; United Nations, 2022). The situation regarding Industry, Innovation, and Infrastructure (SDG 9), for example, seems slightly more favorable, although progress after the pandemic has been uneven. Climate Action (SDG 13) demands urgent initiatives to combat global changes. After the decline in emissions observed during the first pandemic year, emissions from energy combustion and industrial processes reached their highest level, with the global average temperature continuing to increase, along with more frequent and intense extreme weather events. Amid challenging scenarios, one particularly positive trend is seen in Sustainable Cities and Communities (SDG 11). Between 2015 and 2021, the number of countries with disaster risk reduction strategies doubled, reaching a total of 98 (Shulla & Leal-Filho, 2023).

The current Latin American context has been significantly influenced by the repercussions of the recent neoliberal wave and the ongoing challenges posed by the COVID-19 pandemic. Consequently, any collaborative efforts between these regions appear to underscore priorities such as poverty eradication (SDG 1), health and well-being (SDG 3), reduction of inequalities (SDG 10), sustainable industry, innovation, and infrastructure (SDG 9), sustainable cities and communities (SDG 11), and climate action (SDG 13). This multifaceted approach reflects a comprehensive commitment to address diverse sustainable development goals, acknowledging the interconnectedness of economic, social, and environmental dimensions.

To promote sustainable development and achieve beneficial results, the BRI can learn from the 2030 Agenda and the SDGs, as shown in Fig. 3. The BRI should incorporate them into its development in the Latin America region, considering China's development experience that can serve as a reference for the countries in the region.

One of the most crucial factors that have contributed to China's economic growth and social development is investment in basic infrastructure and industry. This investment has addressed essential aspects such as transport connectivity, market integration, industrialization, investment opportunities and capital attraction, innovation, mobilization of human resources, job creation, and improved public services. The BRI should provide additional opportunities to promote beneficial outcomes and strengthen South-South and multilateral cooperation that can benefit countries along

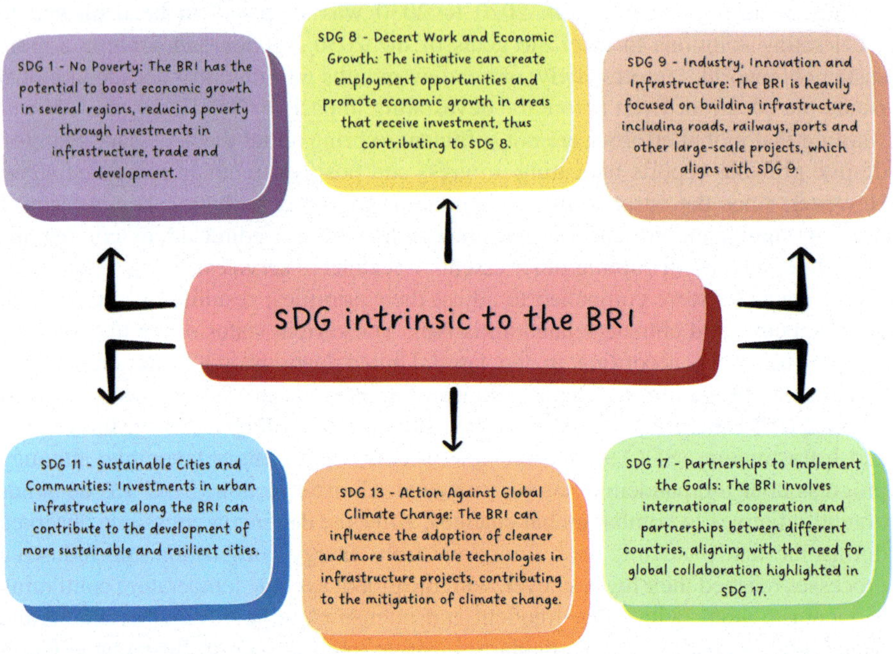

Fig. 3 SDGs that can be strengthened by promoting the BRI

the BRI, including those in the Latin America region. It should focus on connecting unconnected road segments, building a community with shared interests, responsibilities, and destinies. This way, countries along the BRI can strengthen their political trust and economic integration and provide a platform for coordinating the development strategies of diverse stakeholders, including major powers and international and regional institutions (Li & Zhu, 2019).

7 Conclusion

The BRI in Latin America presents a complex array of opportunities and challenges. On the one hand, Chinese investments and the infrastructure brought about by the initiative have the potential to boost economic development in the region, improve connectivity, and reduce logistical costs. These positive outcomes could yield benefits for Latin American countries and contribute to the achievement of Sustainable Development Goals. Additionally, China's non-interventionist approach to its relations with Latin America stands in stark contrast to the history of foreign interference in the region. However, the Chinese presence also raises concerns, particularly regarding extractive projects that could have negative impacts on the environment and local communities.

During the 15th summit of the BRICS bloc held in August 2023 in South Africa, the expansion of the group was formalized. The bloc, which includes emerging economies such as Brazil and China, plans to increase its membership from 5 to 11. Six countries, namely Argentina, Egypt, Iran, Ethiopia, Saudi Arabia, and the United Arab Emirates, have been invited to join the group from January 1, 2024, subject to meeting the founders' conditions. The new configuration has geopolitical and economic implications, aiming to foster access to capital, crucial elements for development strategies. The planning project outlines guidelines to expand trade agreements, secure routes, and promote investments in industrial production strategies, prioritized during the last summit in August.

The implications of this new arrangement are intrinsically linked to China's strategy with the BRI. The expansion broadens China's sphere of influence in the Middle East and North Africa, making it a significant geopolitical player in a key region from a logistical standpoint, given the location of relevant transit points such as the Suez Canal and the Strait of Hormuz. This is evidenced by the promotion of a historic meeting, led by the Chinese, between Saudi Arabia and Iran. In the Latin American context, the initially planned inclusion of Argentina would have expanded the continent's influence in the bloc. However, with the rise of a new government, the country made the surprising decision not to be part of the bloc.

Beyond the geopolitical landscape, a nuanced perspective on progress toward the SDGs was presented, and as observed, we can conclude that despite significant advancements in the initial 5 years, the profound impacts of the pandemic exposed vulnerabilities and underscored the need for more resilient actions to successfully achieve the SDGs. Looking ahead, the pursuit of effective implementation demands not only a resurgence of pre-pandemic momentum but also the integration of innovative and adaptive strategies to address emerging challenges. It is imperative to maintain an unwavering focus on building a more sustainable future, considering lessons learned and adapting strategies for a post-pandemic world. This holistic approach will be instrumental in overcoming obstacles and realizing the vision outlined in the SDGs for a globally sustainable and equitable future.

This evolving scenario remains dynamic, with diverse theories and projections offering a rich foundation for discussions and numerous opportunities for further study, analysis, and research. This holds true for both the BRICS perspective and the potential impacts on BRI initiatives. In conclusion, ongoing research and vigilant monitoring of the unfolding dynamics in Sino-Latin American relations are pivotal for shaping future policies, decisions, and investments. Latin America occupies a strategic position, not only logistically but also environmentally, presenting a unique opportunity for reaping benefits from its engagement with China. However, strategic alignment with its own interests and sustainable development goals is essential for maximizing the positive outcomes of this engagement.

References

Abdenur, A. E., & Levaggi, A. G. (2018). Trans-regional cooperation in a multipolar world: How is the belt and road initiative relevant to Latin America? *London School of Economics Global South Unit: Working Paper Series, 1*, 3–18.

Artmann, M., & Sartison, K. (2018). The role of urban agriculture as a nature-based solution: A review for developing a systemic assessment framework. *Sustainability, 10*(6), 1937. https://doi.org/10.3390/su10061937

Barbosa, G. N., & Santos, B. S. F. (2021). Is there a place for Latin America on the Silk Road: Perspectives of Chinese foreign policy. *6th Undergraduate and Postgraduate Seminar of the Brazilian Association of International Relations.*

Cañón, L. N. C. (2022). *China-financed highway eases Colombia's infrastructure bottlenecks.* Diálogo Chino.

CCICED – China Council for International Cooperation on Environment and Development. (2019). *Strengthening the new green consensus and promoting China's 14th Five-Year Plan.* Quality Development: CCICED policy recommendation to the Chinese Government 2019.

CCICED – China Council for International Cooperation on Environment and Development. (2020). *From recovery to green prosperity—accelerating the promotion of green high-quality development during the 14th Five-Year Plan.*

Chrzanowski, M., & Zawada, P. (2023). Fraction separation potential in the recycling process of photovoltaic panels at the installation site—A conceptual framework from an economic and ecological safety perspective. *Energies, 16*, 2084. https://doi.org/10.3390/en16052084

Convention on Biological Diversity—CBD. (2021). *Decision V/6 Ecosystem Approach. Decision adopted by the Conference of Parties to the Convention on Biological Diversity at its fifth meeting, 15–26* May 2000*, Nairobi, Kenya.* Available at www.cbd.int/decision/cop/default.shtml?id=7148

Economic Commission for Latin America and the Caribbean – ECLAC. (2018). *Plan de Accion Regional para la implementacion de la nueva agenda urbana en America Latina y el Caribe, 2016–2036.* Available at: https://onuhabitat.org.mx/index.php/plan-de-accion-regional-para-la-implementacion-de-la-nau-en-lac

Fan, P., Ouyang, Z., Basnou, C., Pino, J., Prak, H., & Chen, J. (2017). Nature-based solutions for urban landscapes under post-industrialization and globalization: Barcelona versus Shanghai. *Environmental Research, 156*, 272–283. https://doi.org/10.1016/j.envres.2017.03.045

Gabriele, A., & Jabbour, E. (2021). *China: The socialism of the 21st century* (1st ed.). Boitempo Editorial.

Harlan, T. (2021). Green development or greenwashing? A political ecology perspective on China's green Belt and Road. *Eurasian Geography and Economics, 62*(2), 202–226. https://doi.org/10.1080/15387216.2020.1795700

Hogenboom, B., Baud, M., Gonzalez, R., & Steinhöfel, D. (2022). *China's economic and political role in Latin America.* University of Birmingham.

Jenkins, R. (2017). China and Latin America. In *American hegemony and the rise of emerging powers* (pp. 182–197). Routledge.

Klein, R. J. T., Schipper, E. L. F., & Dessai, S. (2005). Integrating mitigation and adaptation into climate and development policy: Three research questions. *Environmental Science & Policy, 8*(6), 579–588. https://doi.org/10.1016/j.envsci.2005.06.010

Larraín, F., & Zhang, P. (2021). How Latin America can navigate the China-US trade wars. *Americas Quarterly.*

Li, Y., & Zhu, X. (2019). The 2030 agenda for sustainable development and China's Belt and Road initiative in Latin America and the Caribbean. *Sustainability, 11*(8), 2297. https://doi.org/10.3390/su11082297

Liang, W. (2019). Pulling the region into its orbit? China's economic statecraft in Latin America. *Journal of Chinese Political Science, 24*(3), 433–449. https://doi.org/10.1007/s11366-018-09603-w

Liévano, A. B. (2019). *How is the belt and road progressing in the Andean region?* Diálogo Chino.
Liu, J. (2021). Analysis: Nine key moments that changed China's mind about climate change. *CarbonBrief*.
Locatelli, B. (2010). *Local, global: Integrating mitigation and adaptation*. Perspective forests/ climate change no. 3. CIRAD, Paris.
Locatelli, B., Evans, V., Wardell, A., Andrade, A., & Vignola, R. (2011). Forests and climate change in Latin America: Linking adaptation and mitigation. *Forests, 2*(1), 431–450. https://doi.org/10.3390/f2010431
McKibbin, W. J., & Wilcoxen, P. J. (2004). *Climate policy and uncertainty: The roles of adaptation versus mitigation*. Economics and environment network working paper EEN0306, The Brookings Institution, Washington, DC.
MEE – Ministry of Ecology and Environment of China. (2020). BRI International Green Development Coalition (BRIGC).
MEE – Ministry of Ecology and Environment of China. (2021). *Guiding opinions on coordinating and strengthening the work related to climate change and ecological environmental protection*.
Narain, D., Maron, M., Teo, H. C., Hussey, K., & Lechner, A. M. (2020). Best-practice biodiversity safeguards for belt and road Initiative's financiers. *Nature Sustainability, 3*(8), 650–657. https://doi.org/10.1038/s41893-020-0528-3
Niu, H. (2018). A strategic analysis of Chinese infrastructure projects in Latin America and the Caribbean. Building Development for a New Era. In *Chinas Infrastructure Projects in Latin America and the Caribbean* (pp. 180–193).
Oliveira, M. (2023). *Brazil and China: On the agenda, Belt and Road, and partnerships for reindustrialization*. Monitor Mercantil.
Pautasso, D., Nogara, T. S., Ungaretti, C. R., & Doria, G. (2020). The Belt and Road initiative and the dilemmas of Latin America. *Tempo do Mundo Magazine*.
Qi, J. J., & Dauvergne, P. (2022). China and the global politics of nature-based solutions. *Environmental Science & Policy, 137*, 1–11. https://doi.org/10.1016/j.envsci.2022.08.008
Qiu, B., Chen, G., Tang, Z., Lu, D., Wang, Z., & Chen, C. (2017). Assessing the Three-North Shelter Forest Program in China by a novel framework for characterizing vegetation changes. *ISPRS Journal of Photogrammetry and Remote Sensing, 133*, 75–88. https://doi.org/10.1016/j.isprsjprs.2017.10.003
Renwick, N., Gu, J., & Gong, S. (2018). *The impact of BRI investment in infrastructure on achieving the sustainable development goals. K4D Emerging Issues Report*. Institute of Development Studies.
Sachs, J. D., Schmidt-Traub, G., Mazzucato, M., Messner, D., Nakicenovic, N., & Rockström, J. (2019). Six transformations to achieve the sustainable development goals. *Nature Sustainability, 2*(9), 805–814. https://doi.org/10.1038/s41893-019-0352-9
Shulla, K., & Leal-Filho, W. (2023). *Achieving the UN agenda 2030: Overall actions for the successful implementation of the sustainable development goals before and after the 2030 deadline*. https://doi.org/10.2861/295344
Swart, R., & Raes, F. (2008). Making integration of adaptation and mitigation work: Mainstreaming into sustainable development policies? *Climate Policy, 7*(4), 288–303.
Temmerman, S., Meire, P., & Bouma, T. J. (2013). Ecosystem-based coastal defence in the face of global change. *Nature, 504*, 79–83. https://doi.org/10.1038/nature12859
UN-Habitat. (2020). *La Nueva Agenda Urbana*. Available at: https://publicacionesonuhabitat.org/onuhabitatmexico/Nueva-Agenda-Urbana-Ilustrada.pdf
United Nations. (2019). *Human development and planetary health:The role of nature-based solutions, United Nations*. Available at: https://hdr.undp.org/content/human-development-and-planetary-healththe-role-nature-basedsolutions
United Nations. (2022). *Young people key for COVID recovery and a sustainable future*.
United Nations Framework Convention on Climate Change – UNFCCC. (2015). *Paris Agreement*. Available at: https://unfccc.int/sites/default/files/spanish_paris_agreement.pdf

Wise, C., & Chonn Ching, V. (2018). Conceptualizing China–Latin America relations in the twenty-first century: The boom, the bust, and the aftermath. *The Pacific Review, 31*(5), 553–572. https://doi.org/10.1080/09512748.2017.1408675

World Bank. (2008). *Biodiversity, climate change, and adaptation: Nature-based solutions from the world bank portfolio.* Washington, DC.

Xu, H., Chen, S., Guo, Y., Jiang, J., Wang, B., Xu, J., et al. (2023). Searching for the nano-hertz stochastic gravitational wave background with the chinese pulsar timing array data release. *Research in Astronomy and Astrophysics, 23*(7), 075024. https://doi.org/10.1088/1674-4527/acdfa5

Zhang, X., Du, J., Huang, T., Zhang, L., Gao, H., Zhao, Y., & Ma, J. (2017). Atmospheric removal of PM2.5 by man-made Three Northern Regions Shelter Forest in Northern China estimated using satellite retrieved PM2.5 concentration. *The Science of the Total Environment, 593-594*, 713–721. https://doi.org/10.1016/j.scitotenv.2017.03.206

Implementing Circular Economy Principles in Road Transport: Enhancing Efficiency to Mitigate Resource Depletion and Carbon Footprint

Victor Hugo Souza de Abreu

1 Introduction

Currently, growing concerns about climate change and excessive consumption of natural resources have put sustainable development under the spotlight (Litman, 2007). Urban planning for sustainable development focuses on the balance between economic, social, and environmental activities (Chen et al., 2006) and develops and anticipates the needs of cities, identifying the most pressing issues and ensuring that the measures needed to solve their problems are implemented as efficiently as possible (Santos et al., 2021; De Assis et al., 2022c).

In this context, road transport in the urban environment plays a key role in efficient sustainable development (De Assis et al., 2023) because it promotes economic growth, social inclusion, urban expansion, the distribution of activities, and equity in the use of urban space and all utilities (De Abreu et al., 2022, 2023). However, the transportation sector is a major consumer of fossil fuels (for example, to fuel traditional cars with internal combustion engines) (Hickmana et al., 2010; Santos et al., 2021), an important contributor to the use of urban space and all utilities (De Abreu et al., 2023), an important contributor to the increase in Greenhouse Gas (GHG) emissions, which leads to global warming and consequently climate change (Zhao et al., 2019; Bonsu, 2020), and air pollution, which is recognized as the main global risk factor for health problems related to the respiratory system and death (Apte et al., 2017).

In addition, the transport sector is a major generator of end-of-life waste from the constituent components of vehicles (Zhou et al., 2019), such as end-of-life tires

V. H. S. de Abreu (✉)
Federal University of Rio de Janeiro (Universidade Federal do Rio de Janeiro – UFRJ), Transport Engineering Program (Programa de Engenharia de Transportes – PET), COPPE-UFRJ, Rio de Janeiro, Brazil
e-mail: victor@pet.coppe.ufrj.br

S. S. Muthu (ed.), *Carbon Footprint Assessments*, Environmental Footprints and Eco-design of Products and Processes,
https://doi.org/10.1007/978-3-031-70262-4_12

(Gigli et al., 2019) and electric vehicle batteries. To make matters worse, End-of-Life Vehicles are classified as hazardous waste and have the potential to pollute the environment if not managed properly (Simic, 2016), representing a category of waste whose processing is especially difficult because of its complex structure and varied composition (Karagoz et al., 2019). Thus, one response to contemporary threats to the environment and the quality of life of the urban population is the Circular Economy (CE) (Ghisellini et al., 2016. Saidani et al., 2018; De Jesus & Mendonça, 2018), which can be understood as a system solution that aims to mitigate the adverse environmental impact of the production and consumption of goods and services, especially in the context of reducing GHG emissions and waste generation (Seroka-Stolka & Ociepa-Kubicka, 2019).

The reason is that CE is an economic system that supports more sustainable management of resources throughout the life cycle of systems and is characterized by closed loops rather than a linear system (De Abreu et al., 2022a, b), promoting more sustainable practices such as reuse, remanufacturing, and recycling (Saidani et al., 2019; Albuquerque et al., 2019). For example, remanufacturing costs to restore a product to its original condition are usually lower compared to the material and production costs of the new product (Prendeville et al., 2016). This raises the following question: what is the current state of research applying CE to minimize the impacts of the urban road transport sector on the environment? Therefore, the aim of this chapter is to present a general overview of studies applicable to the topic, through a systematic review, which uses careful filters to include and qualify studies, with a bibliometric approach.

This chapter seeks to provide a more general overview of the application of CE in the urban road transport sector, which is still lacking in the literature, presenting a broader scope than the studies currently published, which direct their efforts to a specific field of transport engineering. This is necessary in order to understand which are the main areas of application on the subject, as well as to direct learning efforts on the subject. To achieve its objectives, this study is structured as follows. Section 1 presents a contextualization of the subject, the problem, and the research objectives. Section 2 deals with the importance of using CE concepts in the transportation sector. Section 3 presents and describes the methodological procedure. Section 4 presents and discusses the bibliometric and systematic results. Finally, Sect. 5 contains the concluding remarks.

2 The Transportation Sector and the Circular Economy

Transportation heralds the development of a region because it is a logistical element that allows the mobility of people, goods, and services in a greater geographical dispersion between productive activities (De Assis et al., 2022b). Transport, despite being considered a middle activity from an economic point of view, is currently seen as an influential factor in regional development, proving to be an important axis for the planning, development, and management of cities (De Assis et al., 2022a), being considered the lifeblood of cities in recognition of this critical role (Vuchic, 1999). However, much of the energy currently consumed in the transportation sector comes from non-renewable fossil fuels, with high levels of GHG emissions and air pollutants (Bonsu, 2020). Globally, the transportation sector is responsible for 23% of total energy-related GHG emissions and 13% of global GHG emissions (UN-Habitat, 2011) from fossil fuel consumption, with road transport accounting for the majority of these emissions (OECD, 2010).

The road transport sector also remains a major source of air pollution through exposure to particulate matter, which contributed to 4.1 million deaths from heart disease, stroke, lung cancer, chronic lung disease, and respiratory infections in 2016 (HEI, 2018). However, despite these negative externalities, it has been observed that the efforts made by the global transport industries in recent years to gradually decarbonize the sector remain modest and are not sufficient (Zhang et al. 2019; Hasan et al., 2019). Therefore, without the implementation of aggressive GHG mitigation policies, transport emissions could increase at a faster rate than emissions from any other energy end-use sector and reach around 12 Gigatons (Gt) of carbon dioxide equivalent per year (CO_2e/year) by 2050 (IPCC, 2014).

In addition, the scarcity of consolidated vehicle recycling systems and the large volume of End-of-Life Vehicles are a problem for efficient resource management in the transportation sector (Karagoz et al., 2019). End-of-life vehicles consist of a vehicle that is discarded by its owner and registered as waste (Simic, 2013). As the automotive sector generates around 5% of industrial waste worldwide (Simic, 2013), recycling End-of-Life Vehicles is not only an environmental issue to be addressed, but also a source of funding for industries (Karagoz et al., 2019).

There are two losses involved in the lack of proper recycling and the placement of all waste in landfills: (i) the cost of disposal and (ii) the cost of not selling this waste (Da Silva, 2018). For the disposal management of End-of-Life Vehicles and waste oil, national guidelines and regulations can be developed and enforced as a means of reducing environmental impacts (Dri et al., 2018). Around 75% of vehicle components are recyclable, especially metal and aluminum, while 25% are characterized as hazardous waste (Simic, 2013).

Thus, one of the most effective ways to mitigate these negative effects of transportation, without inhibiting economic development, is to implement CE concepts in this sector (Moore et al., 2020). The reason is that CE presents a direct response to growing concerns about environmental impacts associated with GHG emissions and waste generation (Seroka-Stolka & Ociepa-Kubicka, 2019; Geerken et al., 2019), which in the context of transportation comes mainly from fuel consumption and excessive vehicle production. Unlike the Linear Economy (characterized by producing, consuming, and discarding), CE offers an alternative perspective for recovering materials and returning them to productivity as part of a sustainable future (Tam et al., 2019).

CE, in general, promotes cyclical flows of resources and services in the production-consumption system, as illustrated in Fig. 1. The system is designed to be restorative and regenerative in itself, as in the cycle shown, and can be applied at different urban scales (Charonis, 2012). An extended service life leads to longer

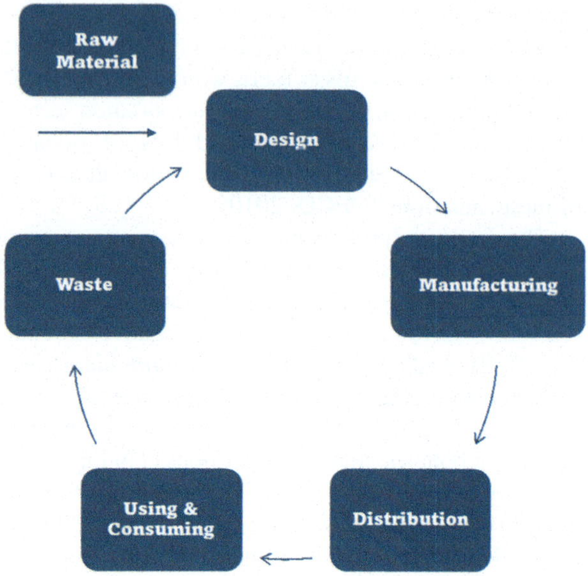

Fig. 1 Cycle of CE. (Source: Based on Olabi 2019)

replacement cycles and results in the reduction of newly manufactured products (Kamper et al., 2019). In addition, for CE to work, waste needs to be efficiently returned to production (Tam et al., 2019).

CE aims to mimic the functioning of ecosystems, transforming waste into resources, establishing a closed material cycle by minimizing dissipative emissions, dematerializing products and economic activities, and decarbonizing energy (Erkman, 1998; Dente and Tavasszy, 2018). The concept of CE aims to decouple economic growth from the use of finite resources (Kamper et al., 2019).

It is also worth noting that although the “3Rs” (reuse, recovery, and recycling) represent concepts that have been used for years, they have perhaps never been more valued than now in CE because of the emphasis on the design and integration of decision and material return (Tam et al., 2019). Thus, to support economic growth and the efficient and sustainable management of resources, the CE paradigm appears to be a great opportunity (Saidani et al., 2018). Indeed, the promises and benefits of circular practices have been widely discussed in the literature in recent years (Lacy, 2015; Ghisellini et al., 2016; Saidani et al., 2018).

3 Methodological Procedure

The methodological procedure adopted in this chapter consists of a systematic review, with a bibliometric approach, to map the main studies on the application of CE in the transportation sector. During this review, sources that support the presence of the problem under investigation should be used (Barnes, 2005). This

Table 1 Description of search strategies

Criteria	Description
Topic	*TS*[1] = (*"circular economy" AND "urban" AND "transport*") OR TS* = (*"circular economy" AND "road") OR TS* = (*"circular economy" AND "vehicle"*)
Database	*Web of Science*
Indexes	SCI-EXPANDED, SSCI, A&HCI, CPCI-S, CPCI-SSH, ESCI
Inclusion	(I) Time of coverage: all years of the database (1945–2020); (II) Fit with the proposed objective; (III) Impact factor of the journal; and (IV) Types of documents: papers only
Qualifications	(I) Does the research present a well-founded literature review? (II) Does the study present technical innovation? (III) Are the contributions discussed? (IV) Are the limitations explicitly stated? and (V) Are the results and conclusions consistent with the pre-established objectives?
Search date	January 22, 2024, at 10:00 p.m.

activity allows the researcher to provide a strong argument related to the need for the study, as well as the point at which the literature fits the stipulated objective (Levy & Ellis, 2006).

It should be noted that the most appropriate search terms were defined through preliminary research in primary sources. The selection of these terms can influence the search results and is therefore a crucial activity. In addition, we sought to define the inclusion and qualification criteria for the papers, which were necessary for the proper conduct of the research. These criteria were based on the work of De Abreu et al. (2022a). Therefore, the terms used (search topics), inclusion and qualification criteria, as well as the details for the search and database extraction are shown in Table 1. It should be noted that TS = Topic, which means the words that will be searched in the titles, abstracts, and keywords of the papers.

As a limitation, it should be mentioned that the study used only one database (Web of Science)—although this database has satisfactory reach and coverage—and did not consider publications contained in other databases such as ScienceDirect, Scielo, and Scopus. It is also important to note that search terms can influence the results, so the studies included in the research repository are limited by the database and search terms used.

4 Results

A search of the Web of Science database revealed that only 155 publications were suitable for inclusion in the research repository, i.e. they met the inclusion and qualification criteria (quality and applicability). The process of selecting and discarding studies is exemplified in Fig. 2.

Figure 3 illustrates the trajectory of publications related to the topic over the years, making it possible to identify the status of the subject, whether it is out of date or constantly evolving, evidenced by an upward curve that opens doors to various

Fig. 2 Database manipulation process

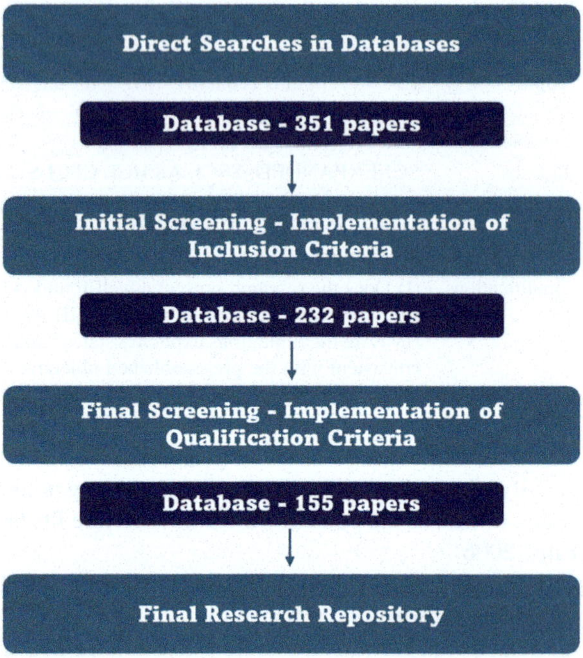

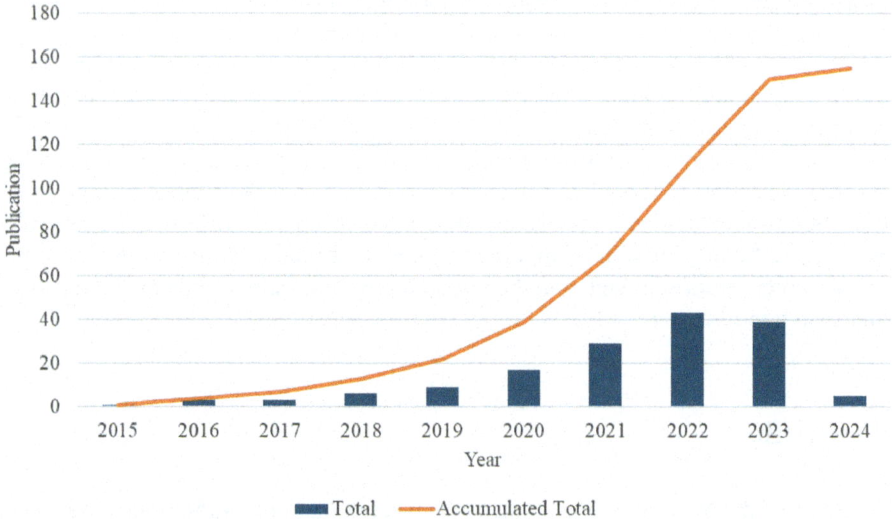

Fig. 3 Evolution of publications on the subject

Table 2 Main journals on the subject

Source title	P (%)	IF (2022)
Sustainability	12.3	3.9
Journal of Cleaner Production	11.0	11.1
Resources Conservation and Recycling	7.7	13.2
Construction and Building Materials	3.2	7.4
Materials	3.2	0.59
Applied Sciences-Basel	1.9	0.57
Environmental Science & Technology	1.9	11.4
Infrastructures	1.9	2.6
Journal of Energy Storage	1.9	9.4
Journal of Industrial Ecology	1.9	5.9
Polymers	1.9	5.0
Sustainable Production and Consumption	1.9	12.1
Waste Management	1.9	8.1

study opportunities. Figure 3 shows that the first publication dates back to 2015, indicating the contemporary nature of the subject and the potential for new research. In addition, the topic continues to expand, with the last 5 years (2020–2024) concentrating 86% of publications, while the accumulated curve reflects the growing interest in the topic over time.

It is also pertinent to evaluate the papers by journal of publication, in order to identify which journals are most interested in the subject, as well as the impact factor of each one. This helps researchers to target their publication efforts to journals that are directly interested in a given topic, preventing them from wasting time sending studies to journals that would probably reject their studies for not fitting the established scope. In addition, as the impact factor is evaluated, it is possible to identify the most relevant journals on the subject. Table 2 shows the journals in which the articles included in the research repository were published. It should be noted that P (Publications) refers to the percentage of articles published in the journal on the area of interest investigated and IF (Impact Factor) assesses the importance of scientific journals in their respective areas. The values shown in Column IF correspond to the year 2022.

Table 2 shows that the pre-eminent journals dealing with the subject are *Sustainability*, *Journal of Cleaner Production*, and *Resources Conservation and Recycling*, representing approximately 31% of publications. The subsequent analysis, classifying the journals according to impact factor, highlights the importance of *Sustainable Production and Consumption*, *Environmental Science & Technology*, and *Journal of Cleaner Production*, all with an impact factor of more than eleven. These results indicate that researchers interested in conducting studies on the subject are more likely to be accepted by these journals, pointing to the most relevant publications for this area of research. It is worth noting that the database contains a total of 75 publication sources.

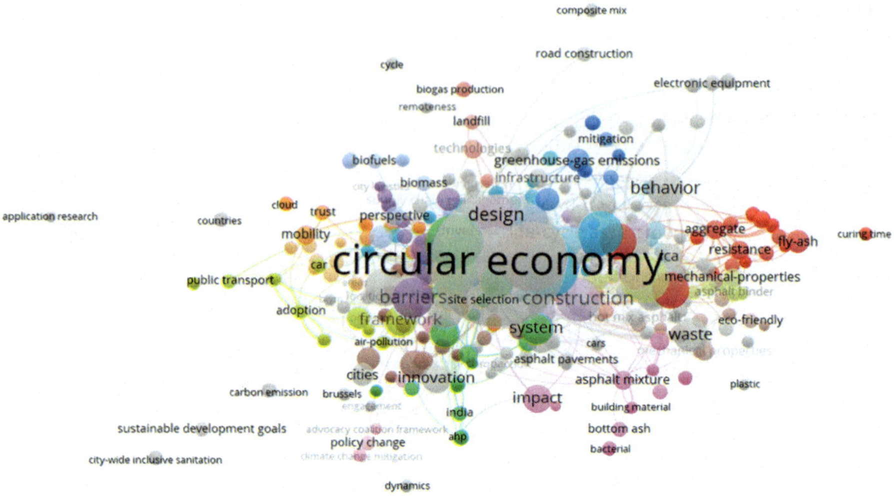

Fig. 4 Interconnection network between the top keywords

It is also important to evaluate the main keywords found in the studies included in the research repository, as identified in the interconnection network between keywords shown in Fig. 4, developed with the help of the VOSviewer software (Centre for Science and Technology Studies, Leiden University, Netherlands), which builds and visualizes bibliometric networks. This strategy allows researchers to more easily find studies directly related to the subject under investigation, as well as to identify new research directions by determining the factors, key dimensions, and main areas.

Figure 4 shows a complex network made up of 986 elements distributed in 38 clusters, totaling 8898 interrelated connections. In this graphic representation, the size of the spheres reflects the frequency of the keywords, while the lines connecting them indicate their interrelationships. Terms with high recurrence stand out, such as "Circular Economy" (92 occurrences), "Sustainability" (29 occurrences), "Management" (25 occurrences), "Recycling" (21 occurrences), "Performance" (15 occurrences), and "Barriers" (11 occurrences).

Also noteworthy is the presence of terms associated with road infrastructure, such as "Road Construction," "Concrete," "Asphalt," and "Pavement." Terms related to Urban Planning and Urban Mobility, with a focus on sustainable transport, also stand out, including "Electric Vehicles," "Policy Change," and "Public Transport." In the context of Life Cycle or End-of-Life Cycle Analysis, terms such as "Life Cycle Assessment"and "End-of-life batteries" are identified. Finally, aspects pertinent to the Supply Chain and Logistics are represented by terms such as "Sustainable Supply Chain."

This analysis provides a comprehensive view of the landscape in question, revealing not only the most prominent keywords but also the thematic interconnections between different areas of research. This in-depth approach contributes to a more holistic understanding of the research context, offering valuable insights for researchers and professionals working in this field.

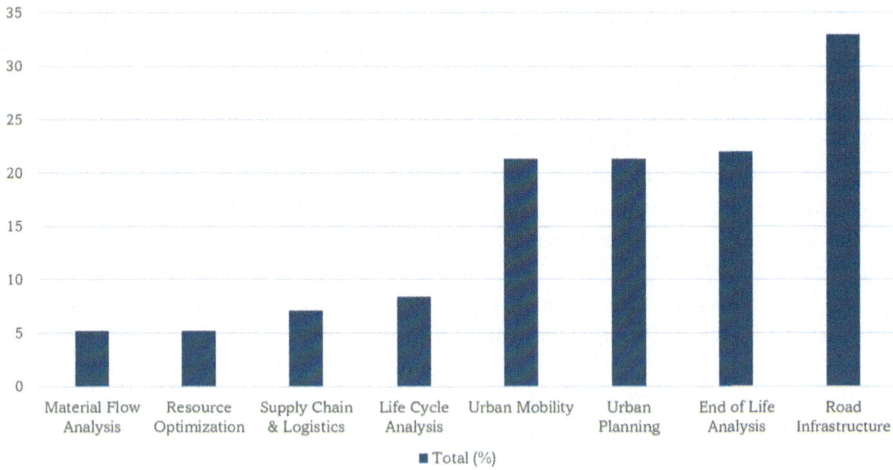

Fig. 5 Categorization of papers by area of study

Evaluating the research by area of study proved to be a relevant step in gaining an in-depth understanding of the various analytical approaches present in the studies that apply the principles of CE in the context of urban road transport. This approach was based on the inputs provided by the keyword network, as shown graphically in Fig. 4.

Figure 5 shows the segmentation of the studies included in the research repository, stratified by area of application in the field of transport. Analysis of Fig. 5 shows that the predominant areas, in terms of volume of publications, are as follows: Road Infrastructure, accounting for 33% of the studies; End-of-Life Analysis, with 22% of the studies; and Urban Planning and Urban Mobility, both contributing 21% of the studies each. In addition, there were studies in the areas of Logistics and Supply Chain, Resource Optimization, Life Cycle Assessment, and Material Flow Analysis. It should be noted that the sum of the percentages attributed to the areas of application exceeds 100%, since the same study can be classified in more than one category.

Another point worth highlighting is the distinction between Urban Planning and Urban Mobility, because although both are intrinsically linked to the sustainable development of cities, they have specific approaches and focuses. Urban Planning, more broadly, addresses critical urban infrastructure, including not only the transport sector but also urban design, land use, and distribution of public services, among others. It aims to shape the growth and expansion of urban areas in an integrated way, considering the present and future needs of the population. On the other hand, Urban Mobility focuses more directly on the transport sector, presenting specific strategies to optimize movement within cities. This includes the development of efficient public transport systems, the promotion of sustainable means of locomotion, such as walking and cycling, and the implementation of policies to reduce the negative environmental impacts associated with urban transportation.

Furthermore, a distinction that requires attention is between Life Cycle Assessment and End-of-Life Analysis. The reason is that Life Cycle Assessment focuses on assessing the environmental impact of a product or process throughout all its phases, from the extraction of raw materials to final disposal. On the other hand, End-of-Life Analysis specifically examines the treatment and destination of waste after the end of the product's useful life, considering aspects such as recycling, reuse, and proper disposal. These complementary approaches are essential for a holistic understanding of the environmental impacts associated with products and processes throughout their entire life cycle.

Having delimited these areas of application, the next step involves a detailed discussion of each of them in the following subsections. This systematic approach enables a more in-depth analysis of the nuances and complexities inherent in the application of CE in urban road transport, contributing to a more holistic and comprehensive understanding of the implications and benefits associated with each specific area.

4.1 Road Infrastructure

These studies intricately examine the implementation of cutting-edge technologies in road paving. For instance, they explore the use of bio-recycled asphalt mixtures (Blanc et al., 2019), the reclamation of asphalt (Mantalovas & Di Mino, 2019), the incorporation of coal mine waste for sub-base construction in low-traffic roads (Suescum-Morales et al., 2019), and the utilization of phosphogypsum waste as an additive for lime-stabilized bentonite in soil reinforcement for road construction (Oumnih et al., 2019). Additionally, they investigate the recycling and application of used tires (Wang et al., 2019) or plastic bottles (Anwar et al., 2021) and the substitution of natural aggregates with diverse solid mineral waste (Mkahal et al., 2022). These advancements hold immense promise for enhancing sustainability across paving practices and other components of road infrastructure. Adopting these technologies not only aims to improve efficiency and extend the life of roads but also actively seeks to integrate CE concepts to promote more responsible and environmentally friendly practices in road infrastructure.

In addition, studies highlight the relevance of these innovations not only in the technical context of paving but also in environmental and economic terms (Wang et al., 2019). The use of bio-recycled asphalt mixes and reclaimed asphalt, for example, not only reduces dependence on non-renewable natural resources but also minimizes the carbon footprint associated with the production of asphalt materials. Incorporating coal mine waste as sub-base and phosphogypsum waste into soil stabilization not only offers a sustainable solution to the disposal of these materials but also contributes to responsible environmental management. In addition, recycling used tires not only solves the problem of tire waste but also introduces an economically viable and ecologically friendly alternative to road construction.

These approaches not only reflect the ongoing evolution of pavement engineering but also indicate a growing commitment to sustainable practices and the effective integration of CE principles (Kowalski et al., 2022). By understanding and applying these innovations, it is possible not only to improve road infrastructure but also to contribute significantly to building a more sustainable and resilient environment (Almokdad & Zentar, 2023; Connolly et al., 2023).

It is pertinent to note a significant limitation within this category of studies, which pertains to the prevailing tendency to concentrate solely on specific practices when applying circular economy principles to pavement infrastructures. This limitation is underscored by the observation that current research often lacks a holistic and systemic perspective, instead opting for fragmented analyses of isolated practices (Liu & Kringos, 2024).

4.2 Urban Planning

This research is not just restricted to the field of road infrastructure, but extends into the sphere of urban planning, incorporating the principles of CE to shape the sustainable development of cities. A clear example is the analysis of how urban planning, grounded in CE, can positively influence energy consumption patterns during the urbanization process (Zhou et al., 2012). In addition, these studies explore ways to promote urban sustainability by optimizing flow systems within the city (Lee et al., 2016). It is worth noting that although these studies do not focus primarily on the transportation sector, they do consider this sector as critical infrastructure that is fundamental to achieving strategic urban planning.

By adopting an ecocentric approach, urban planning seeks not only to manage urban growth but also to transform and develop cities based on industrial resources, thus promoting a more efficient and sustainable integration of available resources (Pang, 2017). In addition, the consideration of CE in urban planning emerges as a strategic tool to mitigate the impacts of climate change (Christis et al., 2019), strengthening the resilience of cities in the face of global environmental challenges. Therefore, these studies not only enrich the understanding of the practical application of CE principles in urban planning but also highlight its potential to create more sustainable, resilient, and energy-efficient cities, including the large-scale implementation of electric mobility.

4.3 Life Cycle Assessment

This research is not only limited to innovation in city design and the application of materials in paving but also extends to a comprehensive analysis of life cycle impact (Da Costa et al., 2022). The life cycle, which comprises the various phases that a product or service goes through, from its production to its disposal, is evaluated as an integral part of sustainable resource management (Guo et al., 2017; Folega & Burchart-Korol, 2017). A concrete example of this approach is the investigation of the life cycle of lithium-ion batteries, which are widely used in electric vehicles (Grandjean et al., 2019).

In this context, Iturrondobeitia et al. (2021) developed a life cycle assessment methodology to analyze and compare how different types of batteries and their components influence environmental footprints, distributed into 18 categories. For a comprehensive perspective, the results were compared with the environmental loads of a standard lithium-ion battery, a reference sodium-ion battery, and the average lithium-sulfur battery. The results revealed that, in general, Li-O_2 batteries have lower environmental loads in nine impact categories, while exhibiting similar impacts in five categories when compared to lithium-sulfur and lithium-ion batteries. With an average of 55.76 kg-CO_2 equiv. in Global Warming Potential (GWP) for the Li-O_2 battery as a whole, it stands out that the cathode is the main contributor, accounting for 44.5% of the relative weight.

With a different approach, Balaguera et al. (2021) used Environmental Life Cycle Assessment on segments of two low-traffic roads located in different regions of Colombia: one in Urrao (Antioquia), located in the center of the country, and the other in La Paz (Cesar), located in the northeast. Each stretch was stabilized with alternative materials such as brick dust, fly ash, sulfonated oil, and polymer. The study comprised three phases: the first covered the manufacture of the stabilizer; the second encompassed preliminary actions from locating the material to its application on site; and the third consisted of the stabilization process, covering all stages from the application of the stabilizer to the completion of the road. The environmental impacts were predominantly associated with the manufacture of the stabilizers (60% of the total), especially for sulfonated oil or polymer, due to the different compounds used during production, before being used as stabilizers. The most significant impact categories were Abiotic Depletion Potential (ADP), GWP, and Terrestrial Ecotoxicity Potential (TETP).

By considering the life cycle of these technologies, studies seek not only to assess their effectiveness during application but also to understand the environmental impact throughout all stages, from the extraction of raw materials to disposal. This holistic approach contributes to a more complete view of the sustainable landscape, enabling the implementation of practices that optimize not only operational efficiency but also responsible resource management, in line with the principles of the Circular Economy.

4.4 Material Flow Analysis

This approach seeks to deepen the understanding of materials accumulated in inventories by exploring broader contexts, such as in a road network (Miatto et al., 2017). By adopting CE in material flow accounting, these studies aim to provide a comprehensive analysis of the resources used and accumulated throughout the life cycle of road infrastructures, highlighting opportunities to improve efficiency and sustainability.

In addition, these investigations delve into more specific contexts, such as the consumption of platinum in catalytic converters, especially in diesel-powered vehicles, where platinum shows particular effectiveness under conditions of excess oxygen (Saidani et al., 2019). Another noteworthy study is that developed by Nurdiawati and Agrawal (2022) which seeks to present a scenario-based material flow analysis to estimate the future volume of waste electric vehicle batteries to be potentially generated in Sweden and the future demand for essential battery materials, considering the potential fleet of electric vehicles, the chemical developments of batteries, and the end of life. Each scenario investigated highlights the need for effective policy frameworks to promote a circular value chain for vehicle batteries.

By taking a closer look at the specific use of materials in automotive technologies, these studies contribute to optimizing resource management and identifying opportunities for circularity in key sectors. In this way, the application of CE in material flow accounting not only offers crucial insights into inventory management on a road network but also improves efficiency and sustainability in more specific applications, promoting a more holistic approach to resource management (Miatto et al., 2017).

4.5 Logistics and Supply Chain

Expanding the areas of application, we move on to logistics and the supply chain, where the implementation of CE principles demonstrates significant benefits (Demartini et al., 2023). An illustrative example is the improvement of eco-efficiency in the closed-loop supply chain for agricultural products, particularly in road initiatives (Zhao et al., 2018). By integrating CE concepts, these studies seek not only to optimize logistics but also to reduce the environmental impact associated with the transport and distribution of agricultural products.

On another research front, Demartini et al. (2023) seek to create a model based on system dynamics and agent-driven approaches to assess the impacts of transitions toward a net zero and electric economy on automotive supply chains and associated stakeholders. The results indicate the crucial need for companies and policymakers to carry out in-depth studies to mitigate the adverse effects on employment in the automotive industry resulting from increased automation. In addition, the importance of planning specific programs and policies to support the development of new skills for workers and facilitate their transition along the supply chain (for example, by phasing out certain activities) is emphasized.

The application of advanced route optimization methods using Geographic Information Systems (GIS) data (Laasasenaho et al., 2019) highlights another dimension of this research. These methods aim not only to increase efficiency in the movement of goods but also to minimize the environmental footprint associated with transport, resulting in more sustainable and effective logistics operations. In addition, the studies are dedicated to developing sustainable methods for collecting and transporting electronic waste, in both urban and rural environments, within the context of the circular economy (Nowakowski & Mrowczynska, 2018).

It is also worth mentioning the study by Nguyen-Tien et al. (2022) who developed a supply chain model based on the Geospatial Supply Chain (GSC) for the electric vehicle and battery recycling industry in the UK. The study highlights the UK's current lag behind other nations in the manufacture and recycling of lithium-ion batteries. However, this gap offers a unique opportunity to provide valuable evidence that can guide industry and government in strategic decisions about the planning and spatial distribution of the recycling industry. Although the scenarios presented are idealized, they point to significant potential for translating the lessons of the analysis into an implementable reality, especially considering the early stage of development of the lithium-ion battery recycling industry in the UK.

These approaches not only efficiently manage waste but also promote reuse and recycling, thus contributing to reducing the environmental impact associated with the inappropriate disposal of these materials. In this way, these studies expand our understanding of how the application of CE in logistics and the supply chain not only optimizes operations but also drives more sustainable and responsible practices in line with CE objectives.

4.6 End-of-Life Analysis

This research specifically addresses the reuse of vehicle components at the end of their life cycle, a crucial phase that deserves special attention (Casals et al., 2019; Fallah & Fitzpatrick, 2022), even though these studies also fall within the analysis of the life cycle as a whole. An illustrative example is the study conducted by Landi et al. (2018), which investigates the feasibility of a reuse scenario for textile fibers recovered from tires at the end of their useful life. This type of approach not only contributes to waste reduction but also highlights the importance of considering reuse options at advanced stages of the vehicle life cycle.

Additionally, the work of Saidani et al. (2018) highlights the end-of-life management of heavy-duty vehicles as an integral part of the path to CE. This study not only explores sustainable practices for dealing with the disposal of heavy-duty vehicles but also highlights the need for effective strategies for the reuse of their components, thus aligning with the fundamental principles of CE. Thus, by focusing on the final stage of the vehicle life cycle, this research provides valuable insights into how to maximize the reuse of components, promoting more circular and sustainable practices in the management of waste from the automotive sector.

An additional case in point is the work carried out by Bonsu (2020), whose purpose is to investigate the end-of-life applications of electric vehicle batteries in a low-carbon circular economy, following the global transition to electric vehicles and a net-zero economy. The study highlights the importance of defining commitments within the framework of political objectives, considering global value chains, and exploring synergies between social, economic, and environmental goals.

In a complementary way, Tripathy et al. (2022) analyze the critical enablers of adopting electric vehicle battery recycling to reduce the unprecedented accumulation of battery wastes—minimizing environmental problems and raw material shortages as batteries approach the end of their useful life. This article highlights the significant influence of organizational factors on decision-making compared to technological and environmental factors in the context of electric vehicle recycling. Professionals showed a preference for technological enablers, while policy planners considered all dimensions equally. Finally, academic researchers adopt an intermediate perspective on innovations in electric vehicle battery recycling.

Still on electric vehicle batteries, Dunn et al. (2021), understanding the concerns regarding the environmental and social impacts of the production of lithium-ion batteries, especially cathode materials, seek to implement CE strategies, such as reuse and recycling, to reduce these impacts and ensure regional supply. The results highlight the importance of incorporating the second life of batteries, indicating that in order to take full advantage of the recovery of secondary materials, the development of robust recycling and production infrastructures is imperative.

4.7 Resource Optimization

These studies aim not only to minimize costs but also to promote organizational profitability by adopting an approach that avoids waste. A notable example is the study carried out by Crisan et al. (2019), which conducts a case analysis on the opportunities for optimizing the technical and economic activities of a car wash company. This study addresses not only the economic aspects but also the environmental impacts, such as excessive water consumption during vehicle washing.

An additional example is the study conducted by Peyravi and Jakubavičius (2022), which aims to examine organizational capacity and exploitation strategies to promote a circular economy and overcome obstacles to sustainability. The practical contribution of this study consists of a consolidated and comprehensive eco-innovation framework, providing significant insights for practitioners and facilitating the dissemination of innovative practices. This study will be beneficial for policymakers, prominent institutions, and companies, enabling them to strengthen organizational capacity through the effective implementation of eco-innovation processes toward a circular economy, where structural change and transformative eco-innovation play a central role.

As a further example in this area of study, Szabo et al. (2023) highlight possibilities for strengthening the protection of the integrity of the cells that make up a battery in situations of impact or road accident. They present a numerical analysis of a topographically optimized battery module casing. The proposed method is innovative and demonstrates a cell protection efficiency ranging from 16.6% to 60%

(with averages of 19.7% to 40.7% considering all three impact speeds). Cell protection efficiency decreases as impact velocity increases, providing the basis for reintegrating many salvaged cells into other energy storage systems (such as photovoltaic and/or wind), preventing possible future problems related to environmental pollution.

By highlighting sustainable practices and strategies for resource protection, this study serves as a practical model for implementing the theoretical concept of CE. The aim is to go beyond financial efficiency, also considering environmental responsibility and effective resource management. In this way, by adopting measures that minimize waste and promote a more circular approach, this research contributes not only to the profitability of organizations but also to building more sustainable business practices in line with the principles of CE.

4.8 Urban Mobility

This research not only promotes sustainable urban mobility with a focus mainly on electric mobility (Song & Zhou, 2023; López et al., 2023) but also explores the crucial role of developing new business models as essential catalysts for the transition toward a more comprehensive CE. An emblematic example is the study conducted by Sarasini and Linder (2018), which analyzes a new business model called Mobility as a Service (MaaS). This innovative model has the potential to radically transform the road transport landscape.

MaaS aims to offer an alternative to private vehicle ownership, proposing an innovative approach that integrates various mobility services into a single, cohesive offer. This offer is made available to users via subscription-based smartphone applications (Beutel et al., 2014; Goldman and Gorham, 2006; Sochor et al., 2015). By exploring and implementing business models such as MaaS, these studies not only foster a more sustainable approach to urban mobility but also lay the foundation for an effective transition toward CE.

An additional study, conducted by Jalas and Numminen (2022), seeks to examine equity in the rhythms and dynamic pricing of infrastructure services in transportation and electricity. The concept of the circular economy promotes the sharing and efficient use of capital assets. For these strategies to be effective, it is crucial to manage the timing of demand. Dynamic pricing is often used to manage demand for electricity services, for example, by aligning it with available production capacity. Dynamic pricing helps establish new daily rhythms and facilitates "consumption shifting" by restricting access during periods of high demand.

By considering the economic and behavioral implications, this research enriches our understanding of how these models can not only improve the efficiency of urban transport but also positively influence consumption patterns, promoting a more conscious and integrated approach to mobility in the cities of the future, thus reducing the depletion of natural resources and the carbon footprint.

4.9 Final Discussion

Considering all that has been discussed, CE can play a crucial role in promoting sustainable transportation, permeating various areas of study and mitigating the associated environmental impacts. In this context, it becomes essential to recognize the importance of incorporating CE principles into future policies and research, providing a systemic approach to sustainable practices (De Jesus & Mendonça, 2018; De Abreu et al., 2022a, b), including those related to the transport sector.

It is worth noting that although electric mobility or the study of electric vehicles and their components, especially batteries, was not mentioned as a specific area, it is present in several facets of research (Da Costa et al., 2022). The presence of electric mobility in these different areas indicates its relevance and widespread impact on sustainable practices related to road transport (De Abreu et al., 2023). Electric mobility emerges as a cross-cutting element, influencing studies ranging from route optimization and waste management to the reuse of vehicle components at the end of their life cycle, for example batteries. Furthermore, when considering MaaS, the inclusion of electric vehicles in the available transportation options can be crucial to the model's effectiveness and sustainability.

Implementing CE concepts in the transport sector not only offers an opportunity to rethink and redesign conventional road transport practices but also contributes to building more effective and eco-efficient solutions. It is essential to highlight that the global transition toward cleaner air, low-carbon economies, and the reduction of excessive consumption of natural resources are intrinsically linked to the Sustainable Development Goals (SDGs) and their targets, as outlined in the United Nations (UN) 2030 Agenda.

Meanwhile, despite the progress made, the main gaps in studies on the implementation of the Circular Economy in the urban road transport sector are the lack of studies on:

- Integration of renewable energy sources: Although the transition to electric vehicles is a trend, there is a lack of studies exploring the integration of renewable energy sources, such as recharging through solar energy, in urban road transport systems.
- Life Cycle of Renewable Hydrogen in Transportation: Few studies have comprehensively addressed the life cycle of renewable hydrogen in the specific context of urban road transportation. A complete analysis, from production to consumption in vehicles, taking into account the principles of the Circular Economy, is absent. This includes assessing the origin of raw materials, efficiency in production processes, and sustainable disposal methods at the end of life.
- Evaluation of social and economic impacts: Many studies focus on the environmental aspects of the Circular Economy, but there is a lack of evaluation of the social and economic impacts of the transition to more circular practices in urban road transport. This includes analysis of jobs, equity in access to transportation, and associated costs.

- Circular policies and regulations: Research on policies and regulations that encourage the adoption of circular practices in urban road transport is limited. Understanding how public policies can boost the Circular Economy in this sector is essential for an effective transition.
- Education and awareness: Studies investigating the awareness and education of all stakeholders involved in urban road transport about the benefits of the Circular Economy are needed. This includes transport operators, vehicle manufacturers, city managers, and, above all, users.

Addressing these gaps can make a significant contribution to promoting more sustainable and circular practices in the urban road transport sector.

5 Final Considerations

Various sectors are seeking to apply the concepts of the Circular Economy (CE) to mitigate the adverse environmental impact of the production and consumption of goods and services, especially in the context of reducing Greenhouse Gas (GHG) emissions and waste generation and the natural resources depletion. One of these sectors is transportation, which, despite playing a key role in efficient sustainable urban development, is a major consumer of fossil fuels, an important contributor to the increase in GHG emissions. The aim of this study was to collect, collate, and evaluate the main relevant studies on the application of CE in the urban road transport sector.

The results of the analysis revealed that the first scientific contribution on the application of CE to urban road transport dates back to 2015. Since then, a steady upward trend has been identified, culminating in its peak in 2022, both in terms of the volume of publications and citations. This phenomenon indicates a growing attention from the academic and scientific community to the challenges and opportunities related to the integration of CE in this specific sector.

It is worth noting that high-impact journals such as *Sustainable Production and Consumption*, *Environmental Science & Technology*, and *Journal of Cleaner Production*, with impact factors greater than 11, have played a crucial role in advancing knowledge in this area. The quality and relevance of these publications highlight the importance attributed to the topic by the scientific community and indicate the need for a multidisciplinary approach to understanding and solving the challenges faced by the urban road transport sector in the light of CE principles.

In addition, it was possible to identify eight distinct areas of application of CE in this sector, outlining the broad scope of the scientific contributions. These areas include Road Infrastructure, Urban Planning, Life Cycle Assessment, Material Flow Analysis, Logistics and Supply Chain, End-of-Life Analysis, Resource Optimization, and Urban Mobility. Each of these areas represents a specific domain in which CE principles are applied to promote sustainability and efficiency in urban road transport.

These findings not only consolidate the growing interest in CE integration in the urban road transport sector but also provide a solid basis to guide future research. To advance this field, future studies could delve deeper into each specific area identified, conducting comprehensive analyzes to gain a more nuanced understanding of the practices and challenges involved. Furthermore, expanding the scope of research to cover additional relevant databases such as ScienceDirect, Scielo, and Scopus could significantly increase knowledge on the application of the circular economy in this context.

Furthermore, investigating the practical implementation of circular economy initiatives in real urban road transport contexts could provide valuable insights into their feasibility, effectiveness, and impact on sustainability metrics. Another possible front consisting of the socio-economic implications and stakeholder perspectives around the adoption of circular economy practices in road transport could offer a comprehensive understanding of the broader implications of such initiatives.

References

Albuquerque, T. L. M., Mattos, C. A., Scur, G., & Kissimoto, K. (2019). Life cycle costing and externalities to analyze circular economy strategy: Comparison between aluminum packaging and tinplate. *Journal of Cleaner Production, 234*, 477. https://doi.org/10.1016/j.jclepro.2019.06.091

Almokdad, M., & Zentar, R. (2023). Characterization of recycled dredged sediments: Toward circular economy in road construction. *Construction and Building Materials, 402*, 132974. https://doi.org/10.1016/j.conbuildmat.2023.132974

Anwar, M. K., Shah, S. A. R., & Alhazmi, H. (2021). Recycling and utilization of polymers for road construction projects: An application of the circular economy concept. *Polymers, 13*(8), 1330. https://doi.org/10.3390/polym13081330

Apte, J. S., Messier, K. P., Gani, S., Brauer, M., Kirchstetter, T. W., Lunden, M. M., Marshall, J. D., Portier, C. J., Vermeulen, R. C. H., & Hamburg, S. P. (2017). High-resolution air pollution mapping with Google street view cars: Exploiting big data. *Environmental Science and Technology, 51*(12), 6999–7008. https://doi.org/10.1021/acs.est.7b00891

Balaguera, A., Alberti, J., Carvajal, G. I., & Fullana-i-Palmer, P. (2021). Stabilising rural roads with waste streams in Colombia as an environmental strategy based on a life cycle assessment methodology. *Sustainability, 13*(5), 2458. https://doi.org/10.3390/su13052458

Barnes, S. J. (2005). Assessing the value of IS journals. *Communications of the ACM, 48*(1), 110–112.

Beutel, M. C., Gokay, S., Kluth, W., Krempels, K.-H., Samsel, C., & Terwelp, C. (2014). Product oriented integration of heterogeneous mobility services. In *17th IEEE international conference on intelligent transportation systems, ITSC 2014* (pp. 1529–1534). https://doi.org/10.1109/ITSC.2014.6957650

Blanc, J., Hornych, P., Sotoodeh-Nia, Z., Williams, C., Porot, L., Pouget, S., & Chailleux, E. (2019). Full-scale validation of bio-recycled asphalt mixtures for road pavements. *Journal of Cleaner Production, 227*, 1068. https://doi.org/10.1016/j.jclepro.2019.04.273

Bonsu, N. O. (2020). Towards a circular and low-carbon economy: Insights from the transitioning to electric vehicles and net zero economy. *Journal of Cleaner Production, 256*, 120659. https://doi.org/10.1016/j.jclepro.2020.120659

Casals, L. C., Barbero, M., & Corchero, C. (2019). Reused second life batteries for aggregated demand response services. *Journal of Cleaner Production, 212*, 99–108. https://doi.org/10.1016/j.jclepro.2018.12.005

Charonis, G. K. (2012). Degrowth, steady state economics and the circular economy: Three distinct yet increasingly converging alternative discourses to economic growth for achieving environmental sustainability and social equity. In World Economics Association Conference, 2012.

Chen, M. C., Ho, T. P., & Jan, C. G. (2006). A system dynamics model of sustainable urban development: Assessing air purification policies at Taipei City. *Asian Pacific Planning Review, 4*(1), 29–52.

Christis, M., Athanassiadis, A., & Vercalsteren, A. (2019). Implementation at a City level of circular economy strategies and climate change mitigation—The case of BRUSSELS. *Journal of Cleaner Production, 218*, 511. https://doi.org/10.1016/j.jclepro.2019.01.180

Connolly, L., Sheils, E., Lamb, M., O'Connor, A., OBrien, E., Viner, H., et al. (2023). Circular economy in road construction and maintenance. *Transportation Research Procedia, 72*, 3458–3465.

Crisan, O. A., Crisan, H. G., Dan, V., & Serdean, F. M. (2019). Approaching the circular economy in Romania – A higher alternative to the linear economy. ACTA TECHNICA NAPOCENSIS. Series: Applied mathematics, mechanics, and engineering, 62, I.

Da Costa, M. G., de Abreu, V. H. S., de Assis, T. F., da Costa, V. X., de Almeida D'Agosto, M., & Santos, A. S. (2022). Life cycle assessment and circular economy strategies for electric vehicle: A systematic review on mitigating climate change and reducing resource depletion in road transportation. In *Carbon footprints of manufacturing and transportation industries* (pp. 113–137). https://doi.org/10.1007/978-981-19-7226-3_5

Da Silva, C. L. (2018). Proposal of a dynamic model to evaluate public policies for the circular economy: Scenarios applied to the municipality of Curitiba. *Waste Management, 78*, 456–466. https://doi.org/10.1016/j.wasman.2018.06.007

De Abreu, V. H. S., Da Costa, M. G., Da Costa, V. X., De Assis, T. F., Santos, A. S., & D'Agosto, M. D. A. (2022a). The role of the circular economy in road transport to mitigate climate change and reduce resource depletion. *Sustainability, 14*(14), 8951. https://doi.org/10.3390/su14148951

De Abreu, V. H. S., Santos, A. S., & Monteiro, T. G. M. (2022b). Climate change impacts on the road transport infrastructure: A systematic review on adaptation measures. *Sustainability, 14*(14), 8864. https://doi.org/10.3390/su14148864

De Abreu, V. H. S., D'Agosto, M. D. A., Angelo, A. C. M., Marujo, L. G., & Carneiro, P. J. P. (2023). Action plan focused on electric mobility (APOEM): A tool for assessment of the potential environmental benefits of urban mobility. *Sustainability, 15*(13), 10218. https://doi.org/10.3390/su151310218

De Assis, T. F., Monteiro, T. G. M., de Abreu, V. H. S., D'Agosto, M. D. A., & Santos, A. S. (2022a). Enabling the green bonds market for sustainable transport projects based on the measure/monitoring, reporting and verification method. In *Carbon footprints of manufacturing and transportation industries* (pp. 1–24). Springer Nature Singapore. https://doi.org/10.1007/978-981-19-7226-3_1

De Assis, T. F., de Abreu, V. H. S., da Costa, M. G., & Marcio de Almeida, D. A. (2022b). Methodology for prioritizing best practices applied to the sustainable last mile—The case of a Brazilian parcel delivery service company. *Sustainability, 14*(7), 3812. https://doi.org/10.3390/su14073812

De Assis, T. F., Ricci, L. M., Monteiro, T. G. M., de Abreu, V. H. S., D'Agosto, M. D. A., & Santos, A. S. (2022c). Sustainable transport indicators and mitigation actions applied to the green bond principles. In *Carbon footprints of manufacturing and transportation industries* (pp. 139–169). Springer Nature Singapore. https://doi.org/10.1007/978-981-19-7226-3_6

De Assis, T. F. D., Marujo, L. G., de Abreu, V. H. S., da Costa, M. G., Rodrigues, L. M., & D'Agosto, M. D. A. (2023). Best practices to support the transition towards sustainable logistics from the perspective of Brazilian carriers. *Sustainability, 15*(18), 13434. https://doi.org/10.3390/su151813434

De Jesus, A., & Mendonça, S. (2018). Lost in transition? Drivers and barriers in the eco-innovation road to the circular economy. *Ecological Economics, 145*, 75–89. https://doi.org/10.1016/j.ecolecon.2017.08.001

Demartini, M., Ferrari, M., Govindan, K., & Tonelli, F. (2023). The transition to electric vehicles and a net zero economy: A model based on circular economy, stakeholder theory, and system thinking approach. *Journal of Cleaner Production, 410*, 137031. https://doi.org/10.1016/j.jclepro.2023.137031

Dente, S. M. R., & Tavasszy, L. A. (2018). Impacts of trade related sustainability strategies on freight transportation: Modelling framework and application for France. *Transportation Research Part D: Transport and Environment, 58*, 308–319. https://doi.org/10.1016/j.trd.2017.04.006

Dri, M., Canfora, P., & Gaudillat, P. (2018). *Best environmental management practice for the waste management sector*. Publications Office of the European Union.

Dunn, J., Slattery, M., Kendall, A., Ambrose, H., & Shen, S. (2021). Circularity of lithium-ion battery materials in electric vehicles. *Environmental Science & Technology, 55*(8), 5189–5198. https://doi.org/10.1021/acs.est.0c07030

Erkman, S. (1998). Vers une écologie industrielle: comment mettre en pratique le développement durable dans une société hyper – industrielle, 1998th, PARI ed.

Fallah, N., & Fitzpatrick, C. (2022). How will retired electric vehicle batteries perform in grid-based second-life applications? A comparative techno-economic evaluation of used batteries in different scenarios. *Journal of Cleaner Production, 361*, 132281. https://doi.org/10.1016/j.jclepro.2022.132281

Folega, P., & Burchart-Korol, D. (2017). Environmental assessment of road transport in a passenger car using the life cycle approach. *Transport Problems, 12*(2), 147–153. https://doi.org/10.20858/tp.2017.12.2.14

Geerken, T., Schmidt, J., Boonen, K., Christis, M., & Merciai, S. (2019). Assessment of the potential of a circular economy in open economies – Case of Belgium. *Journal of Cleaner Production, 227*, 683. https://doi.org/10.1016/j.jclepro.2019.04.120

Ghisellini, P., Cialani, C., & Ulgiati, S. (2016). A review on circular economy: The expected transition to a balanced interplay of environmental and economic systems. *Journal of Cleaner Production, 114*, 11–32. https://doi.org/10.1016/j.jclepro.2015.09.007

Gigli, S., Landi, D., & Germani, M. (2019). Cost-benefit analysis of a circular economy project: A study on a recycling system for end-of-life tyres. *Journal of Cleaner Production, 229*, 680. https://doi.org/10.1016/j.jclepro.2019.03.223

Goldman, T., & Gorham, R. (2006). Sustainable urban transport: Four innovative directions. *Technology in Society, 28*, 261–273. https://doi.org/10.1016/j.techsoc.2005.10.007

Grandjean, T. R. B., Groenewald, J., McGordon, A., & Marco, J. (2019). Cycle life of lithium ion batteries after flash cryogenic freezing. *Journal of Energy Storage, 24*, 100804. https://doi.org/10.1016/j.est.2019.100804

Guo, Z., Shi, H., Zhang, P., Chi, Y., & Feng, A. (2017). Material metabolism and lifecycle impact assessment towards sustainable resource management: A case study of the highway infrastructural system in Shandong peninsula, China. *Journal of Cleaner Production, 153*, 195–208. https://doi.org/10.1016/j.jclepro.2017.03.194

Hasan, M. A., Frame, D. J., Chapman, R., & Archie, K. M. (2019). Emissions from the road transport sector of New Zealand: key drivers and challenges. In *Environmental science and pollution research 26* (pp. 23937–23957). https://doi.org/10.1007/s11356-019-05734-6

HEI – Health Effects Institute. (2018). *State of Global Air 2018*. Special Report. Boston: Health Effects Institute. Available at: https://www.stateofglobalair.org/sites/default/files/soga-2018-report.pdf

Hickmana, R., Ashirub, O., & Banisterc, D. (2010). Transport and climate change: Simulating the options for carbon reduction in London. *Transport Policy, 17*(2), 110–125. https://doi.org/10.1016/j.tranpol.2009.12.002

IPCC (Intergovernmental Panel on Climate Change). (2014). Summary for policymakers. In O. Edenhofer, R. Pichs-Madruga, Y. Sokona, E. Farahani, S. Kadner, K. Seyboth, A. Adler, I. Baum, S. Brunner, P. Eickemeier, B. Kriemann, J. Savolainen, S. Schlömer, C. von Stechow, T. Zwickel, & J. C. Minx (Eds.), *Climate change 2014, mitigation of climate change. Contribution of Working Group III to the fifth assessment report of the Intergovernmental Panel on Climate Change*. Cambridge University Press.

Iturrondobeitia, M., Akizu-Gardoki, O., Minguez, R., & Lizundia, E. (2021). Environmental impact analysis of aprotic Li–O2 batteries based on life cycle assessment. *ACS Sustainable Chemistry & Engineering, 9*(20), 7139–7153. https://doi.org/10.1021/acssuschemeng.1c01554

Jalas, M., & Numminen, S. (2022). Prime-time access for whom? Rhythms fairness and the dynamic pricing of infrastructure services. *Local Environment, 27*(10–11), 1355–1371. https://doi.org/10.1080/13549839.2022.2040468

Kamper, A., Triebs, J., Hollah, A., & Lienemann, C. (2019). Remanufacturing of electric vehicles: Challenges in production planning and control. *Procedia Manufacturing, 33*, 280–287. https://doi.org/10.1016/j.promfg.2019.04.034

Karagoz, S., Aydin, N., & Simic, V. (2019). End-of-life vehicle management: A comprehensive review. *Journal of Material Cycles and Waste Management, 22*, 416. https://doi.org/10.1007/s10163-019-00945-y

Kowalski, K. J., Bańkowski, W., Król, J. B., Andersen, B. H., Komkova, A., & Casado Barrasa, R. (2022). Practical application of sustainable road structure: Mechanical and environmental approach. *Applied Sciences, 12*(23), 11914. https://doi.org/10.3390/app122311914

Laasasenaho, K., Lensu, A., Lauhanen, R., & Rintala, J. (2019). GIS-data related route optimization, hierarchical clustering, location optimization, and kernel density methods are useful for promoting distributed bioenergy plant planning in rural areas. *Sustainable Energy Technologies and Assessments, 32*, 47–57. https://doi.org/10.1016/j.seta.2019.01.006

Lacy, P. (2015). *Growth, innovation and customer value though the circular economy.*

Landi, D., Gigli, S., Germani, M., & Marconi, M. (2018). Investigating the feasibility of a reuse scenario for textile fibres recovered from end-of-life tyres. *Waste Management, 75*, 187–204. https://doi.org/10.1016/j.wasman.2018.02.018

Lee, S., Quinn, A., & Rogers, C. (2016). Advancing city sustainability via its systems of flows: The urban metabolism of Birmingham and its hinterland. *Sustainability, 8*(3), 220. https://doi.org/10.3390/su8030220

Levy, Y., & Ellis, T. J. (2006). A systems approach to conduct an effective literature review in support of information systems research. *Informing Science: The International Journal of an Emerging Transdiscipline, 9*, 181–212. https://doi.org/10.28945/479

Litman, T. (2007). Developing indicators for comprehensive and sustainable transport planning. *Transportation Research Record: Journal of the Transportation Research Board, 1*, 10–15. https://doi.org/10.3141/2017-02

Liu, Z., & Kringos, N. (2024). Transition from linear to circular economy in pavement engineering: A historical review. *Journal of Cleaner Production, 141809*, 141809. https://doi.org/10.1016/j.jclepro.2024.141809

López, A. I., Ramírez-Díaz, A., Castilla-Rodríguez, I., Gurriarán, J., & Mendez-Perez, J. A. (2023). Wind farm energy surplus storage solution with second-life vehicle batteries in isolated grids. *Energy Policy, 173*, 113373. https://doi.org/10.1016/j.enpol.2022.113373

Mantalovas, K., & Di Mino, G. (2019). The sustainability of reclaimed asphalt as a resource for road pavement management through a circular economic model. *Sustainability, 11*(8), 2234. https://doi.org/10.3390/su11082234

Miatto, A., Schandl, H., Wiedenhofer, D., Krausmann, F., & Tanikawa, H. (2017). Modeling material flows and stocks of the road network in the United States 1905–2015. *Resources, Conservation and Recycling, 127*, 168–178. https://doi.org/10.1016/j.resconrec.2017.08.024

Mkahal, Z., Mamindy-Pajany, Y., Maherzii, W., & Abriak, N. E. (2022). Recycling of mineral solid wastes in backfill road materials: Technical and environmental investigations. *Waste and Biomass Valorization, 13*(1), 667–687. https://doi.org/10.1007/s12649-021-01544-5

Moore, E. A., Russell, J. D., Babbitt, C. W., Tomaszewski, B., & Clark, S. S. (2020). Spatial modeling of a second-use strategy for electric vehicle batteries to improve disaster resilience and circular economy. *Resources, Conservation and Recycling, 160*, 104889. https://doi.org/10.1016/j.resconrec.2020.104889

Nguyen-Tien, V., Dai, Q., Harper, G. D., Anderson, P. A., & Elliott, R. J. (2022). Optimising the geospatial configuration of a future lithium ion battery recycling industry in the transition to electric vehicles and a circular economy. *Applied Energy, 321*, 119230. https://doi.org/10.1016/j.apenergy.2022.119230

Nowakowski, P., & Mrówczyńska, B. (2018). Towards sustainable WEEE collection and transportation methods in circular economy – Comparative study for rural and urban settlements. *Resources, Conservation and Recycling, 135*, 93–107. https://doi.org/10.1016/j.resconrec.2017.12.016

Nurdiawati, A., & Agrawal, T. K. (2022). Creating a circular EV battery value chain: End-of-life strategies and future perspective. *Resources, Conservation and Recycling, 185*, 106484. https://doi.org/10.1016/j.resconrec.2022.106484

OECD (Organization for Economic Co-Operation and Development). (2010). *Reducing transport greenhouse gas emissions – Trends & data*. Organization for Economic Co-operation and Development.

Olabi, A. G. (2019). Circular economy and renewable energy. *Energy, 181*, 450. https://doi.org/10.1016/j.energy.2019.05.196

Oumnih, S., Bekkouch, N., Gharibi, E. K., Fagel, N., Elhamouti, K., & El Ouahabi, M. (2019). Phosphogypsum waste as additives to lime stabilization of bentonite. *Sustainable Environment Research, 29*, 1. https://doi.org/10.1186/s42834-019-0038-z

Pang, M. (2017). Planning, transformation and development of resource based industrial cities. *Open House International, 42*(3), 88–92. https://doi.org/10.1108/OHI-03-2017-B0018

Peyravi, B., & Jakubavičius, A. (2022). Drivers in the eco-innovation road to the circular economy: Organiational capabilities and exploitative strategies. *Sustainability, 14*(17), 10748. https://doi.org/10.3390/su141710748

Prendeville, S., Peck, D., Balkenende, R., Cor, E., Jansson, K., & Karvonen, I. (2016). *Map of remanufacturing product design landscape*. Available at: https://www.remanufacturing.eu/assets/pdfs/D3.2_FINAL.pdf

Saidani, M., Yannou, B., Leroy, Y., & Cluzel, F. (2018). Heavy vehicles on the road towards the circular economy: Analysis and comparison with the automotive industry. *Resources, Conservation and Recycling, 135*, 108–122. https://doi.org/10.1016/j.resconrec.2017.06.017

Saidani, M., Kendall, A., Yannou, B., Leroy, Y., & Cluzel, F. (2019). Management of the end-of-life of light and heavy vehicles in the U.S.: Comparison with the European Union in a circular economy perspective. *Journal of Material Cycles and Waste Management, 21*(6), 1449–1461. https://doi.org/10.1007/s10163-019-00897-3

Santos, A. S., de Abreu, V. H. S., de Assis, T. F., Ribeiro, S. K., & Ribeiro, G. M. (2021). An overview on costs of shifting to sustainable road transport: A challenge for cities worldwide. In S. S. Muthu (Ed.), *Carbon footprint case studies. Environmental footprints and eco-design of products and processes*. Springer. https://doi.org/10.1007/978-981-15-9577-6_4

Sarasini, S., & Linder, M. (2018). Integrating a business model perspective into transition theory: The example of new mobility services. *Environmental Innovation and Societal Transitions, 27*, 16–31. https://doi.org/10.1016/j.eist.2017.09.004

Seroka-Stolka, O., & Ociepa-Kubicka, A. (2019). Green logistics and circular economy. *Transportation Research Procedia, 39*, 471–479. https://doi.org/10.1016/j.trpro.2019.06.049

Simic, V. (2013). End-of-life vehicle recycling – A review of the state-of-the-art. Teh. Vjesn., v. 20, n. 2, p. 371–380. ISSN 1330-3651 (Print), ISSN 1848-6339 (Online). UDC/UDK 504.064.4:658.567.3]:629.33.

Simic, V. (2016). Interval-parameter chance-constraint programming model for end-of-life vehicles management under rigorous environmental regulations. *Waste Management, 52*, 180–192. https://doi.org/10.1016/j.wasman.2016.03.044

Sochor, J., Strömberg, H., & Karlsson, I. C. M. (2015). Challenges in integrating user, commercial, and societal perspectives in an innovative mobility service. In *94th annual meeting of the Transportation Research Board, Washington DC* (pp. 1–16). https://doi.org/10.1104/pp.109.13885

Song, A., & Zhou, Y. (2023). Advanced cycling ageing-driven circular economy with E-mobility-based energy sharing and lithium battery cascade utilisation in a district community. *Journal of Cleaner Production, 137797*, 137797. https://doi.org/10.1016/j.jclepro.2023.137797

Suescum-Morales, D., Romero-Esquinas, Á., Fernández-Ledesma, E., Fernández, J. M., & Jiménez, J. R. (2019). Feasible use of colliery spoils as subbase layer for low-traffic roads. *Construction and Building Materials, 229*, 116910. https://doi.org/10.1016/j.conbuildmat.2019.116910

Szabo, I., Scurtu, L. I., Raboca, H., & Mariasiu, F. (2023). Topographical optimization of a battery module case that equips an electric vehicle. *Batteries, 9*(2), 77. https://doi.org/10.3390/batteries9020077

Tam, E., Soulliere, K., & Sawyer-Beaulieu, S. (2019). Managing complex products to support the circular economy. *Resources, Conservation and Recycling, 145*, 124–125. https://doi.org/10.1016/j.resconrec.2018.12.030

Tripathy, A., Bhuyan, A., Padhy, R., & Corazza, L. (2022). Technological, organizational, and environmental factors affecting the adoption of electric vehicle battery recycling. *IEEE Transactions on Engineering Management., 71*, 12992. https://doi.org/10.1109/TEM.2022.3164288

Un-Habitat. (2011). *Cities and climate change: Global report on human settlements* (p. 279). Earthscan. ISBN. 9781849713719.

Vuchic, V. R. (1999). *Transportation for livable cities*. Centre for Urban Policy Research.

Wang, Q.-Z., Wang, N.-N., Tseng, M.-L., Huang, Y.-M., & Li, N.-L. (2019). Waste tire recycling assessment: Road application potential and carbon emissions reduction analysis of crumb rubber modified asphalt in China. *Journal of Cleaner Production, 119411*, 119411. https://doi.org/10.1016/j.jclepro.2019.119411

Zhang, L., Long, R., Chen, H., & Geng, J. (2019). A review of China's road traffic carbon emissions. *Journal of Cleaner Production, 207*, 569–581. https://doi.org/10.1016/j.jclepro.2018.10.003

Zhao, R., Liu, Y., Zhang, Z., Guo, S., Tseng, M.-L., & Wu, K.-J. (2018). Enhancing eco-efficiency of agro-products' closed-loop supply chain under the belt and road initiatives: A system dynamics approach. *Sustainability, 10*(3), 668. https://doi.org/10.3390/su10030668

Zhao, S., Li, G., He, W., Huang, J., & Zhu, H. (2019). Recovery methods and regulation status of waste lithium-ion batteries in China: A mini review. *Waste Management and Research, 37*(11), 1142–1152. https://doi.org/10.1177/0734242X19857130

Zhou, W., Zhu, B., Chen, D., Griffy-Brown, C., Ma, Y., & Fei, W. (2012). Energy consumption patterns in the process of China's urbanization. *Population and Environment,* (2/3), Special Issue on Urbanization and Environment in China, 202–220. https://doi.org/10.1007/s11111-011-0133-5

Zhou, F., Lim, M. K., He, Y., Lin, Y., & Chen, S. (2019). End-of-life vehicle (ELV) recycling management: Improving performance using an ISM approach. *Journal of Cleaner Production, 228*, 231–243. https://doi.org/10.1016/j.jclepro.2019.04.182

Best Practices Applied in Logistic Intern: From the Perspective of Reducing the Carbon Footprint

Tássia Faria de Assis, Victor Hugo Souza de Abreu, Lino Guimarães Marujo, and Marcio de Almeida D'Agosto

1 Introduction

The sustainable logistics and the reduction of greenhouse gas (GHG) emissions related to its operations have been increasingly garnering attention from both the scientific and business communities. Society and stakeholders are demanding more conscientious measures from companies, while governments at all levels are adopting more stringent regulatory measures and recommendations, influencing the logistics decision-making process (Perotti et al., 2023; De Assis et al., 2023).

According to McKinnon and Piecyk (2012), there can be various reasons for a company to reduce carbon emissions, with cost savings being a primary and compelling motivation for many companies. Additionally, the pursuit of eco-efficiency driven by financial concerns, as well as environmental efforts for marketing and corporate social responsibility (CSR), plays a significant role.

However, as the global business landscape continues to evolve, environmental, social, and governance (ESG) considerations have taken center stage, transforming how companies operate and are perceived by stakeholders. It is observed that in recent years, the importance of sustainable practices in various sectors has been highlighted due to growing concerns about climate change, resource scarcity, and

T. F. de Assis · V. H. S. de Abreu (✉) · M. de Almeida D'Agosto
Federal University of Rio de Janeiro (Universidade Federal do Rio de Janeiro – UFRJ), Transport Engineering Programme (Programa de Engenharia de Transportes – PET), COPPE-UFRJ, Rio de Janeiro, Brazil
e-mail: victor@pet.coppe.ufrj.br

L. G. Marujo
Federal University of Rio de Janeiro (Universidade Federal do Rio de Janeiro – UFRJ), Production Engineering Programme (Programa de Engenharia de Produção – PEP), COPPE-UFRJ, Rio de Janeiro, Brazil

S. S. Muthu (ed.), *Carbon Footprint Assessments*, Environmental Footprints and Eco-design of Products and Processes,
https://doi.org/10.1007/978-3-031-70262-4_13

social inequality. This has led companies to increasingly integrate ESG principles and action plans into their strategies to reduce negative impacts and mitigate risks (Deloitte, 2023).

Warehouses are recognized as essential components of supply chains; however, their operations have become more complex over time. They have evolved from simple stock repositories to multifunctional logistics centers, posing significant challenges and requiring improvements in operational efficiency, service level fulfillment, and environmental impact reduction. This has led to a growing search for practices capable of reducing resource consumption and simultaneously decreasing emissions related to logistics facilities, which account for 11% of GHG emissions (Perotti et al., 2023).

To comply with carbon emissions regulations, companies have sought to adopt equipment, facilities, or vehicles with higher energy efficiency. They also aim to optimize operational decisions in production, transportation, and inventory to reduce carbon emissions (Hua et al., 2011). In this context, a considerable amount of energy is required for heating, cooling, lighting, and material handling equipment (MHE) in warehouses, representing about 20% of overall logistics costs. Therefore, reducing warehouse energy consumption would result in a significant environmental benefit. Sustainable strategies that enable energy consumption minimization pose a new challenge in warehouse management (Carli et al., 2020).

The increasing need to study warehouse sustainability has driven new research regarding management concepts, technologies, and techniques to reduce carbon emissions (Ibrahim et al., 2022). This study aims to explore optimal practices for reducing activity, improving operation, and mitigating greenhouse gas emissions in internal logistics, focusing on low-carbon practices within the operational boundaries of an organization and proposing a methodology to help evaluators and decision-makers define evaluation limits for the scopes, categories, activities, and elements of the internal logistics operation.

This encompasses all logistical processes and operations in a warehouse, including receiving, internal transportation, material handling, facility management, warehousing (inventory), and packaging (use and retail) activities (Freis et al., 2016; De Lima et al., 2020). The goal is to reduce energy usage, atmospheric pollutant emissions, and GHG emissions (Ibrahim et al., 2022).

2 Literature

The purpose of this section is to provide a comprehensive description of internal logistics, covering its various constituent elements. Included in this analysis are crucial areas such as Receiving, Internal Transportation, Material Handling, Warehouse (installation), Storage (stock), and Packaging (use and retail). In addition, it is important to pay special attention to the carbon footprint of internal logistics, considering not only its measurement but also the strategies and solutions to reduce it. Understanding and effectively managing the carbon footprint in internal logistics has become a priority as companies seek not only to comply with regulatory requirements but also to take a proactive stance toward environmental sustainability.

In this context, exploring innovative strategies and sustainable solutions is crucial. This involves everything from optimizing routes and using more efficient technologies to adopting sustainable packaging and integrating renewable energy sources into facilities. By addressing these practices, companies not only reduce their environmental impact but also contribute positively to broader sustainability goals by promoting a responsible and conscious approach in their internal logistics operations.

2.1 Internal Logistic

Internal logistics, also known as intralogistics, focuses on organizing and managing the flow of information and materials within a company. The term internal logistics refers to all the internal processes of the supply chain, from receiving raw materials to shipping finished products (MASTERMOVER, 2024).

Internal logistics is the part of logistics that covers all internal supply chain management, which includes the storage, transport, and distribution of goods within the organization, in order to meet its domestic demands, as support for manufacturing. In this way, the support cycle for production activities is considered to be directly related to internal logistics, i.e., production planning and control. Thus, the main objective of production logistics support is to establish and maintain an economical and orderly flow of materials and stocks in process, to meet the production sector's schedules.

Production logistics support has operational responsibility for the following activities: handling and storage of products, materials, components, and semi-finished parts. With changes in the business environment, the concepts of logistics services have evolved and various issues have been added to operational logistics tasks such as packaging, outsourced stock management, barcoding, and information systems, in which these activities must interact with other functional areas of the company.

2.1.1 Receiving

Performs the function of physically receiving the material and performing the initial inspection of the incoming shipment for damage and validation against the purchase order quantity to ensure compliance with the purchase order, identification and delivery to destination, and preparation of receiving reports, usually including some initial data recording (Vitasek, 2013; IWLA, 2020).

2.1.2 Internal Transportation

Internal transportation is a process whose purpose is to transport raw materials, work in progress, and finished products within an industrial plant and is necessary for the smooth running of the production flow in an industry. It can be classified as a support process which, with a few exceptions, consumes/uses a significant part of a company's energy, as the energy use of internal transportation is usually low. Examples of different types of internal transportation are conveyor belts, sleepers, cranes, and other lifting devices and various types of trucks (Thollander et al., 2020).

2.1.3 Materials Handling

Materials handling is a system or combination of methods, facilities, labor, and equipment for moving, packaging, and storing materials to meet specific objectives (Nilsson, 2018). The function of materials handling is to control and protect materials, goods, and products throughout the manufacturing, distribution, consumption, and disposal process (Vieira et al., 2011). Where it involves short-distance movement within the physical spaces of a facility, such as a factory or warehouse and between a facility and a transportation agency, it can create utility of time and place through material handling, storage and control (Kay, 2012). In addition, materials handling is an important logistics activity for industrial organizations, which must be subject to meticulous care. Ineffective material handling can result in product damage, delays, theft, deterioration, etc. (IIMM, 2020).

2.1.4 Warehouse (Facility)

A warehouse can be defined as a physical facility in the supply chain where the product pauses, even if briefly, to be moved and handled, which consumes space and time (man-hours) (Bartholdi & Hackman, 2019). In other words, it is a place for storing products, whose main warehouse activities include product receipt, storage, dispatch, and order picking (IWLA, 2020).

2.1.5 Storage (Stock)

Warehousing is the activity that encompasses inventory management, materials handling, and the temporary storage of goods in facilities, usually warehouses (Satolo et al., 2020). The warehouse's basic function is to serve as a bridge between the production and consumption of goods. It ensures that products remain properly stored until they are sent for collection, facilitating their distribution and contributing to the operational efficiency of the logistics process (IIMM, 2020).

However, these operations generate a significant environmental impact, especially in terms of carbon dioxide (CO_2) emissions. Therefore, warehouse management must look beyond operational efficiency and include environmental impact, seeking to reduce CO_2 emissions to comply with legal requirements and sustainability targets (Sadeghi et al., 2024).

2.1.6 Packaging (Use and Retail)

Packaging is an essential activity related to protection, containment, unification, distribution, and communication (Satolo et al., 2020). It is a coordinated system for preparing goods for movement, transport, distribution, storage, retail, consumption, and recovery, reuse, or disposal in a safe, efficient, and effective manner, combined with maximizing consumer value, sales, and profit (García-Arca et al., 2017). In addition, packaging must fulfill the functions of: (i) protection, adapting to the technical and functional characteristics of the products, as well as their value; (ii) handling and movement, storage, and transport adapted according to the production system adopted, be it manual, automated, or mechanized; (iii) informative; and (iv) recycling (Szymonik, 2016).

According to Saghir (2004), packaging is considered a support activity for other logistics activities, such as transportation, inventory management, order processing, and warehousing. In this case, it involves the transportation of goods and reduces the possibility of damage caused during transportation. For example, if items such as processed food are not packaged correctly, they can perish during transportation (IIMM, 2020). It offers convenience to products, as well as contributes to sustainability in logistical efficiency in the supply chain by reducing environmental impacts, where approximately 75% of packaging is recycled after use. This activity plays a fundamental role in the transportation of goods, mitigating potential damage during the transportation activity.

2.2 Carbon Footprint in Internal Logistics

Climate change, brought about by the increase in GHG emissions, and its social, environmental, economic, and ethical implications are widely recognized as one of the primary interconnected sets of challenges that human societies face (Huisingh et al., 2015; De Abreu et al., 2022a, b; Da Costa et al., 2022). In this context, it is essential to acknowledge that the transportation and logistics sector emerges as one of the main contributors to this issue (De Assis et al., 2022a, b).

Thus, many companies view ecologically sustainable practices, including reducing their carbon footprint, as a crucial component of their logistical operations (Hrasky, 2011; Lewis, 2016; Górny et al., 2021; De Assis et al., 2022c, 2023). This is because they need to promote the efficient use of resources, with minimal

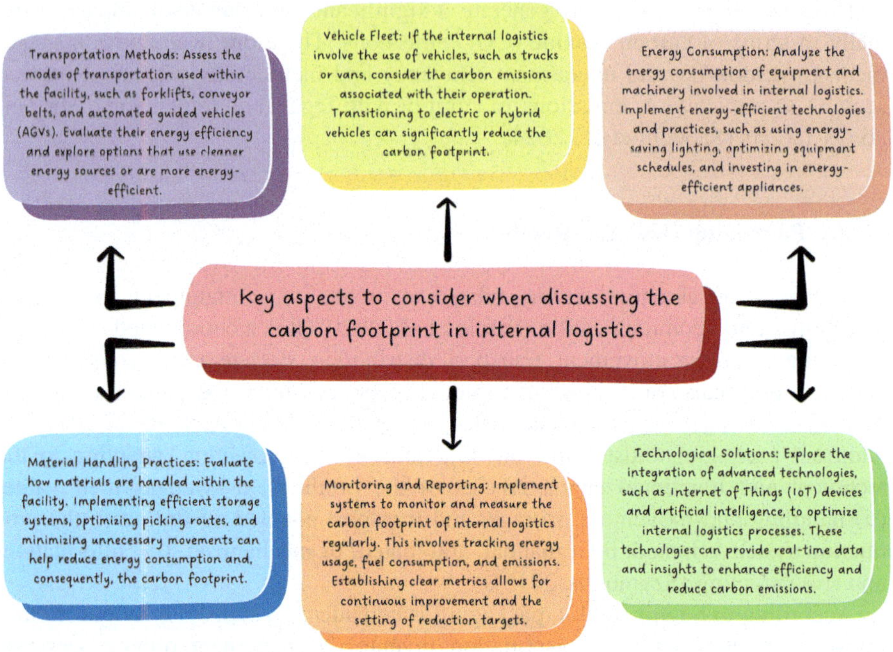

Fig. 1 Key elements of the carbon footprint in internal logistics

impact on the environment and the quality of life of the population, and reflect regional requirements that often do not align with global perspectives (Lagoudis & Shakri, 2015; Liu et al., 2023).

According to the World Economic Forum (2016), the logistics sector, heavily dependent on fossil fuels, is responsible for 13% of all carbon emissions worldwide. In this regard, reducing emissions in warehouses and other facilities that make up companies' internal logistics is becoming increasingly important, both for legal compliance and to meet sustainability goals (Sadeghi et al., 2024). This relevance is even more evident when addressing key aspects related to the carbon footprint in internal logistics, as highlighted in Fig. 1.

Looking at Fig. 1, there is a pressing need to direct more detailed attention to various critical elements in transportation methods, the composition of companies' vehicle fleets, energy consumption, material handling practices, and internal monitoring and reporting activities. In this context, it is crucial to consider not only operational efficiency but also environmental sustainability. This means exploring and implementing innovative technological solutions that not only optimize logistics processes but also significantly reduce the carbon footprint associated with internal logistics activities. Companies can consider introducing low-emission vehicles, using renewable energy sources to power their logistics operations, applying efficient material handling and storage practices, as well as implementing advanced monitoring systems to track and reduce environmental impacts. In addition, reverse logistics strategies can be explored to minimize waste and promote the reuse of materials (Sun, 2017). Partnerships with suppliers committed to sustainable practices and the pursuit of environmental certifications can also play a key role in creating a greener supply chain. Therefore, when considering the complexity of internal logistics, it is crucial to adopt a holistic approach that not only optimizes operational efficiency but also demonstrates a tangible commitment to environmental sustainability and responsible practices.

The issue of packaging is a crucial aspect to consider in the context of sustainability in the production chain. While the primary function of packaging is to contain and protect products during the storage, transportation, and distribution phases, the appropriate choice of packaging system plays a significant role in ensuring the final quality of the product in the destination market (Lo-Iacono-Ferreira et al., 2021). Furthermore, this choice has the potential to improve the sustainability of the entire process. It is important to note that packaging can also have substantial environmental impacts, especially about GHG emissions. These emissions can vary considerably, representing from 7% to 54% of the total emissions associated with the production of fresh fruit and vegetables, depending on factors such as production volume and specific market characteristics (Payen et al., 2015).

Given this scenario, the choice of sustainable packaging becomes crucial. Opting for recyclable or biodegradable materials can not only mitigate environmental impacts but also contribute to building a more eco-efficient production chain. Reducing the excessive use of packaging is another effective strategy, not only to promote sustainability but also to reduce the carbon footprint associated with internal logistics operations. Thus, by considering packaging as an integral part of the product's life cycle, it is possible to implement practices that combine operational efficiency with environmental responsibility. By tackling these aspects, companies can work on minimizing their carbon footprint in internal logistics, thus contributing to the overall goals of sustainability and environmental responsibility (Olatunji et al., 2019). The integration of innovative strategies, such as the adoption of cleaner transport technologies and route optimization, can result in considerable environmental benefits (De Assis et al., 2021).

2.3 Strategies and Solutions to Reduce the Carbon Footprint of Internal Logistics

The complexity of environmental issues imposes significant challenges, underscoring the essential need for research to underpin the adoption of business practices aimed at reducing environmental impact throughout the supply chain (De Assis et al., 2023). This includes the internal logistics of companies, whose impact is substantial in both environmental and social realms (Dey et al., 2011; Bocken & Allwood, 2012). In this context, it becomes imperative to explore strategies and solutions designed to reduce the carbon footprint associated with logistical operations.

To strengthen the credibility and consistency of these initiatives, it is fundamental that carbon reduction goals align with specific principles applicable to the decarbonization of logistics, as outlined by McKinnon and Piecyk (2012):

- Wherever possible, targets should cover the entire logistics operation, recognizing the carbon commitments that permeate the various logistics activities.
- Targets should be expressed in terms of carbon intensity, using logistics production measures as normalizers.
- For target periods longer than 3–4 years, it is necessary to establish "transition" targets for the intervening years, highlighting the trajectory for continuous carbon reduction.
- The scope of the logistics carbon measurement and the associated target must be clearly spelled out, delimiting the relevant organizational, geographical, functional, and hierarchical boundaries.
- Where an industry-wide carbon reduction scheme exists, companies should consider adhering to this scheme to drive logistics decarbonization across the sector.

By adhering to these principles, organizations will be better positioned to address environmental challenges, promote sustainable business practices, and effectively contribute to global-scale logistics decarbonization.

To implement effective and innovative emissions management, it is imperative to establish comprehensive operational limits that encompass both direct and indirect emissions (De Assis et al., 2023). This proactive approach enables a company to more efficiently manage the broad spectrum of risks and opportunities related to GEE across its value chain (Lagoudis & Shakri, 2015). Clearly defining operational limits not only provides a structured framework for emission reduction and monitoring but also allows for the identification of specific areas for improvement (De Assis et al., 2022a, b). Additionally, by encompassing both direct emissions from internal company operations and indirect emissions linked to supply chain activities and product use, effective GEE management becomes more holistic (De Assis et al., 2023).

In this regard, companies can adopt practices such as the implementation of cleaner technologies, the use of renewable energy sources, optimization of operational efficiency, and engagement with suppliers and partners committed to sustainability (Santos et al., 2021). Furthermore, by setting clear emission reduction goals and monitoring progress toward these goals, the organization will be better positioned to identify innovation opportunities and mitigate the negative impacts of GEE on its operations (De Assis et al., 2022a, b).

The engagement of all stakeholders in this process is also crucial. If companies intend to decrease their emissions overall, it is crucial to consider various stakeholders. As highlighted by Henriques and Sadorsky (1999), proactive companies recognize regulators, customers, shareholders, and local communities as significant stakeholders. Governments and NGOs, for example, play important facilitating roles in accelerating the shift toward more sustainable logistics (Bocken & Allwood, 2012).

Furthermore, companies must possess the ability not only to visualize and map their supply chains but also to conduct detailed comparative analyses of their sustainable efforts compared to other organizations in the same industry. This requires a comprehensive understanding of sustainability practices and initiatives adopted by their peers, providing a solid foundation for continuously enhancing their own strategies. Benchmarking within the industry enables companies to identify areas of opportunity, set realistic goals, and promote healthy competition towards higher standards of environmental responsibility and sustainability throughout the supply chain (Dey et al., 2011).

By aligning these practices with overall sustainability goals, companies not only meet regulatory requirements but also strengthen their market position, responding to growing consumer demands for environmental responsibility. Transparency in sustainable initiatives and accountability for environmental impacts can enhance the company's reputation, establishing it as a leader in environmental responsibility. Therefore, by implementing specific strategies to reduce the carbon footprint in internal logistics, companies not only contribute to global sustainability goals but also strategically position themselves to address emerging environmental challenges, building a more sustainable future for future generations (Olatunji et al., 2019).

3 Methodology Applied to Promote Low-Carbon Internal Logistics

After a thorough review of the literature covering the fundamental concepts of internal logistics, carbon footprint, and strategies aimed at mitigating them, the section sets out to present a comprehensive methodology aimed at boosting effective low-carbon internal logistics practices. Inspired by definitions established by the Çelekli & Zariç (2023), which conceptualizes the carbon footprint as a precise measure of the impact of human activities on the amount of carbon dioxide (CO_2) generated by burning fossil fuels, expressed numerically in tons of CO_2 emissions, the intricate steps of this methodology are detailed in a concise and enlightening manner, as presented in a visually elucidating way in Fig. 2.

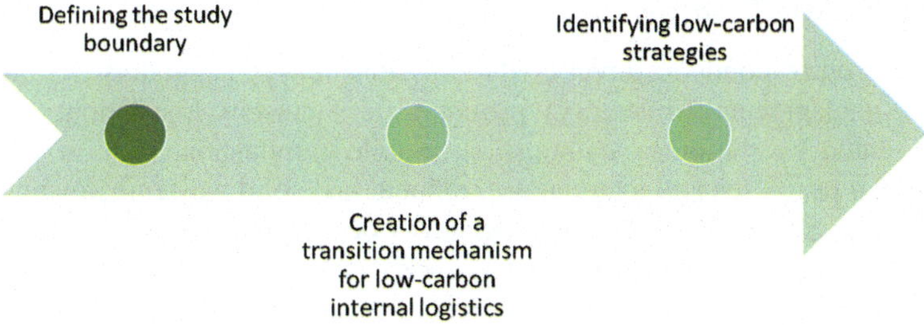

Fig. 2 Steps of methodology proposed

Step 1 is dedicated to meticulously defining the boundaries of the study, a crucial step in establishing clear parameters and ensuring a comprehensive and accurate understanding of the scope involved. This strategic delimitation provides the foundation for an effective and targeted analysis, ensuring that all relevant variables are duly considered. Moving on to Step 2, a proactive approach is taken to creating a transition mechanism aimed at the effective implementation of low-carbon internal logistics practices. This strategic phase involves the careful design of adaptive processes and protocols, aligned with sustainability principles, to facilitate a smooth and efficient transition to low-carbon operations.

Concluding the methodological cycle, Step 3 involves the meticulous identification of specific strategies aimed at tangibly reducing the carbon footprint. At this stage, the aim is not merely to recognize areas for improvement, but to propose innovative and effective solutions, promoting a tangible transformation in internal logistics operations. Thus, the proposed methodology aims to not only offer a practical guide for promoting low-carbon internal logistics but also establish a solid and effective paradigm, outlining clear paths for the sustainable implementation of these practices. These steps are not only sequential, but interconnected, forming a comprehensive and integrated framework to effectively drive change and improve logistics practices in search of a more sustainable future.

4 Defining the Study Boundary

As a first step in the methodology, it is important to limit what will be considered, i.e., the scope of GHG emissions, categories, activities, and elements, as described in the following subsections.

4.1 Scopes

In order to define the limits of the level of detail of the activities subject to carbon emissions accounting that will be assessed, it is necessary to define what to consider as scope 1, scope 2, and scope 3 within the internal logistics approach. Thus, according to ABNT (2015):

- Scope 1 corresponds to direct GHG emissions and is mandatory for almost all GHG management (inventory) programs and current legislation. It can be represented by stationary combustion, mobile combustion, fugitive emissions, industrial processes and material use, effluent and solid waste treatment, agricultural processes, GHG removals (vegetation decomposition, biogas burning and carbon capture and storage), and CH_4 and N_2O emissions from biomass combustion. Table 1 shows examples of operations applied to internal logistics in the context of scope 1.

Table 1 Examples of operations with direct GHG emissions applied to internal logistics

Emission type	Emission source class	Emission source
Stationary combustion	Non-transport	Equipment or machinery (EPA, 2016)
Mobile combustion	On-road vehicles	Cars and trucks that consume fuel (EPA, 2023b)
	Non-road vehicles	Equipment such as forklifts that consume fuel (EPA, 2023b)
	Refrigerants	Mobile air conditioning, transport (internal) (EPA, 2023b)
Fugitive emissions	Refrigerants and air-conditioning equipment	Heat pumps, chillers, retail food refrigeration, cold stores, refrigerators, industrial process refrigeration and commercial unitary air-conditioning systems, fixed and portable fire suppression equipment (EPA, 2023a)

It should be noted that stationary combustion includes the burning of fuels to generate electricity, steam, heat, or energy in stationary equipment such as boilers, furnaces, etc. (Ranganathan et al., 2004). And emissions from the refrigeration and air-conditioning sector result from the manufacturing process, from leaks and maintenance during the operational life of the equipment, and from disposal at the end of the equipment's useful life (EPA, 2023a).

- Scope 2 corresponds to indirect GHG emissions from energy use associated with the production of electricity, heat, or steam purchased by the organization and used within the defined organizational boundary. They are mandatory in almost all GHG management (inventory) programs and current legislation.
- Scope 3 corresponds to all other sources of indirect GHG emissions and can be reported optionally. In other words, other indirect emissions besides Scope 2 (Emissions by others related to the company's activities) (Ministry of the Environment Japan, 2013). Scope 3 emissions are generally classified as upstream or downstream and correspond to the supply chain. Upstream emissions refer to indirect GHG emissions related to goods and services purchased or acquired by organizations, and downstream emissions refer to indirect GHG emissions related to goods and services sold. This classification avoids double counting between organizations in a supply chain.

4.2 Categories

Internal logistics processes are usually grouped into three main categories, as described in the following subsections.

4.2.1 Internal Material Flow

It refers to the flow of materials within a single organization, either at one facility or across multiple facilities. These internal material flows can be categorized into two types: flows between different activities within the system or subsystem (cross-border internal flows) and flows within each individual activity (internal flows) (OECD, 2007). For example, the flow of materials in a facility, from a warehouse to production, or the transport of parts or components from one factory to another in the same organization.

4.2.2 Information Flow and Stock Management

Information flow in inventory management is part of a management information system that deals with the information needed to control inventory, from the acquisition of raw materials to the distribution of finished products. This crucial component covers the collection, analysis, and dissemination of data relating to the quantity of items in stock, movements, delivery times, market demand, seasonality, and other relevant factors.

By covering all stages of the product life cycle, the flow of information in inventory management provides a comprehensive, real-time view of inventory status, enabling more informed and strategic decision-making. The effectiveness of this information flow is essential for optimizing stock levels, minimizing costs associated with storage, efficiently meeting customer demands, and, consequently, improving the operational efficiency of the organization as a whole.

Internal logistics is often associated with high-tech automated equipment and software systems. However, at their core, intralogistics processes revolve around the "flow of information," whether using advanced Industry 4.0 inventory management software or simply providing clear and easy access to the information needed to complete a task.

4.2.3 Warehouse Management and Logistics Flow

The warehouse management system is used to effectively manage warehouse business processes and direct warehouse activities, including receiving, putaway, picking, shipping, and stock cycle counting. It also includes support for radio frequency communications, enabling real-time data transfer between the system and warehouse staff (IWLA, 2020).

This includes all logistics processes and operations within a warehouse (MASTERMOVER, 2024). A warehouse can be defined as a physical facility in the supply chain where products temporarily pause to be moved and handled, consuming both space and time (man-hours) (Bartholdi & Hackman, 2019). In other words, a warehouse is a place for storing products, with key activities such as receiving, storing, shipping, and order picking (IWLA, 2020).

4.3 Activities

The activities of internal logistics within the operational boundaries of an organization are considered to cover all logistics processes and operations in a storage system, covering the following activities: (i) receiving, (ii) internal transport, (iii) materials handling, (iv) warehouse (installation), (v) storage (stock), and (vi) packaging (use and retail) (Freis et al. 2016; de Lima et al., 2020), as defined in the literature, which low-carbon project evaluators and decision-makers should choose.

4.4 Elements

According to Amjed and Harrison (2013) and Klug (2013), internal logistics management can be influenced by different elements, as shown in Fig. 3.

4.4.1 Building

It is important to note that for a building, the size, design, and operation of warehouse buildings are challenges when it comes to energy efficiency, since they can be characterized by: (i) large open spaces, (ii) narrow aisles between high rack

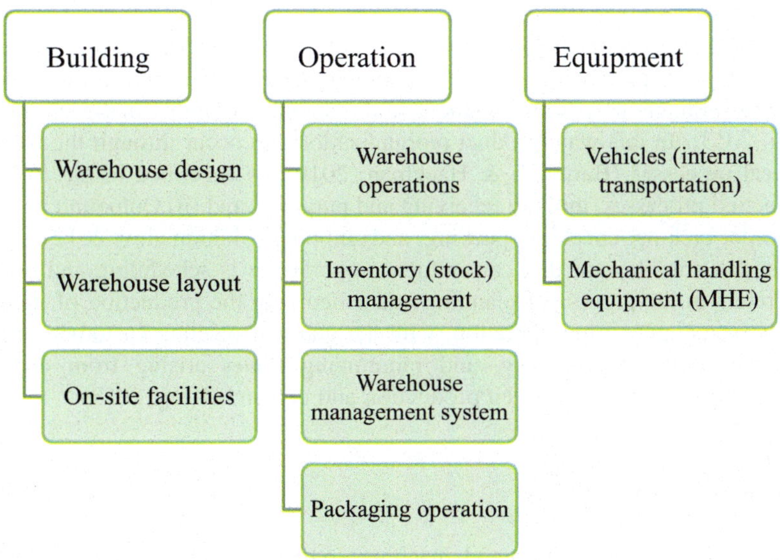

Fig. 3 Classification of elements proposed for low carbon internal logistics

systems, (iii) perishable items that require exact temperatures, (iv) high ceilings and material roofs that can create the risk of high internal temperatures due to solar gain, (v) large external doors that are opened and closed frequently.

It is also important to note that the development of the layout and design of a warehouse is the first stage in the design of a facility, consisting of the areas that must be perfectly defined when designing a warehouse layout: (i) Loading/unloading areas, (ii) Receiving area, (iii) Storage area, (iv) Picking area, (v) Shipping area, and (vi) Service area (Bartholdi & Hackman, 2019). In this context, layout refers to the physical organization of production resources and processes, machines, and workstations in a production or manufacturing environment to optimize efficiency, reduce cycle times, meet customer demands, and minimize variability (production fluctuations, defective products, and process deviations) (Klug, 2013).

4.4.2 Operation

The operation can be defined by the warehouse, stock, and packaging management system, which involves sub-elements such as capacity, batch, and order scheduling. Capacity can therefore be defined as the potential of the value-added system to allow physical materials to be transported and handled within factories, impacting the speed and distance of material movement (Klug, 2013). Material handling capacity includes the availability of equipment, storage facilities, and staging facilities, and it is important that the warehouse capacity accommodates the stock properly and that it fits within the organization (IIMM, 2020). Capacity also generally refers to the maximum output or production capacity of a machine, a person, a process, a factory, a product, or a service (Morgan et al. 2021).

As for the sub-element called batch, this refers to a specific quantity of items or products that are produced, ordered, or handled together for various reasons, such as economic efficiency, cost management, and reducing variability in the supply chain (Klug, 2013). In this way, product reorganization can occur through the following physical processes (Bartholdi & Hackman, 2019), as described in the following: (i) Inbound processes, through receiving and putaway; and (ii) Outbound processes, with order picking, checking, packing, and shipping (Makhmudov, 2012).

Another sub-element that can influence operations is scheduling and ordering. This refers to the process of planning and scheduling the production of items in a factory or manufacturing operation, with the goals of meeting customer demands, optimizing process efficiency, and minimizing issues arising from unplanned changes or mismatches between production and demand (Klug, 2013).

4.4.3 Equipment

Internal vehicles, equipment, and machinery are part of this group of elements, where equipment can be defined as tools used to move, store, control, and protect goods and products in the storage system and can be classified as transportation

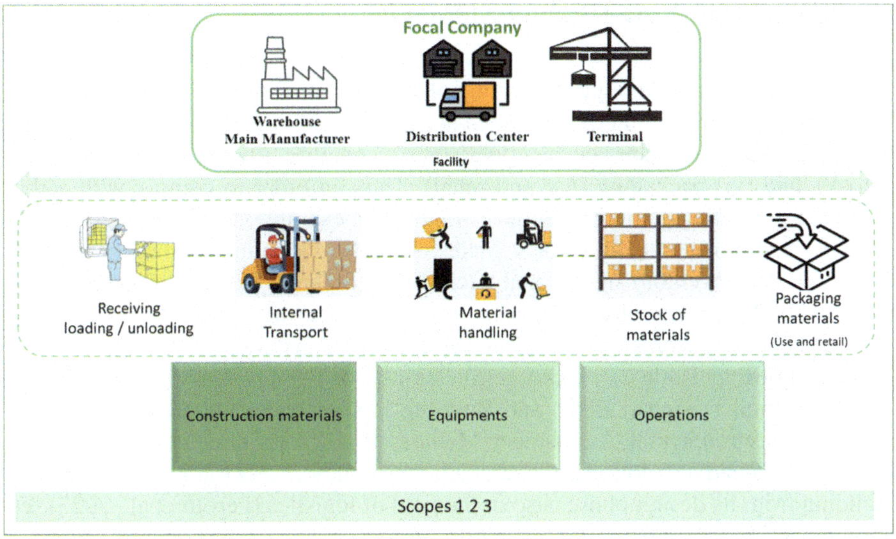

Fig. 4 Study operational boundaries

equipment, positioning equipment, unit load formation equipment, and storage equipment (IIMM, 2020). The limitations suggested for this methodology for assessing low-carbon practices in internal logistics, which involves logistics activities within the operational boundaries of an organization, are shown in Fig. 4.

5 Creation of a Transition Mechanism for Low-Carbon Internal Logistics

The transition mechanism to low-carbon internal logistics has certain challenges, such as understanding where to start, what to do, and when and how to act. In this way, it can be adopting measures and actions that can reduce activity, such as handling distance in the facility (warehouse, terminal, distribution centers, etc.), improve a process (for example, by automating a manual system), and mitigate or eliminate (net zero) emissions.

In climate policy, mitigation measures are technologies, processes, or practices that contribute to human intervention to reduce emissions or increase greenhouse gas sinks, for example renewable energy technologies, waste minimization processes, and public transport practices. While net zero measures are net zero emissions, achieved when anthropogenic emissions of GHG into the atmosphere are balanced by anthropogenic removals over a specific period (IPCC, 2018).

6 Identifying Low-Carbon Strategies

Strategies to comply with a low-carbon policy in internal logistics can include good practices according to internal logistics activities such as (i) receiving, (ii) internal transportation, (iii) materials handling, (iv) warehouse (installation), (v) storage (stock), and (vi) packaging (use and retail). Thus, in order to comply with carbon emission regulations, companies have sought, for example, to adopt more energy-efficient equipment, facilities, or vehicles, as well as optimizing their production, transport, and inventory operational decisions in order to reduce carbon emissions (Hua et al., 2011). Therefore, examples of best practices are presented for receiving, warehousing, and storage activities in Table 2.

In addition to Table 2, various certifications, such as Leadership in Energy and Environmental Design (LEED) and Building Research (Building Research Establishment Environmental Assessment Method, BREEAM), are becoming increasingly widespread as a low-carbon strategy, since they assess the sustainability of a building from its design phase, also in the field of logistics (Perotti et al., 2023). For moving and handling activities, examples of best practices are shown in Table 3.

For inventory management, examples of best practices are shown in Table 4.

In Table 5, exemplary instances of best practices for the packaging activity are elucidated.

7 Final Consideration

Reducing the carbon footprint is still a challenge for all sectors of the economy, including internal logistics, a crucial aspect of supply chain management that encompasses the entire spectrum of activities within an organization. Therefore, to meet the needs of this sector, this study aims to explore optimal practices for reducing activity, improving operation, and mitigating greenhouse gas emissions in internal logistics, focusing on low-carbon practices within the operational boundaries of an organization and proposing a methodology to help evaluators and decision-makers define evaluation limits for the scopes, categories, activities, and elements of the internal logistics operation.

The best practices encompass warehouse, internal transport, movement, handling, and storage activities. Therefore, for the warehouse, it was possible to observe strategies aimed at the construction element, which seeks to improve the layout and design of the installation, as well as reduce energy consumption by adopting techniques that promote clearer installations in terms of lighting. According to the best practices that correspond to warehouses, we can see the influence of other activities such as receiving, warehousing, and storage activities, which make up a group that complements each other. For the handling and movement activity, it is important to adopt strategies that reduce activities and movements in the internal environment to reduce energy consumption, as well as strategies aimed at stock

Table 2 Best practices associated with low-carbon warehouse management

Best practices	Description
5S or Housekeeping methodology	The 5S stands for the five Japanese words Seiri, Seiton, Seiso, Seiketsu, and Shitsuke, and comprises a set of practices aimed at generating productivity improvements by creating and maintaining clean and well-organized workplaces. Implementing 5S can significantly reduce the area required for operations by organizing and disposing of unused equipment and supplies. Reducing storage space reduces the energy needed to heat and light the space
Improving Warehouse Design (layout)	Strategies for designing the most efficient layout for material handling focus on reducing costs, minimizing travel distances, and thereby decreasing emissions and carbon footprint
Use of energy-saving and ecological lighting	Use of certified energy-saving lighting (e.g., LED) with high intensity; reduction of lamps containing mercury compounds
Use of automatic lighting control devices	Application of motion sensors and clock-control lighting help reduce energy consumption
Roof surface covered by skylights/light transmission through skylights	The use of highly transparent glass (according to the requirements of the stored products) helps to reduce energy consumption
Use of renewable energy	Installation of solar panels; use of wind energy to reduce fossil fuel consumption
Level of insulation in the building	Use of building materials for thermal insulation and anti-condensation of warehouse walls, floors, and ceilings to reduce the need for energy use
Use of automatic temperature control devices	Installation of air conditioning with thermostats; installation of temperature sensors and weather regulators to reduce the need for energy use
Use of air purification methods and tools	Installation of respirators and filters; installation of actuators for automatic opening of skylights to reduce the use of energy
Use of ecological materials in building structures	Use of certified low-emission paints, varnishes, silicones and other sealants, concrete floors, and floor coverings; use of soundproofing materials; use of recycled materials
Improving the level of automation	Use of robots, automatic conveyors, self-propelled trolleys, automatic sheet presses, etc.; use of mobile terminals and sorting zones increase energy efficiency
Improving the IT level	Applying IT systems to optimize warehouse work (e.g., WMS, WCS, ERP, SCM, etc.)
Effective stock control	Optimizing storage strategy; optimizing stock levels; maximizing the use of storage space
Reduced distance during warehouse operations	Optimizing collection routes; optimizing the distribution of storage areas and storage equipment; creating picking zones, picking paths, consolidating orders

(continued)

Table 2 (continued)

Best practices	Description
Training employees in materials handling equipment	Cyclical employee training to reduce equipment energy use and improve equipment energy efficiency
The use of low-emission moving through the warehouse	Purchase of low-emission internal transport equipment

Source: Elaborated by the author and adapted from Facchini et al. (2018), Malinowska et al. (2018), Gutierrez et al. (2020), Carli (2020), and Oloruntobi et al. (2023)

Table 3 Best practices associated with low-carbon handling and movement

Best practices	Description
Use of equipment with an alternative energy source (electricity, LPG)	Use of cleaner energy to minimize GHG emissions
Use of picker-to-part system	It is based on the adoption of fixed equipment for material handling, featuring only stable belts, such as belts or sorters, whose energy consumption is only related to the size (length) of the system
Use of part-to-picker	It is based on the adoption of mobile materials handling equipment, considering only mobile equipment such as forklifts, pick-ups, and automated storage/retrieval systems, whose energy consumption depends on the specifications and movements of the equipment
Use of order batching system	Strategies for classifying orders from a group of customers. The order sequencing problem plays a central role in reducing energy consumption due to material handling in warehouses. Reduces warehouse carbon emissions by reducing useless works in order picking
Use of picker routing system	Strategies to identify routing in order to reduce the distance traveled by order pickers
Automated guided vehicles (AGVs)	AGVs can theoretically move freely in the zone in front of the warehouse and can deliver and retrieve goods on pallets to dedicated transfer stations in an energy-efficient manner
Use of remanufactured pallets	The use of remanufactured pallets helps to reduce GHG emissions
Maximizing the fill rate in transport and storage	It is important that packaging contributes to minimizing transport along the supply chain because the environmental impact of transport is as high or higher for many products than the direct environmental impact of packaging
Optimizing unitization	The packaging unitization function contributes to efficient materials handling, as the number of discrete packages/loads is reduced

Source: Elaborated by the author and adapted from Facchini et al. (2018), Molina-Besch e Pålsson (2016), Witczak et al. (2019), and Carli et al. (2020)

Table 4 Best practices associated with low-carbon stock management

Best practices	Description
Sustainable integrated optimization of transport and stock	It seeks optimal decisions to achieve sustainable levels of emissions associated with factors such as sales optimization, carbon tax rate, and investment sharing value derived from transport and inventory
Optimizing stock management	It seeks to design sustainable inventory management in order to reduce environmental impact by optimizing batch size through cycle time optimization with a reduction in rework due to a reduction in defective products
Stock routing	The main objective of the traditional stock routing problem model is to minimize total stock and transport costs and determine the best vehicle route and quantity of products to be delivered in each period, as well as finding the best stock control strategy, taking into account social issues such as reducing accident rates, the number of expired products, and vehicle noise emissions. It also generates opportunities to improve energy efficiency and reduce emissions resulting from the process
Just in sequence	The just-in-sequence supply strategy is based on the just-in-time philosophy, where the aim is to reduce stock but also to guarantee the sequence of products requested. Its main benefits are to eliminate idle raw materials, work in process, and finished goods inventory reserves to increase cash flow and reduce the need for space in the plant to ensure more efficient optimization
Additive manufacturing	Additive manufacturing, often associated with 3D printing, is a manufacturing technology present in industry 4.0. It contributes to eliminating or at least minimizing inventory waste, reducing inventory risk with unsold finished goods, improving revenue flow by paying for goods before they are manufactured, and reducing inventory waste, including unsold and obsolete products

Source: Elaborated by the author and adapted from Ford and Despeisse (2016), Rahimi et al. (2016), Hove-Sibanda et al. (2018), Sarkar et al. (2018), Soysal et al. (2019), Bányai et al. (2019), Delmia (2019), Tang et al. (2020), and Godina et al. (2020)

management seek to optimize and improve stock management to reduce the size of batches and consequently the number of trips (internal movement). As far as packaging is concerned, there is a concern to reduce material consumption and improve the type of material used in packaging, also helping to promote lower GHG emissions.

The limitations of this methodology for assessing low-carbon practices in internal logistics, which encompass logistics activities within an organization's operational boundaries, are important to consider. While this study highlights several best practices for promoting an operational assessment framework for internal logistics, future research should delve deeper into the relationship between these best practices and the various scopes, categories, activities, and elements outlined in this study. Such an analysis would help clarify the significance and applicability of the

Table 5 Best practices associated with low-carbon packaging

Best practices	Description
Material modification	It applies to the replacement of hazardous or non-renewable inputs with less hazardous or renewable materials, auxiliary materials with a long useful life, or materials with a purification process
Promote recycling	Improved recycling rates for each of the main packaging materials; improved quality of recycled materials; increased use of recycled content
Optimizing the use of packaging (eco-design, reducing packaging, designing for reuse)	It's about using the right amount of packaging for what is needed. It includes eco-design, which is the design of products and packaging systems to ensure that products (including their packaging) can be produced, distributed, used, and recovered with minimal environmental impact
Optimization of material use	This approach is concerned with using as little packaging material as possible, for example by optimizing the dimensions of the packaging or reducing the thickness of the material
Environmentally responsible supply of packaging material	Production processes in the packaging industry have a number of environmental impacts, such as the use of fossil energy, emissions of volatile organic compounds from printing processes, and effluent emissions from chlorine bleaching of paper
Reusable travel containers	The reuse of reusable disposable containers can fundamentally reduce the waste of travel packaging
Minimizing cooling/heating needs (Molina-Besch e Pålsson, 2016)	Packaging influences the product's sensitivity to temperature, which changes throughout the SC because it acts as an insulating barrier to the environment, where temperature-sensitive products, such as fresh food, must be stored and transported under specific temperature conditions

Source: Elaborated by the author and adapted from Molina-Besch and Pålsson (2016), Hove-Sibanda et al. (2018), De Assis et al. (2021), and Jiang et al. (2020)

proposed practices, ultimately providing a more comprehensive understanding of how to effectively implement low-carbon strategies in internal logistics operations. Additionally, future studies could explore potential barriers to implementation, assess the long-term impact of these practices, and examine how they align with broader sustainability goals within the organization.

References

ABNT. (2015) Guia de Implementação: Gestão de Emissões e Remoções de Gases de Efeito Estufa (GEE). ABNT NBR ISO 14064. Obtido de http://abnt.org.br/paginampe/biblioteca/files/upload/anexos/pdf/4ee5b810af4a3aee073ab89f0a573a1a.pdf

Amjed, T. W., & Harrison, N. J. (2013, July). A model for sustainable warehousing: From theory to best practices. In *Proceedings of the International Decision Sciences Institute and Asia Pacific DSI conference* (pp. 1–28).

Bányai, T., Illés, B., Gubán, M., Gubán, Á., Schenk, F., & Bányai, Á. (2019). Optimization of just-in- sequence supply: A flower pollination algorithm-based approach. *Sustainability, 11*(14), 3850. https://doi.org/10.3390/su11143850

Bartholdi, J., & Hackman, S. (2019). Warehouse & distribution science (Release 0.98. 1). URL https://www.warehouse-science.com/book/editions/wh-sci-0.98, 1.

Bocken, N. M., & Allwood, J. M. (2012). Strategies to reduce the carbon footprint of consumer goods by influencing stakeholders. *Journal of Cleaner Production, 35*, 118–129.

Carli, R., Dotoli, M., Digiesi, S., et al. (2020). Sustainable scheduling of material handling activities in labor-intensive warehouses: A decision and control model. *Sustainability (Switzerland), 12*(8). https://doi.org/10.3390/SU12083111

Çelekli, A., & Zariç, Ö. E. (2023). From emissions to environmental impact: understanding the carbon footprint. *International Journal of Environment and Geoinformatics, 10*(4), 146–156. https://doi.org/10.30897/ijegeo.1383311

da Costa, M. G., de Abreu, V. H. S., de Assis, T. F., da Costa, V. X., de Almeida D'Agosto, M., & Santos, A. S. (2022). Life cycle assessment and circular economy strategies for electric vehicle: A systematic review on mitigating climate change and reducing resource depletion in road transportation. In *Carbon footprints of manufacturing and transportation industries* (pp. 113–137).

De Abreu, V. H. S., Da Costa, M. G., Da Costa, V. X., De Assis, T. F., Santos, A. S., & D'Agosto, M. D. A. (2022a). The role of the circular economy in road transport to mitigate climate change and reduce resource depletion. *Sustainability, 14*(14), 8951.

de Abreu, V. H. S., Santos, A. S., & Monteiro, T. G. M. (2022b). Climate change impacts on the road transport infrastructure: A systematic review on adaptation measures. *Sustainability, 14*(14), 8864.

De Assis, T. F., De Abreu, V. H. S., Carneiro, P. J. P., & Marcio De Almeida, D. A. (2021). Contribution of best practices to promote sustainable urban freight transport. *Revista Geintec-Gestao Inovacao e Tecnologias, 11*(4), 5928–5949. https://doi.org/10.47059/revistageintec.v11i4.2658

de Assis, T. F., Monteiro, T. G. M., de Abreu, V. H. S., D'Agosto, M. D. A., & Santos, A. S. (2022a). Enabling the green bonds market for sustainable transport projects based on the measure/monitoring, reporting and verification method. In *Carbon footprints of manufacturing and transportation industries* (pp. 1–24). Springer Nature Singapore.

de Assis, T. F., Ricci, L. M., Monteiro, T. G. M., de Abreu, V. H. S., D'Agosto, M. D. A., & Santos, A. S. (2022b). Sustainable transport indicators and mitigation actions applied to the green bond principles. In *Carbon footprints of manufacturing and transportation industries* (pp. 139–169). Springer Nature Singapore.

de Assis, T. F., de Abreu, V. H. S., da Costa, M. G., & Marcio de Almeida, D. A. (2022c). Methodology for prioritizing best practices applied to the sustainable last mile—The case of a Brazilian parcel delivery service company. *Sustainability, 14*(7), 3812.

De Assis, T. F. D., Marujo, L. G., de Abreu, V. H. S., da Costa, M. G., Rodrigues, L. M., & D'Agosto, M. D. A. (2023). Best practices to support the transition towards sustainable logistics from the perspective of Brazilian carriers. *Sustainability, 15*(18), 13434.

de Lima, O. P., Santiago, S. B., Taboada, C. M. R., Rodríguez, J. L. M., Rodríguez, M. B. R., Maduro, M. R., & da Silva Lima, J. C. (2020). Conceptualization, definition and assessment of internal logistics through different approaches using artificial intelligence. In *Operations management-emerging trend in the digital era*. IntechOpen.

Delmia. (2019). Just-in-sequence get real-time visibility & control. Available at: https://discover.3ds.com/sites/default/files/2020-03/delmiaapriso-just-in-sequence-ds.pdf

Deloitte. (2023). ESG in Warehousing. https://www2.deloitte.com/content/dam/Deloitte/be/Documents/climate-and-sustainability/be_ESG-in-warehousing.pdf

Dey, A., LaGuardia, P., & Srinivasan, M. (2011). Building sustainability in logistics operations: A research agenda. *Management Research Review, 34*(11), 1237–1259.

EPA. Guidance, G. G. I. (2016). *Direct emissions from stationary combustion sources*. US EPA.

EPA. Guidance, G. G. I. (2023a). *Direct fugitive emissions from refrigeration, air conditioning, fire suppression, and industrial gases*. US EPA.

EPA. Guidance, G. G. I. (2023b). *Direct emissions from Mobile combustion source*. US EPA.

Facchini, F., De Pascale, G., & Faccilongo, N. (2018). Pallet picking strategy in food collecting center. *Applied Sciences, 8*(9), 1503.

Ford, S., & Despeisse, M. (2016). Additive manufacturing and sustainability: an exploratory study of the advantages and challenges. *Journal of Cleaner Production, 137*, 1573–1587. https://doi.org/10.1016/j.jclepro.2016.04.150

Freis, J., Vohlidka, P., & Günthner, W. A. (2016). Low-carbon warehousing: Examining impacts of building and intra-logistics design options on energy demand and the CO_2 emissions of logistics centers. *Sustainability, 8*(5), 448.

García-Arca, J., González-Portela Garrido, A. T., & Prado-Prado, J. C. (2017). "Sustainable packaging logistics". The link between sustainability and competitiveness in supply chains. *Sustainability, 9*(7), 1098.

Godina, R., Ribeiro, I., Matos, F., T. Ferreira, B., Carvalho, H., & Peças, P. (2020). Impact assessment of additive manufacturing on sustainable business models in industry 4.0 context. *Sustainability, 12*(17), 7066. https://doi.org/10.3390/su12177066

Górny, K., Idaszewska, N., Sydow, Z., & Bieńczak, K. (2021). Modelling the carbon footprint of various fruit and vegetable products based on a company's internal transport data. *Sustainability, 13*(14), 7579.

Gutierrez, J., Santaolalla, A., Tercjak, A., Rojo, N., Encinas, D., Gomez-de-Balugera, Z., & Gallastegui, G. (2020). Creating a green chemistry lab: Towards sustainable resource management and responsible purchasing. *Sustainability, 12*(21), 8934.

Henriques, I., & Sadorsky, P. (1999). The relationship between environmental commitment and managerial perceptions of stakeholder importance. *Academy of Management Journal, 42*(1), 87–99.

Hrasky, S. (2011). Carbon footprints and legitimation strategies: Symbolism or action? *Accounting, Auditing & Accountability Journal, 25*(1), 174–198.

Hua, G., Cheng, T. C. E., & Wang, S. (2011). Managing carbon footprints in inventory management. *International Journal of Production Economics, 132*(2), 178–185.

Huisingh, D., Zhang, Z., Moore, J. C., Qiao, Q., & Li, Q. (2015). Recent advances in carbon emissions reduction: Policies, technologies, monitoring, assessment and modeling. *Journal of Cleaner Production, 103*, 1–12.

Ibrahim, A., Fernando, Y., Tseng, M.-L., & Lim, M. K. (2022). Low-carbon warehousing practices and challenges: Insights from emerging country. *International Journal of Logistics Research and Applications*. https://doi.org/10.1080/13675567.2022.2145276

IIMM. (2020). Logistics and warehousing management, Indian Institute of Materials Management, Navi Mumbai, 1–220. 15.Sisinni, Emiliano & Saifullah. ISBN: 978-93- 89795-50-9.

IPCC. (2018). Annex I: Glossary [Matthews, J.B.R. (ed.)]. In: Global Warming of 1.5 °C. An IPCC Special Report on the impacts of global warming of 1.5 °C above pre-industrial levels and related global greenhouse gas emission pathways, in the context of strengthening the global response to the threat of climate change, sustainable development, and efforts to eradicate poverty [Masson-Delmotte, V., P. Zhai, H.-O. Pörtner, D. Roberts, J. Skea, P.R. Shukla, A. Pirani, W. Moufouma-Okia, C. Péan, R. Pidcock, S. Connors, J.B.R. Matthews, Y. Chen, X. Zhou, M.I. Gomis, E. Lonnoy, T. Maycock, M. Tignor, and T. Waterfield (eds.)]. Cambridge University Press, Cambridge, UK and New York, NY, USA, pp. 541–562. https://doi.org/10.1017/9781009157940.008

IWLA. (2020). *Warehouse supply chain & logistics terms and glossary*. https://www.iwla.com/wp-content/uploads/2020/09/IWLA-Glossary-and-Terminology.pdf

Jiang, X., Dong, M., He, Y., Shen, J., Jing, W., Yang, N., & Guo, X. (2020). Research on the design of and preference for collection modes of reusable takeaway containers to promote sustainable consumption. *International Journal of Environmental Research and Public Health, 17*(13), 4764.

Kay, M. G. (2012). *Material handling equipment*. North Carolina, [s.n.]. https://people.engr.ncsu.edu/kay/Material_Handling_Equipment.pdf

Klug, F. (2013). The internal bullwhip effect in car manufacturing. *International Journal of Production Research, 51*(1), 303–322.

Lagoudis, I. N., & Shakri, A. R. (2015). A framework for measuring carbon emissions for inbound transportation and distribution networks. *Research in Transportation Business & Management, 17*, 53–64.

Lewis, A. (2016). Towards a harmonized framework for calculating logistics carbon footprint. In *Sustainable logistics and supply chains: innovations and integral approaches* (pp. 163–181).

Liu, L., Long, Z., Kou, C., Guo, H., & Li, X. (2023). Evaluation of the environmental cost of integrated inbound logistics: A case study of a gigafactory of a Chinese logistics firm. *Sustainability, 15*(15), 11520.

Lo-Iacono-Ferreira, V. G., Viñoles-Cebolla, R., Bastante-Ceca, M. J., & Capuz-Rizo, S. F. (2021). Carbon footprint comparative analysis of cardboard and plastic containers used for the international transport of Spanish tomatoes. *Sustainability, 13*(5), 2552.

Makhmudov, I. (2012). Design and implementation of the storage layout and internal material flow at the new General Motors Powertrain Uzbekistan Greenfield manufacturing plant.

Malinowska, M., Rzeczycki, A., & Sowa, M. (2018). Roadmap to sustainable warehouse. In *SHS web of conferences* (Vol. 57, p. 01028). EDP Sciences.

MASTERMOVER. (2024). What is intralogistic. https://www.mastermover.com/en-us/material-handling/what-is-intralogistics

McKinnon, A. C., & Piecyk, M. I. (2012). Setting targets for reducing carbon emissions from logistics: Current practice and guiding principles. *Carbon Management, 3*(6), 629–639.

Ministry of the Environment Japan. (2013). Explanations by Industry (Logistics Industry) for the Basic Guidelines on Accounting for Greenhouse Gas Emissions Throughout the Supply Chain, Ver. 1.0 (Draft).

Molina-Besch, K., & Pålsson, H. (2016). A supply chain perspective on green packaging development-theory versus practice. *Packaging Technology and Science, 29*(1), 45–63.

Morgan, J., Halton, M., Qiao, Y., & Breslin, J. G. (2021). Industry 4.0 smart reconfigurable manufacturing machines. *Journal of Manufacturing Systems, 59*, 481–506. https://doi.org/10.1016/j.jmsy.2021.03.001

Nilsson, E. (2018). Improving material flow and production layout using value stream mapping: A case study in a manufacturing company. In *Production systems: Production development and management.*

OECD. (2007). Measuring Material Flows and Resource productivity – OECD guidance manual Volume Ii. A Theoretical Framework For Material Flow Accounts And Their Applications At National Level. 11th Meeting of the London Group on Environmental Accounting Johannesburg, 26–30 March 2007.

Olatunji, O. O., Akinlabi, S. A., Ayo, O. O., Madushele, N., Adedeji, P. A., & Fatoba, S. O. (2019). Drivers and barriers to competitive carbon footprint reduction in manufacturing supply chain: A brief review. *Procedia Manufacturing, 35*, 992–1000.

Oloruntobi, O., Mokhtar, K., Rozar, N. M., Gohari, A., Asif, S., & Chuah, L. F. (2023). Effective technologies and practices for reducing pollution in warehouses-a review. *Cleaner Engineering and Technology, 13*, 100622.

Payen, S., Basset-Mens, C., & Perret, S. (2015). LCA of local and imported tomato: An energy and water trade-off. *Journal of Cleaner Production, 87*, 139–148.

Perotti, S., Coslovich, M., & Granata, E. (2023, June). Transitioning towards net-zero warehouses: Empirical insights and best practices in Italy. In *Proceedings of the 2023 9th international conference on e-society, e-learning and e-technologies* (pp. 82–87).

Rahimi, M., Baboli, A., Rekik, Y. (2016). Sustainable inventory routing problem for perishable products by considering reverse logistic. *Ifac-Papersonline, 49*(12), 949–954. https://doi.org/10.1016/j.ifacol.2016.07.898

Ranganathan, J., et al. (2004). Greenhouse gas protocol. In *A corporate accounting and reporting standard* (Revised edition).

Sadeghi, M., Nikfar, M., & Rad, F. (2024). Optimizing warehouse operations for environmental sustainability: A simulation study for reducing carbon emissions and maximizing space utilization. *Journal of Future Sustainability, 4*(1), 35–44.

Saghir, M. (2004). A Platform for Packaging Logistics Development—A Systems Approach. PhD dissertation, Packaging Logistics, Lund University, Lund.

Santos, A. S., de Abreu, V. H. S., de Assis, T. F., Ribeiro, S. K., & Ribeiro, G. M. (2021). An overview on costs of shifting to sustainable road transport: A challenge for cities worldwide. In *Carbon footprint case studies: Municipal solid waste management, sustainable road transport and carbon sequestration* (pp. 93–121).

Sarkar, B., Ahmed, W., Choi, S.B., et al. (2018). et al. Sustainable inventory management for environmental impact through partial backordering and multi-trade-creditperiod, Sustainability (Switzerland), v. 10, n. 12, 2018. https://doi.org/10.3390/su10124761

Satolo, E. G., Campos, R. S. D., Ussuna, G. D. A., Simon, A. T., Mac-Lean, P. A. B., & Braga Júnior, S. S. (2020). Sustainability assessment of logistics activities in a dairy: An example of an emerging economy. *Production, 30*.

Sibanda, K., Hove-Sibanda, P., & Mukarumbwa, P. (2018). Greening up in logistics: Managerial perceptions of small and medium-sized enterprises on sustainability in Zimbabwe. *TD: The Journal for Transdisciplinary Research in Southern Africa, 14*(1), 1–13.

Soysal, M., Çimen, M., Belbağ, S., & Toğrul, E. (2019). A review on sustainable inventory routing. *Computers & Industrial Engineering, 132*, 395–411. https://doi.org/10.1016/j.cie.2019.04.026

Sun, Q. (2017). Research on the influencing factors of reverse logistics carbon footprint under sustainable development. *Environmental Science and Pollution Research, 24*, 22790–22798.

Szymonik, A. (2016). *Packaging in logistics*. URL: https://www.researchgate.net/publication/297368732_packaging_in_logistics

Tang, Z., Liu, X., & Wang, Y. (2020). Integrated optimization of sustainable transportation and inventory with multiplayer dynamic game under carbon tax policy. *Mathematical Problems in Engineering, 2020*, 1–16. https://doi.org/10.1155/2020/4948383

Thollander, P., Karlsson, M., Rohdin, P., Wollin, J., & Rosenqvist, J. (2020). *Introduction to industrial energy efficiency: Energy auditing, energy management, and policy issues*. Academic.

Vieira, B., Bergmann, G., Pasa, G. S., et al. (2011). Materials handling management:A case study. *Journal of Operations and Supply Chain Management, 4*(2), 19. https://doi.org/10.12660/joscmv4n2p19-30

Vitasek, K. (2013). Supply chain management. Terms and glossary.

Witczak, M., Majdzik, P., Stetter, R., & Lipiec, B. (2019). Multiple AGV fault-tolerant within an agile manufacturing warehouse. *IFAC-PapersOnLine, 52*(13), 1914–1919. https://doi.org/10.1016/j.ifacol.2019.11.482

World Economic Forum and Nike. (2016). The World Economic Forum and Nike: Emerging 'shared responsibility' and institutional control models for achieving a socially responsible global supply chain?. [Hemphill, T. A., & WHITE III, G. O.]. *Business and Human Rights Journal, 1*(2), 307–313.

Sowing the Future: The Vital Role of Green Bonds in Sustainable Agribusiness to Reduce the Carbon Footprint

Thaís Guedes Máximo Monteiro, Renata Angeli, and Victor Hugo Souza de Abreu

1 Introduction

In the year 2022, the Intergovernmental Panel on Climate Change (IPCC) unveiled its sixth assessment report, a comprehensive synthesis of new insights derived from scientific, technological, environmental, economic, and social facets related to climate change mitigation. This report builds upon and reinforces the findings of previous IPCC assessments. As of 2019, global net anthropogenic greenhouse gas (GHG) emissions stood at a substantial 59 gigatons of CO2-eq, marking a notable 12% increase compared to 2010 and a staggering 54% surge from 1990 levels—an alarming upward trajectory.

The decade spanning 2010 to 2019 witnessed an annual average of emissions surpassing that of the preceding 2000–2009 period, resulting in the most significant 10-year average increase ever documented. Notably, in 2019, the Agriculture, Forestry, and Other Land Use (AFOLU) sector contributed approximately 22% (equivalent to 13 gigatons of CO2-eq) of the total net anthropogenic GHG emissions. A pivotal revelation from the IPCC's 2023 report is that nearly half of these AFOLU emissions were predominantly attributed to deforestation—a stark reminder of the urgent need to address land-use practices to mitigate the escalating impact of climate change (IPCC, 2023).

T. G. M. Monteiro · R. Angeli
State University of Rio de Janeiro (Universidade Estadual do Rio de Janeiro—UERJ), Postgraduate Program in Environmental Science and Technology (Programa de Pós-Graduação em Ciência e Tecnologia Ambiental—PPGCTA), Rio de Janeiro, Brazil

V. H. S. de Abreu (✉)
Federal University of Rio de Janeiro (Universidade Federal do Rio de Janeiro—UFRJ), Transport Engineering Programme (Programa de Engenharia de Transportes—PET), COPPE-UFRJ, Rio de Janeiro, Brazil
e-mail: victor@pet.coppe.ufrj.br

S. S. Muthu (ed.), *Carbon Footprint Assessments*, Environmental Footprints and Eco-design of Products and Processes,
https://doi.org/10.1007/978-3-031-70262-4_14

With the commitment of the signatories to the Paris Agreement, policies aimed at keeping the rise in temperatures within 1.5 °C of pre-industrial levels were publicly ratified, which means an annual reduction of at least 7.6% in GHG emissions by 2030 (United Nations Environmental Programme, 2019). Currently, two options are considered capable of fulfilling this promise: (i) a drastic reduction in the consumption of goods, whose entire production chain is associated with GHG emissions; (ii) a "green transition," with the consumption of goods based on the production of low-carbon technologies. Given the homeric challenges of the first option, public and private policymakers have focused on the transition to low-carbon technologies. As an example of tangible change, based on serious policies, in the 1980s and 1990s chlorofluorocarbons in industrial processes were replaced by technological alternatives that do not degrade the ozone layer (Dugoua, 2023).

As a result, concern about environmental issues has led investors around the world to seek investments that pay the expected return and are linked to effective green projects (Monteiro et al., 2023). As a result, green bonds have come onto the scene to fulfill this role, as they are fixed-income security instruments with funds invested exclusively in financing or refinancing eligible new and/or existing green projects, which must comply with the four main components of the Green Bonds Principles (GBP) (ICMA, 2021). Thus, in recent years, central banks have assimilated the potential risks to the financial market of climate change and abandoned the use of fossil fuels. In September 2015, the Governor of the Bank of England and Chairman of the G20 Financial Stability Board, Mark Carney, highlighted the threats posed by climate change to financial resilience, recognizing the importance of structured action for the efficient transition to a low-carbon economy (Sartzetakis, 2021). With global warming comes significant economic impacts, such as lower productivity and profitability due to higher temperatures, where the risk resulting from carbon emissions begins to affect investment decisions, mergers and acquisitions, and financing costs (Wang et al., 2022).

Consequently, in response to the need to mitigate and adapt to climate change, the use of green finance has expanded rapidly around the world, attracting the attention of investors interested in the transition to a low-carbon economy (De Abreu et al., 2022a, b; Da Costa et al., 2022). This interest has grown even more since the environmental commitments signed by countries at the United Nations Framework Convention on Climate Change (UNFCCC) in Paris and due to the investment needs associated with the 2030 Agenda, as well as the implementation of the SDGs resulting from it. The practice of financing investments that have environmental benefits focuses mainly on renewable energy projects that aim to reduce carbon emissions, promote climate-resilient infrastructure, and environmental sustainability (Taghizadeh-Hesary & Yoshino, 2019). As the green bond market only covers investors focused on the environment, social, and governance, as well as investors who aim to green their portfolios (Nanayakkara & Colombage, 2019), it still represents a small percentage of the overall bond market (Ketterer et al., 2019).

The agricultural sector plays a crucial role in reducing the carbon footprint to meet global decarbonization targets. Nonetheless, as a result of the expansion of agricultural land and pastures into forest areas, agriculture was one of the main drivers of deforestation, contributing at least 2.6 $GtCO_2$ to annual emissions between 2010 and 2014 (Pendrill et al., 2019), denoting the need to invest financially in projects to reduce the carbon footprint.

Over the past decade, groundbreaking mechanisms have steered private financial resources toward initiatives addressing climate change (Santos et al., 2021; De Abreu et al., 2022b). Notably, the emergence of innovative instruments such as green debt has harnessed investors' eagerness to pay a premium for engaging in climate-friendly activities (Ketterer et al., 2019). Green bonds, in particular, have played a pivotal role in financing projects associated with climate change mitigation or adaptation, bolstering resilience, combating natural resource depletion, addressing biodiversity loss, and promoting soil protection and other related endeavors (Sustainable Banking Network, 2018; De Assis et al., 2022a).

The reason is that, since climate change has become an exponentially significant challenge for the sustainable development of society as a whole, it has become equally essential to direct investments toward a low-carbon, climate-resilient economy (Santos et al., 2021). It is also estimated that US$6 trillion a year will need to be invested worldwide in sustainable infrastructure over the next 15 years, almost double the current figure (Xia et al., 2022). Thus, with increasingly recurrent and catastrophic impacts around the world resulting from accelerated global warming, the implementation of sustainable agricultural practices has become crucial. As the sector is under pressure related to food security, it is essential to adapt to changes in climate patterns in order to withstand the pressures related to the supply of food and other inputs for an accelerating world population (Climate Bonds Initiative, 2021b).

Due to its substantial contribution to GHG emissions and, consequently, the carbon footprint, the agribusiness sector assumes particular importance in the context of environmental impact. Green bonds, therefore, emerge as instrumental tools intricately linked to multiple Sustainable Development Goals (SDGs). Directing financial resources toward activities and projects that align with these goals, such as SDG 12 emphasizing Sustainable Consumption and Production, underscores the imperative to channel increased financial flows to address climate impacts within the agribusiness sector. Consequently, a comprehensive literature review has explored the trajectory of agribusiness and underscored the imperative for green bonds to facilitate the sector's decarbonization. This underscores the crucial role of the financial market as an indispensable actor in realizing the ambitious objectives outlined in the 2030 Agenda.

Thus, this chapter focuses on presenting a literature review of studies on the impacts of agribusiness on the environment and discussing possible strategies for reducing the carbon footprint through the issuance of green bonds, which, in line with the SDGs, continue with the participation of private companies (Taghizadeh-Hesary & Yoshino, 2019), public companies, and governments in financing and refinancing green projects. Furthermore, there is the objective of alerting agribusiness decision-makers and scholars on the subject about the importance of more robust analyses, as the subject requires more in-depth studies. The relevance of this study lies in addressing a topic of global scope, directly linked to the 2030 Agenda and the SDGs, but which still lacks related studies. In addition to this introductory section, the rest of the document is organized as follows: Section 2 deals with the methodological procedure developed to efficiently conduct the research; Section 3 discusses the main impacts of agribusiness on the environment related to soil management; Section 4 presents an overview of the green bond market; Section 5 correlates the importance of the green bond market for agribusiness; and, finally, Sect. 6 concludes the research and proposes guidelines for future studies on the topic.

2 Methodology

This study was meticulously developed through a literature review, inspired by the work of Santos et al. (2021). This approach consists of a critical and systematized analysis of the existing literature on the subject, covering the collection of relevant studies, reports, and scientific publications. The main objective is to investigate the trends and practices associated with green bonds in the context of agribusiness.

The comprehensive selection of sources included scientific articles, book chapters, reports, and technical documents, obtained through searches on the free Google Scholar platform. Differentiating itself from multidisciplinary databases such as WoS and Scopus, Google Scholar offers access to a wide range of academic works, including conference papers, theses and dissertations, as well as open access or paid journals (Caregnato, 2011).

Searches on Google Scholar were conducted using strategically chosen keywords such as "green bonds," "agribusiness," "sustainable agriculture," "climate change," "carbon footprint," and "green bonds." In addition, documentary searches were carried out on renowned official websites, such as the Climate Bonds Initiative and ICMA, to enrich the database with information from reliable sources. The texts resulting from these investigations were subjected to careful analysis and synthesis. This process made it possible to carefully filter the information and conclusions presented by the various authors, identifying the most relevant and robust results for building knowledge.

At the same time, a complementary approach was adopted through the Scopus paid database. Using specific keywords, such as "(KEY(green bonds AND agriculture) OR KEY(green bonds AND climate change)", 75 studies carried out between 2017 and 2023 were identified. Of these, those selected for analysis and inclusion in the study are listed in detail in Table 1.

3 Agribusiness and the Carbon Footprint

According to The Economics of Land Degradation Initiative (ELD Initiative, 2013), approximately one-third of the planet's arable land is suffering from the process of degradation and desertification, which continues to grow. The processes of soil

Table 1 Studies used from the Scopus database

Authors	Title	Document type
Andersen et al. (2020)	Resolving intergenerational conflict over the environment under the pareto criterion	Article
Gibon et al. (2020)	Shades of green: Life cycle assessment of renewable energy projects financed through green bonds	Article
Sartzetakis (2021)	Green bonds as an instrument to finance low carbon transition	Article
Dent and Boincean (2021)	An investable proposal for regenerative agriculture across the steppes	Book Chapter
de Assis et al. (2022a)	Enabling the Green Bonds Market for Sustainable Transport Projects Based on the Measure/Monitoring, Reporting and Verification Method	Book Chapter
Pombo-Romero and Rúas-Barrosa (2022)	A Blockchain-Based Financial Instrument for the Decarbonization of Irrigated Agriculture	Article
Wang et al. (2022)	Does green bond issuance have an impact on climate risk concerns?	Article
Xia et al. (2022)	Forecasting green bond volatility via novel heterogeneous ensemble approaches	Article
Hsu (2023)	Influence of climate finance and natural resource consumption on the mitigation of climate change in developed countries in the Pre-COP26 era	Article

erosion, compaction, and salinization, which are also increasing, signal that strategies for the sustainability of agricultural production must undoubtedly be a priority.

As well as contributing to the loss of biodiversity, deforestation also releases considerable amounts of carbon into the atmosphere, exacerbating the impacts of climate change. Intensive agriculture, a common practice in the sector, soil erosion, and nutrient loss are examples of consequences that compromise agricultural productivity in the long term. In addition, the excessive use of pesticides and fertilizers also ends up contaminating soil, water, and aquatic ecosystems, damaging human and animal health, as well as biodiversity. About the carbon footprint, soil is considered the largest carbon reservoir in nature (Tordin, 2022).

With increasingly recurrent extreme climate events, such as hurricanes, prolonged droughts, and floods, efforts to limit the increase in global temperature to 1.5 °C become urgent. In this way, agricultural production is capable of mitigating climate change by optimizing practices that in themselves present high levels of greenhouse gas (GHG) emissions. As an example, it is possible to mention mineral fertilization, which acts as a carbon footprint forming agent, making it possible to replace it by implementing sustainable alternatives such as crop rotation, which maintains soil health and enables the reduction of GHG emissions arising from the application of these fertilizers (Holka et al., 2022).

According to Olsson et al. (2019), the soil is of great importance in the biogeochemical carbon cycle, as it stores approximately three times more carbon than the atmosphere and, once released, this carbon contributes to the increase in the

Table 2 Main sources of increased carbon footprint in agriculture

Activity	Consequence
Soil cultivation	Whether planting annual or perennial crops, it can lead to CO_2 emissions, especially in areas where vegetation cover is removed
Animal management	Enteric fermentation of ruminant animals such as cattle and digestion of pigs and poultry releases CH_4, while the decomposition of animal waste, such as feces and urine, can also lead to the emission of the GHG CH_4 and N_2O
Use of fertilizers and pesticides	Due to the intense use of synthetic nitrogen fertilizers around the world, agricultural soil is mainly responsible for the release of N_2O—around 300 times more powerful than CO_2 in warming the atmosphere

Source: Based on Chrobak (2021), de Lima et al. (2010) and Tordin (2022)

greenhouse effect. Changes in land use are among the basic sources of anthropogenic carbon emissions into the atmosphere, as cultivation and management alter the efflux and influx of CO2 at the soil-atmosphere interface, especially through plowing and harrowing, as these activities promote greater oxidation of organic matter when in contact with the air (Tordin, 2022). For this reason, the increased carbon footprint in agriculture occurs in various ways, with soil cultivation, animal management, and the use of fertilizers and pesticides being the main sources, as can be seen in Table 2.

Studies have demonstrated the significant impact of fertilization on the size of the carbon footprint (Gasol et al., 2012). This occurs because nitrogen fertilizers used in cultivation emit gaseous nitrogen compounds, with nitrogen being a chemically reactive compound that plays a role in global warming. In the form of nitrous oxide (N_2O), it is one of the biggest GHG emitters in agriculture (Hergoualc'h et al., 2019).

Erosion, as the most widespread form of soil degradation, removes topsoil material, which usually holds the largest stocks of organic carbon, favoring mineralization and release as CO2. The cascading effects of erosion on other environmental processes can, in affected locations, also cause net CO2 emissions through its indirect influence on soil fertility, interacting with degradation processes such as nutrient depletion, compaction, and salinization (van de Koppel et al., 1997). It is worth noting that, as both natural and human-induced erosion can lead to net carbon deposition in very stable pools, degradation at these deposition sites has a high potential for carbon release (Olsson et al., 2019).

Land degradation is therefore a driver of climate change because of GHG emissions and reduced carbon absorption rates. Since 1990, forest area has decreased globally with lower carbon density in regenerating forests compared to carbon stocks before deforestation, resulting in net emissions from land use change (agricultural soils from conventional agriculture continue to be a source of GHGs). Concerning soil, deforestation, the advance of fires, the degradation of peat soils, and the thawing of permafrost all contribute significantly to climate change through the release of GHGs and the reduction of carbon sinks, which are extinguished by deforestation. Agricultural practices, such as the high concentration of fertilizers, also exacerbate GHG emissions which are intensified by climate change. In addition, primary forests

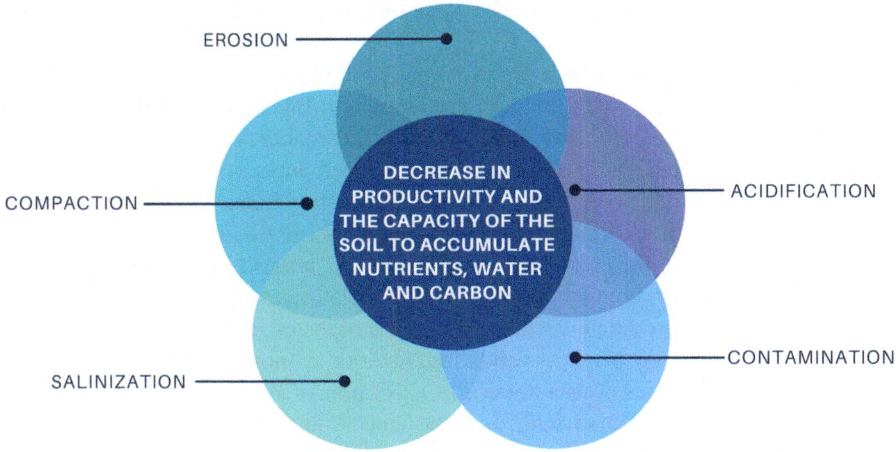

Fig. 1 Aspects that directly influence the availability of natural resources for food production. (*Source:* Based on Montanarella et al. (2015))

transformed into managed forests, illegal logging, and unsustainable forest management result in more GHG emissions, with additional physical effects on the regional climate (Olsson et al., 2019).

Globally, a staggering 33% of arable land faces degradation, resulting in annual losses of approximately 25 to 40 billion tons of soil. This substantial degradation has profound implications, significantly diminishing agricultural productivity and the soil's capacity to retain essential elements such as nutrients, water, and carbon (Montanarella et al., 2015). Figure 1 visually illustrates the magnitude of this issue, emphasizing the urgent need for sustainable soil management practices and concerted efforts to address the escalating challenges facing our valuable arable lands.

It is important to note that any loss of soil productivity would cause serious damage to food production and security, making food prices more volatile and possibly driving an unprecedented number of people into hunger and poverty. Since the agricultural sector is responsible for approximately one-third of GHG emissions into the atmosphere, the lack of a plan to address this issue is worrying given that, since 1970, the amount of soil carbon in agricultural areas has been substantially reduced. The impact on land and water resources and the impairment of soil fertility, as well as the reduction in its capacity to retain water and sustain crops, must be considered as a matter of urgency (Dent & Boincean, 2021).

4 The Green Bond Market

GHG emissions involve the concept of carbon footprint, which has become an inherent part of the economy (Bajan & Mrówczyńska-Kamińska, 2020). The global green bond market began in 2007, when multilateral development banks raised funds for climate-related projects. The Scandinavian Stock Exchange in Oslo and Stockholm and the London Stock Exchange were the pioneers in creating specific lists of green bonds, and the first green exchange took place in Luxembourg during the issuance of the first green bond by the European Investment Bank. During the issuance of the first green bond, there was still no market or regulation for these issues, and European stock exchanges and asset managers saw this gap as an opportunity to respond to growing demand by raising funds for climate change-related solutions with environmental benefits (Filkova et al., 2018).

With the growth of the market, the issuance of self-labeled green bonds increased considerably, but such bonds had no external regulation, and it was enough for the issuer (borrower) to provide investors with details of the eligibility criteria for the use of the funds, without any need for validation. Over time, more cases of greenwashing have emerged, raising concerns about the allocation of funds and the actual environmental benefits associated with these issuances, thus highlighting the need for regulation (World Bank, 2019) through established taxonomies.

Thus, in 2014 the GBP were developed by Climate Bonds Initiative (CBI), promoting security and credibility in "green" operations and projects/assets. From then on, a movement began in the financial market regarding sustainable guidelines, and in the period from 2017 to 2018, the financing of actions related to climate change already had an increase of 25%, compared to previous years, with Europeans leading emissions with a representativeness of approximately 37% and 40%, respectively (EU TEG, 2019).

This growth has been driven by the advancement of energy production from renewable sources in China and India, as well as the commitment of several countries to adopt sustainable practices in land management and energy efficiency, in a progressive and accelerated manner (de Assis et al., 2022a). As a result of this accelerated growth in the green bond market, by June 2019 there was a total of approximately 550 billion euros in debt (100 billion euros accumulated in the year), of which the main issues came from Europe, North America, and Asia-Pacific (EU TEG, 2019).

However, it is equally important to contextualize the participation of developing countries in the green bond financial market. A 2019 report by the Economic Commission for Latin America and the Caribbean (ECLAC) underscored the region's asymmetrical position, noting that while less than 10% of global greenhouse gas emissions originate from LAC, the region remains highly vulnerable to the impacts of climate change (Sánchez et al., 2019). LAC also has a vast biodiversity that is particularly vulnerable to climate change, making the transition to a green and resilient economy vital for long-term development through tools such as green bonds. As such, the allocation of resources to Land Use and Industry becomes more frequently found in LAC than in other parts of the world, signaling a poignant concern associated with agribusiness, with a considerably high share in relation to the global market (Almeida & Filkova, 2019).

It is also worth noting that, in order to issue green bonds, it is crucial to present studies to financiers and certifiers that justify and make feasible the financial contribution to a given project, as well as monitor the actions implemented. As an

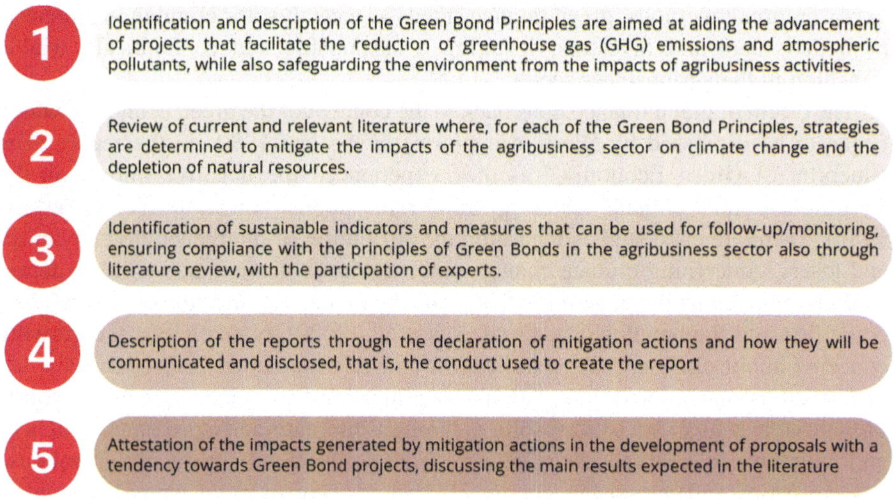

Fig. 2 Steps of roadmap applied to MRV mitigation actions in agribusiness projects. (*Source:* Based on de Assis et al. (2022b))

example, we can mention the creation of a Roadmap divided into five stages, which are broken down into (i) identifying and describing the principles of green bonds; (ii) identifying the mitigation actions applied to the project; (iii) identifying monitoring measures/indicators; (iv) describing the reports; and (v) verifying the impacts, as shown in Fig. 2.

5 Agribusiness and the Issuance of Green Bonds

Whether it's due to environmental awareness or the need to adapt to the impacts of climate change, it's a fact that in different parts of the world, producers have been adopting agroecological practices, such as agroforestry and low-carbon agriculture, to integrate agricultural production with the conservation of natural resources. With a long list of environmental impacts, especially related to deforestation, soil erosion, environmental contamination, and GHG emissions, agribusiness with a focus on sustainability seeks to make agricultural production less harmful, to guarantee the long-term conservation of natural resources (Alves et al., 2015).

Nevertheless, the shift toward sustainable agribusiness occasionally poses financial challenges. The insufficient awareness of environmental responsibilities within agribusiness, coupled with budget constraints, often renders the available funds incompatible with the necessary actions to yield substantial benefits. In addressing this dilemma, green finance emerges as a strategic tool for the sector's decarbonization. By offering financing options with distinct advantages compared to conventional methods, green finance stands as a catalyst for aligning the interests of both issuers and investors, fostering a sustainable and economically viable transition in the agribusiness sector.

The origin of green finance goes back to the concept of the green economy, which began to be discussed by Western countries in the late 1980s in the report "The Blueprint of Green Economy," as they experienced the negative impact on the environment and climate resulting from rapid industrialization (Pearce et al., 1989). There are even studies regarding the attribution of intergenerational winners and losers, concerning climate change, since the investments made will not necessarily be enjoyed by the generation that made the financial contribution—as the costs come first and the benefits later (Andersen et al., 2020). Nonetheless, the growing concerns about climate change mitigation and environmental protection have led developed countries and international organizations to pursue green economic growth with a focus on sustainability, through internationally promoted practices and initiatives such as reducing air, water, and soil pollution, reducing greenhouse gas emissions and increasing energy efficiency using existing natural resources, among others (Platform, 2013).

Green finance has a double impact by incorporating environmental preservation and sustainability objectives into financial decisions by (i) influencing the demand for and supply of environmental protection practices, encouraging companies to adopt sustainable practices and directing consumer interest towards green products and (ii) encouraging the financial sector to avoid excessive speculation with a focus on long-term interests, considering the environment and assessing risk factors to meet sustainability criteria in investment and lending decisions—while also focusing on insurance practices to protect against climate risks (Hsu, 2023).

However, the transition to a low-carbon economy that is resistant to climate change requires a shift in financial capital toward greener solutions. Thus, green bonds are financial instruments that can support this transition and offer an opportunity for investors to minimize their climate risks, provided they have high environmental integrity, since the capital is invested in green solutions or projects related to climate change (Clapp et al., 2016). In the field of finance, this approach, known as "Sustainable Finance," seeks to invest in development projects with a focus on the environment through bonds, creating opportunities by taking into account the various roles of the global financial system (Park, 2019). But to understand how green bonds work, it is important first of all to understand that investing in the financial market is nothing more than investing resources in the present with the prospect of a positive harvest in the future.

A study conducted by Zhang and Du (2020) in the Chinese context evaluated how environmentally friendly financing impacts company performance. This is because, due to pollution, companies in China have been forced to implement green management and, as a result, many of them have modified their production patterns to meet environmental targets. The study also highlighted the importance of funding for the successful implementation of green management programs. However, the scarcity of capital turns out to be one of the limitations to meeting the goals of the 2015 Paris Agreement. Green bonds have therefore become one of the sources of funding for projects and their use can help overcome problems of financial support.

Fig. 3 The SDGs which are an urgent call for action by all countries. (*Source:* Based on United Nations (2015))

In recent years, green bonds, using a taxonomy established by the GBP, have grown exponentially. Although various political perspectives from institutions, governments, and environmental policymakers have recently emerged with the aim of greening the economy, many of the measures introduced have not been able to produce the desired results. Thus, green finance has gained increasing prominence as it aims to combine environmental security with economic profit (Stojanovic & Ilic, 2018). In order to achieve the goals set by the Paris Agreement, the UN estimates that between US$5 and 7 trillion will be needed annually to finance the infrastructure needed to achieve the 17 SDGs, shown in Fig. 3, by 2030. To this end, the mobilization of private capital is crucial, playing an increasing role in financing investments for a green transition, driven by demand from investors seeking to engage in sustainability (Gibon et al., 2020).

The international community understood that there was no alternative but to mobilize both public and private capital in considerable amounts to meet sustainability strategies. Given that the Sustainable Development Goals (SDGs) for achieving the 2030 Agenda and the Paris Agreement explicitly address the inherent tensions between economic growth and environmental stewardship, it is crucial to highlight that the SDGs aim, among other goals, to eliminate poverty and hunger while promoting sustainable economic growth and universal prosperity. At the same time, the Paris Agreement aims to ensure that global temperature increases in the 21st century remain well below 2 degrees Celsius above pre-industrial levels, with efforts to limit the temperature increase to 1.5 degrees Celsius, through low-carbon and climate-resilient global development (de Assis et al., 2022a, b).

In the context of agribusiness, the central theme of this study, among many examples, is sustainability in irrigated agriculture, which addresses significant social challenges such as climate change, human development, and sustainable growth, using distributed renewable energies and decentralized financial solutions to create effective and economically advantageous financial instruments (Pombo-Romero & Rúas-Barrosa, 2022). For the agriculture, food, and forestry sectors, it is estimated that around US$7 billion a year is needed to implement adaptation measures (World Bank, 2019A). However, despite the inevitability of a consistent financial flow directed at climate impacts in the sectors mentioned, investment still amounts to a negligible percentage, around 3% (Climate Bonds Initiative, 2018).

To curb the progression of deforestation and forest degradation, it is imperative to maintain forests in their original configuration, functioning as net carbon sinks. Taxonomies, such as those outlined by the Climate Bonds Initiative, play a pivotal role in financing initiatives aligned with these objectives by categorizing emitters based on sectors. While considerable progress has been made in developing criteria for various sectors, including land use and waste, encompassing agriculture, it is noteworthy that further refinements are still in progress within the criteria framework (Climate Bonds Initiative, 2021a, b). This ongoing development underscores the dynamic nature of efforts to channel financial resources effectively toward activities that contribute to the preservation of forests and the mitigation of climate change impacts.

6 Final Considerations

It is important to remember that climate change is a market failure that can be considered the greatest externality in history, due to its intergenerational nature. With the long lifespan of GHGs and their effects, the measures adopted in the present will, in theory, generate benefits to be enjoyed mostly by future generations, far beyond the lifetime of those who finance them. However, with the increasingly recurrent catastrophic events resulting from climate change, and with biophysical impacts that are sometimes irreversible, it is clear that financing is not just about guaranteeing the future, but resilience in the present. It is therefore essential to raise awareness of the importance of adopting sustainable practices, especially in a sector that is inherent to every human being, which is agribusiness. Every experience takes into account some aspects resulting from an agroindustry activity and, for this reason, financial investment in the sector should be taken more seriously by investors.

Through financing environmentally responsible projects and initiatives, green bonds not only meet the growing demand for sustainable investments but also play a crucial role in transforming the agroindustry. This is because, in addition to mitigating environmental liabilities arising from agribusiness and financing technologies for soil carbon sequestration, green bonds bring economic and reputational advantages to companies. As a result, they contribute to global decarbonization targets and measures to mitigate and adapt to climate change, financing projects that value the responsible use of natural resources, helping to build a greener and more prosperous agroindustrial future.

Therefore, it is crucial to note that the effectiveness of green bonds in reducing the carbon footprint of agribusiness is intrinsically linked to transparency, accountability, and integrity in the execution of the financed projects. Green bond issuers seek to ensure that their projects are aligned with the SDGs as part of their commitment to sustainability. This means that the financing of projects is in line with specific targets of the 2030 Agenda. Rigorous supervision and accurate disclosure of the use of Green Bond resources are essential to maintain the confidence of investors and the general public. In addition, the commitment of all stakeholders, including governments, companies, and financial institutions, is needed to ensure that the environmental benefits are achieved concretely.

As the world faces increasingly urgent environmental challenges, the role of agribusiness in the search for sustainable solutions is undeniable. Green Bonds represent an innovative and effective approach to financing and accelerating the transition of agribusiness toward a more responsible and environmentally friendly model. It is hoped that this study has provided a comprehensive overview of the benefits of these financial instruments while highlighting the continued importance of their responsible use and global collaboration in the pursuit of a truly green agricultural future. Furthermore, due to the need for new material on the subject, it is advisable that the study be used by other critical agribusiness sectors, as well as in studies of the financial sector.

References

Almeida, M., & Filkova, M. (2019). *América Latina e Caribe Análise de mercado das finanças verdes 2019*. Climate bonds initiative. https://www.climatebonds.net/files/reports/latam_sotm_19_por_final_01_web.pdf

Alves, F. V., Laura, V. A., & de Almeida, R. G. (2015). *Sistemas agroflorestais: A agropecuária sustentável*. ISBN 978-85-7035-420-4. http://www.infoteca.cnptia.embrapa.br/infoteca/handle/doc/1010949

Andersen, T. M., Bhattacharya, J., & Liu, P. (2020). Resolving intergenerational conflict over the environment under the Pareto criterion. *Journal of Environmental Economics and Management, 100*, 102290. https://doi.org/10.1016/j.jeem.2019.102290

Bajan, B., & Mrówczyńska-Kamińska, A. (2020). Carbon footprint and environmental performance of agribusiness production in selected countries around the world. *Journal of Cleaner Production, 276*, 123389. https://doi.org/10.1016/j.jclepro.2020.123389

Caregnato, S. E. (2011). Google Acadêmico como ferramenta para os estudos de citações: Avaliação da precisão das buscas por autor. *Pontodeacesso, 5*(3), 72–86. https://periodicos.ufba.br/index.php/revistaici/article/view/5682/4106

Chrobak, U. (2021). Fighting climate change means taking laughing gas seriously. Agriculture researchers seek ways to reduce nitrous oxide's impact on warming. *Knowable Magazine*. https://knowablemagazine.org/article/food-environment/2021/nitrous-oxide-greenhouse-gas-agriculture

Clapp, C., Alfsen, K. H., Lund, H. F., & Cícero, K. P. (2016). *Green bonds and environmental integrity: Insight from CICERO second opinions*. Cícero. https://core.ac.uk/download/pdf/154672377.pdf

Climate Bonds Initiative. (2018). *Títulos de Dívida e Mudanças Climáticas Análise do Mercado 2018*. https://www.climatebonds.net/files/files/CBI_SotM_2018_POR_Final_02G-web.pdf

Climate Bonds Initiative. (2021a). *Record $269.5bn green issuance for 2020: Late surge sees pandemic year pip 2019 total by $3bn*. Disponível em: https://www.climatebonds.net/2021/01/record-2695bn-green-issuance-2020-late-surge-sees-pandemic-year-pip-2019-total-3bn

Climate Bonds Initiative. (2021b). Critérios de Agricultura The Climate Bonds Standard & Certification Scheme (Sistema de Normas e Certificação). Documento de Critérios – Junho de 2021. Versão 1.0. https://www.climatebonds.net/files/files/CBI_Crite%CC%81rio_de_Agricultura_01B.pdf

da Costa, M. G., de Abreu, V. H. S., de Assis, T. F., da Costa, V. X., de Almeida D'Agosto, M., & Santos, A. S. (2022). Life cycle assessment and circular economy strategies for electric vehicle: A systematic review on mitigating climate change and reducing resource depletion in road transportation. In *Carbon footprints of manufacturing and transportation industries* (pp. 113–137). https://doi.org/10.1007/978-981-19-7226-3_5

de Abreu, V. H. S., Da Costa, M. G., Da Costa, V. X., De Assis, T. F., Santos, A. S., & D'Agosto, M. D. A. (2022a). The role of the circular economy in road transport to mitigate climate change and reduce resource depletion. *Sustainability, 14*, 8951. https://doi.org/10.3390/su14148951

de Abreu, V. H. S., Santos, A. S., & Monteiro, T. G. M. (2022b). Climate change impacts on the road transport infrastructure: A systematic review on adaptation measures. *Sustainability, 14*(14), 8864. https://doi.org/10.3390/su14148864

de Assis, T. F., Ricci, L. M., Monteiro, T. G. M., de Abreu, V. H. S., D'Agosto, M. A., & Santos, A. S. (2022a). Sustainable transport indicators and mitigation actions applied to the green bond principles. In S. S. Muthu (Ed.), *Carbon footprints of manufacturing and transportation industries. Environmental footprints and eco-design of products and processes*. Springer. https://doi.org/10.1007/978-981-19-7226-3_6

de Assis, T. F., Monteiro, T. G. M., de Abreu, V. H. S., D'Agosto, M. D. A., & Santos, A. S. (2022b). Enabling the green bonds market for sustainable transport projects based on the measure/monitoring, reporting and verification method. In *Carbon footprints of manufacturing and transportation industries* (pp. 1–24). Springer Nature Singapore. https://doi.org/10.1007/978-981-19-7226-3_1

de Lima, M. A., Pessoa, M. C. P. Y., Neves, M. C., & de Carvalho, E. C. (2010). *Emissões de metano por fermentação entérica e manejo de dejetos de animais. 2° Inventário Brasileiro de Emissões e Remoções Antrópicas de Gases de Efeito Estufa*. Ministério da Ciência e Tecnologia. https://cetesb.sp.gov.br/inventario-gee-sp/wp-content/uploads/sites/34/2014/04/brasil_mcti_fermentacao_enterica.pdf

Dent, D., & Boincean, B. (2021). An investable proposal for regenerative agriculture across the steppes. In *Regenerative agriculture: What's missing? What do we still need to know?* (pp. 341–348). Springer International Publishing. https://doi.org/10.1007/978-3-030-72224-1_31

Dugoua, E. (2023). Induced innovation and international environmental agreements: Evidence from the ozone regime. *Review of Economics and Statistics*, 1–45. https://doi.org/10.2139/ssrn.4592956

ELD Initiative. (2013). The rewards of investing in sustainable land management. Interim Report for the Economics of Land Degradation Initiative: A global strategy for sustainable land management. https://www.eld-initiative.org/fileadmin/pdf/ELD-Interim_Report_web.pdf

EU TEG. (2019). EU Technical Expert Group on Sustainable Finance. Report on EU Green Bond Standard. TEG report. Proposal for an EU Green Bond standard. https://finance.ec.europa.eu/system/files/2019-06/190618-sustainable-finance-teg-report-green-bond-standard_en.pdf

Filkova, M., Frandon-Martinez, C., Meng A., & Rado, G. (2018). *The green bond market in Europe*. Climate Bonds Initiative. https://www.climatebonds.net/files/reports/the_green_bond_market_in_europe.pdf

Gasol, C. M., Salvia, J., Serra, J., Antón, A., Sevigne, E., Rieradevall, J., & Gabarrell, X. (2012). A life cycle assessment of biodiesel production from winter rape grown in Southern Europe. *Biomass and Bioenergy, 40*, 71–81. https://doi.org/10.1016/j.biombioe.2012.02.003

Gibon, T., Popescu, I. Ş., Hitaj, C., Petucco, C., & Benetto, E. (2020). Shades of green: Life cycle assessment of renewable energy projects financed through green bonds. *Environmental Research Letters, 15*(10), 104045. https://doi.org/10.1088/1748-9326/abaa0c

Hergoualc'h, K., Akiyama, H., Bernoux, M., Chirinda, N., Prado, A. D., Kasimir, Å., et al. (2019). *N2O emissions from managed soils, and CO2 emissions from lime and urea application*. https://www.ipcc-nggip.iges.or.jp/public/2019rf/pdf/4_Volume4/19R_V4_Ch11_Soils_N2O_CO2.pdf

Holka, M., Kowalska, J., & Jakubowska, M. (2022). Reducing carbon footprint of agriculture—Can organic farming help to mitigate climate change? *Agriculture, 12*(9), 1383. https://doi.org/10.3390/agriculture12091383

Hsu, C. C. (2023). Influence of climate finance and natural resource consumption on the mitigation of climate change in developed countries in the Pre-COP26 era. *Resources Policy, 83*, 103714. https://doi.org/10.1016/j.resourpol.2023.103714

ICMA. (2021, Junho). *Princípios para Títulos Verdes. Diretrizes Voluntárias para Emissão de Títulos Verdes*. https://www.icmagroup.org/assets/documents/Sustainable-finance/Translations/Portuguese-GBP-2021_06.pdf

IPCC. (2023). *Climate change 2022—Impacts, Adaptation and Vulnerability: Working Group II Contribution to the Sixth Assessment Report of the Intergovernmental Panel on Climate Change. Intergovernmental Panel on Climate Change*. Cambridge University Press. https://doi.org/10.1017/9781009325844

Ketterer, J. A., Andrade, G., Netto, M., & Haro, M. I. (2019). *Transforming green bond markets: Using financial innovation and technology to expand green bond issuance in Latin America and the Caribbean* (Vol. 751). Inter-American Development Bank. https://doi.org/10.18235/0001900

Montanarella, L., Badraoui, M., Chude, V., Baptista Costa, I. D. S., Mamo, T., Yemefack, M., et al. (2015). *Status of the world's soil resources Main Report*. Food and Agriculture Organization of the United Nations and Intergovernmental Technical Panel on Soils, Rome, Italy. https://www.fao.org/3/i5199e/i5199e.pdf

Monteiro, T. G. M., de Abreu, V. H. S., & Santos, A. S. (2023). Uma Visão Geral sobre Os green bonds no Cenário Pandêmico De 2020: An overview of green bonds in the 2020 pandemic scenario. *MIX Sustentável, 9*(3), 171–180. https://doi.org/10.29183/2447-3073.MIX2023.v9.n3.171-180

Nanayakkara, M., & Colombage, S. (2019). Do investors in green bond market pay a premium? Global evidence. *Applied Economics, 51*(40), 4425–4437. https://doi.org/10.1080/00036846.2019.1591611

Olsson, L., Barbosa, H., Bhadwal, S., Cowie, A., Delusca, K., Flores-Renteria, D., Hermans, K., Jobbagy, E., Kurz, W., Li, D., Sonwa, D. J., & Stringer, L. (2019). Land Degradation. In P. R. Shukla, J. Skea, E. C. Buendia, V. Masson-Delmotte, H.-O. Pörtner, D. C. Roberts, P. Zhai, R. Slade, S. Connors, R. van Diemen, M. Ferrat, E. Haughey, S. Luz, S. Neogi, M. Pathak, J. Petzold, J. Portugal Pereira, P. Vyas, E. Huntley, K. Kissick, M. Belkacemi, & J. Malley (Eds.), *Climate Change and Land: An IPCC special report on climate change, desertification, land degradation, sustainable land management, food security, and greenhouse gas fluxes in terrestrial ecosystems*. https://doi.org/10.1017/9781009157988.006

Park, S. (2019). *Green bonds and beyond: Debt financing as a sustainability driver. Cambridge handbook of corporate law, corporate governance and sustainability*. Cambridge University Press. https://ssrn.com/abstract=3383561

Pearce, D., Markandya, A., & Barbier, E. (1989). *Blueprint 1: For a green economy* (1st ed.). Routledge. https://doi.org/10.4324/9781315070223

Pendrill, F., Persson, U. M., Godar, J., Kastner, T., Moran, D., Schmidt, S., & Wood, R. (2019). Agricultural and forestry trade drives large share of tropical deforestation emissions. *Global Environmental Change, 56*, 1–10. https://doi.org/10.1016/j.gloenvcha.2019.03.002

Platform, G. G. K. (2013). Moving towards a common approach on green growth indicators. Green Growth Knowledge Platform Scoping Paper, 1–46. https://www.oecd.org/greengrowth/GGKP%20Moving%20towards%20a%20Common%20Approach%20on%20Green%20Growth%20Indicators[1].pdf

Pombo-Romero, J., & Rúas-Barrosa, O. (2022). A blockchain-based financial instrument for the decarbonization of irrigated agriculture. *Sustainability, 14*(14), 8848. https://doi.org/10.3390/su14148848

Sánchez, L., Bárcena Ibarra, A., Samaniego, J., Galindo, L. M., Ferrer, J., Alatorre, J. E., et al. (2019). *A economia da mudança climática na América Latina e no Caribe: uma visão gráfica*. https://hdl.handle.net/11362/44486

Santos, A. S., de Abreu, V. H. S., de Assis, T. F., Ribeiro, S. K., & Ribeiro, G. M. (2021). An overview on costs of shifting to sustainable road transport: A challenge for cities worldwide. In *Carbon footprint case studies: Municipal solid waste management, sustainable road transport and carbon sequestration* (pp. 93–121). Springer Nature.

Sartzetakis, E. S. (2021). Green bonds as an instrument to finance low carbon transition. *Economic Change and Restructuring, 54*(3), 755–779. https://doi.org/10.1007/s10644-020-09266-9

Stojanovic, D., & Ilic, B. (2018). Green financing in the function of risk management environment and sustainable economic growth. In *Economic and social development: Book of proceedings* (pp. 69–76) https://www.researchgate.net/publication/339740263

Sustainable Banking Network. (2018). *Creating green bond markets: Insights, innovations, and tools from emerging markets*. World Bank. https://doi.org/10.1596/30940

Taghizadeh-Hesary, F., & Yoshino, N. (2019). The way to induce private participation in green finance and investment. *Finance Research Letters, 31*, 98–103. https://doi.org/10.1016/j.frl.2019.04.016

Tordin, C. (2022). *Manejo adequado do solo aumenta sequestro de carbono em áreas irrigadas*. EMBRAPA. https://www.embrapa.br/busca-de-noticias/-/noticia/72379135/manejo-adequado-do-solo-aumenta-sequestro-de-carbono-em-areas-irrigadas

United Nations. (2015). *Do you know all 17 SDGs?* https://sdgs.un.org/goals

United Nations Environmental Programme. (2019). Emissions gap report (2019). Executive summary. https://unepccc.org/publications/emissions-gap-report-2019-executive-summary/

van de Koppel, J., Rietkerk, M., & Weissing, F. J. (1997). Catastrophic vegetation shifts and soil degradation in terrestrial grazing systems. *Trends in Ecology & Evolution, 12*, 352–356. https://doi.org/10.1016/S0169-5347(97)01133-6

Wang, C. W., Wu, Y. C., Hsieh, H. Y., Huang, P. H., & Lin, M. C. (2022). Does green bond issuance have an impact on climate risk concerns? *Energy Economics, 111*, 106066. https://doi.org/10.1016/j.eneco.2022.106066

World Bank. (2019). *10 Years of green bonds: Creating the Blueprint for Sustainability across Capital Markets* [online]. Retrieved from: https://www.worldbank.org/en/news/immersive-story/2019/03/18/10-years-of-green-bonds-creating-the-blueprint-for-sustainability-across-capital-markets

Xia, Y., Ren, H., Li, Y., Xia, J., He, L., & Liu, N. (2022). Forecasting green bond volatility via novel heterogeneous ensemble approaches. *Expert Systems with Applications, 204*, 117580. https://doi.org/10.1016/j.eswa.2022.117580

Zhang, D., & Du, P. (2020). How China "going green" impacts corporate performance? *Journal of Cleaner Production, 258*, 120604. https://doi.org/10.1016/j.jclepro.2020.120604

Green Bonds and the Incentive to Decrease the Carbon Footprint in the Brazilian Agribusiness Sector

Thaís Guedes Máximo Monteiro, Renata Angeli, and Victor Hugo Souza de Abreu

1 Introduction

Agribusiness stands as a cornerstone of the Brazilian economy, playing a pivotal role in contributing significantly to the Gross Domestic Product (GDP) and employment generation. According to CEPEA data from 2023, agribusiness revenue in the first half of the year accounted for a remarkable 50% of the total national exports, reflecting a 3.79% increase compared to the same period in 2022. Brazil achieved the notable status of being the leading exporter in terms of grain value in 2020, reaching a staggering 37 billion US dollars. Key products such as soy, corn, cotton, and various meat categories continue to witness escalating demand in the international market, a trend projected to persist in the foreseeable future. This trajectory is underpinned by factors such as population growth and rising income, acting as propellants for the burgeoning global demand (Aragão & Contini, 2021). However, amid the economic prosperity, the sector's impact on natural resources remains a contentious issue. While the economic benefits are undeniable, they often come hand-in-hand with substantial environmental repercussions, prompting a critical examination of the trade-offs between agribusiness success and environmental sustainability.

T. G. M. Monteiro · R. Angeli
State University of Rio de Janeiro (Universidade Estadual do Rio de Janeiro – UERJ), Postgraduate Program in Environmental Science and Technology (Programa de Pós-Graduação em Ciência e Tecnologia Ambiental – PPGCTA), Rio de Janeiro, Brazil

V. H. S. de Abreu (✉)
Federal University of Rio de Janeiro (Universidade Federal do Rio de Janeiro – UFRJ), Transport Engineering Programme (Programa de Engenharia de Transportes – PET), COPPE-UFRJ, Rio de Janeiro, Brazil
e-mail: victor@pet.coppe.ufrj.br

S. S. Muthu (ed.), *Carbon Footprint Assessments*, Environmental Footprints and Eco-design of Products and Processes,
https://doi.org/10.1007/978-3-031-70262-4_15

Brazil entered the green bond market in 2015, led by the non-financial company BRF S.A., which issued a €500 million bond for project financing—one of the first of its kind to support industry globally. This transaction marked the first large-scale green bond issuance in Latin America and the Caribbean (LAC). Currently, Brazil still stands out as the largest issuer in the region, with 41% of total emissions, among the eight LAC countries that have issued green bonds to date (Almeida & Filkova, 2019). In 2020, at the height of the pandemic, the minister of the Ministry of Agriculture, Livestock and Supply, Tereza Cristina, highlighted the importance of strengthening the green finance market in Brazil, as an agri-environmental power committed to sustainability. All of this corroborates the fact that agribusiness is a significant contributor to Brazilian GDP (MAPA, 2020).

Furthermore, the country is home to the Amazon Basin with the largest tropical forest in the world, with 6 million square kilometers of Amazon Forest, surrounded by the Venezuelan and Colombian savannas and the Cerrado (Duarte, 2020). The importance of taking green finance seriously stems from significant issues related to GHG emissions, such as land use changes driven by deforestation in the Amazon since the 1990s for the cultivation of soybeans, as well as the expansion of areas for pasture (Margulis, 2003). As the suppression of vegetation almost in its entirety accompanies the justification for development, it is important to understand that the result of such decisions is taking us to a point of no return, with irreversible consequences such as the possible savannization of the Brazilian Amazon and triggering of pandemics (Duarte, 2020).

Considering the pivotal role of the agribusiness sector in driving Brazil's economic development, it becomes imperative to strategize approaches that harmonize sustainable growth with environmental preservation. In this context, green bonds emerge as a viable alternative, offering a pathway to incentivize the adoption of cleaner and more sustainable technologies within the sector. The pressing reality of climate change brings direct consequences to financial markets, amplifying the volatility of commodity prices. To fortify green climate policies and expedite the transition toward a low-carbon economy, the role of green bonds becomes indispensable in providing the necessary financial resources (Gozgor & Karakas, 2023). By facilitating targeted investments in eco-friendly practices, these bonds not only contribute to mitigating the environmental impact of agribusiness but also align with broader sustainability goals, fostering a resilient and responsible economic trajectory for Brazil.

The environmental impacts of agribusiness, especially in the Brazilian context, are directly related to several Sustainable Development Goals (SDGs) since Brazilian agribusiness contributes significantly to issues such as biodiversity loss, deforestation, soil degradation, land use of excessive water resources, and GHG emissions. This occurs because the issuance of green bonds aimed at Brazilian agroindustry is aligned, among others, with SDG 2 which deals with Zero Hunger and Sustainable Agriculture and SDG 13 related to Action Against Global Climate Change. Thus, through a bibliographical review, the environmental impacts of agribusiness and the issuance of green bonds aimed at Brazilian agroindustry were analyzed, with the aim of encouraging reflection on the essentiality of green agendas for the development of a nation referred to as one of the main producers of commodities worldwide.

Therefore, this chapter focuses on presenting a bibliographical review of studies on the impacts of Brazilian agribusiness and the search for reducing the carbon footprint through technologies financed by the issuance of green bonds (Sinha et al., 2021), such as Biological Fixation of Nitrogen, which aims to ensure the sequestration of GHG in the soil and the reduction of the carbon footprint. Furthermore, the objective is to demonstrate advances in public policies, such as the ABC Plan, and to raise awareness about the importance of more in-depth analyzes on the topic. The relevance of this study includes the approach to topics linked to the financial market and the environment, denoting the impossibility of talking about limiting the planet's warming to 1.5 °C, not to mention investing in technologies to mitigate the effects of climate change and reduce the carbon footprint. In addition to this introductory section, the rest of the document is organized as follows: Sect. 2 deals with the methodology applied in the study, in order to conduct the considerations efficiently; Sect. 3 contextualizes agroindustry in Brazil; Sect. 4 presents the sector's environmental impacts, such as deforestation and soil erosion; Sect. 5 correlates efforts to make agribusiness sustainable, such as the use of minimum cultivation and integrated pest management; Sect. 6 provides guidelines for issuing green bonds according to globally defined taxonomies; and, finally, Sect. 7 concludes the research with observations about the topic discussed and proposes guidelines for future studies on a topic as comprehensive as agribusiness and the issuance of green bonds.

2 Methodology

The current study is the product of an extensive bibliographic review, involving a meticulous and systematic analysis of the existing literature on the intersection of green bonds and agribusiness. This investigation encompasses a diverse collection of studies, reports, and scientific publications, with a purposeful aim to scrutinize evolving trends and practices in this domain. To gather pertinent information, scientific articles, book chapters, reports, and technical documents were selectively sourced from the Google Scholar online database using the keywords "Brazilian agriculture," "agribusiness," "carbon footprint," "green bonds," and "Brazilian agroindustry."

The obtained texts underwent a rigorous analytical and synthesis process, aimed at distilling essential information and extracting key conclusions presented by various authors. The narrative unfolds by contextualizing the expansion of Brazilian agribusiness, bolstered by illustrative examples from the soybean market and productivity data. This is seamlessly followed by an exploration into the environmental impacts resulting from the conventional agribusiness model, accentuating the

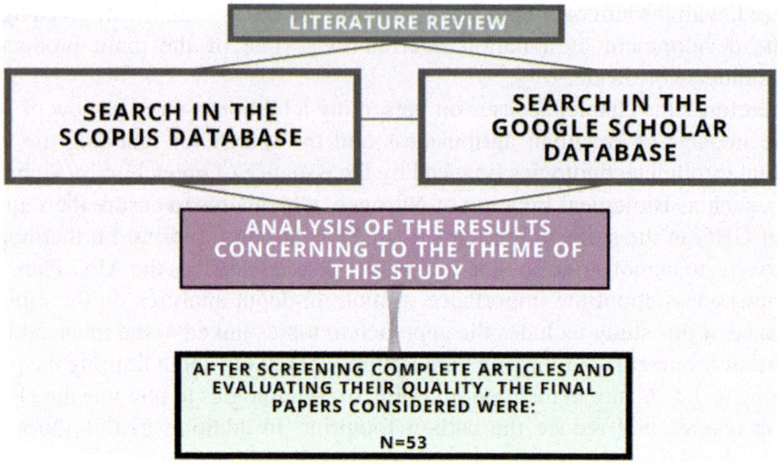

Fig. 1 Methodology used concerns data search. (Source: The Author)

imperative for novel agricultural production approaches. The review delves into alternatives such as Crop-Livestock-Forest integration and the treatment of animal waste.

Concurrently, the study leveraged the paid database Scopus, employing specific keywords such as ("green bonds AND agriculture") OR ("green bonds AND climate change") for a parallel exploration. The findings from both sources were meticulously analyzed to enrich the depth and breadth of this study, as depicted in Fig. 1. This comprehensive approach ensures a well-rounded exploration of the interplay between green bonds and agribusiness, incorporating insights from diverse sources to provide a nuanced understanding of the subject matter.

3 Agroindustry in Brazil

Agricultural activity in Brazil dates back to colonial times and highlights a multitude of product cycles, such as sugar cane, cotton, coffee, cassava, corn, and, more currently, soybeans, so that all governments seek to create specific strategies for the sector. Alongside being the fourth largest grain producer worldwide, Brazil is also the second largest exporter, representing 19% of global grain trade, equivalent to 123 million tons. Over the past 20 years, Brazil has accounted for 12.6% of total global grain exports, amounting to over 1.1 billion tons. In 2020, Brazil reached the position of largest exporter in terms of grain value, which represented 37 billion US dollars. Soy, corn, cotton, and meat in general are the products with growing demand in the foreign market, with the prospect of remaining so in the coming years, considering population growth and rising income as driving forces of global demand (Aragão & Contini, 2021).

The expansion of sugarcane (ethanol) and soybean (biodiesel) production in Brazil has been occurring on a large scale, overloading ecosystems and communities. Advances in technology, as well as the adaptation of sugarcane cultivation to the region, were responsible for transforming São Paulo into the largest producer of sugarcane, with sugar and alcohol as its main focus (Shikida, 2013). This movement boosted the expansion of sugarcane cultivation to the Central-West region, stimulated by the growing production of ethanol, and, in this way, the consolidation of the agroindustry and the considerable growth of the agricultural sector in Brazil resulted in a bonanza. However, new challenges also emerged, such as competition for land and rising prices (Sant'Anna et al., 2016).

Being one of the main pillars of the Brazilian economy, in terms of export trade and internal trade itself, agribusiness has received high investments from the government in recent years to encourage the expansion of monocultures. This is because, as a major producer of commodities on a global scale, Brazil holds a privileged position related to the cultivation of products of primary origin, sold on a large scale, with the price regulated by their market price. As an example, it is worth noting that in the 2022/2023 harvest, soybeans led national production with an unprecedented yield of 154.6 million tons, marking a record volume of exports. These results were obtained due to the excellent climate conditions in most producing regions and advanced technology employed by the farmers (CONAB, 2023).

Nevertheless, while the burgeoning growth of the agribusiness sector contributes substantially to economic development, it concurrently raises escalating apprehensions regarding environmental liabilities. These concerns encompass issues like soil contamination attributed to the use of increasingly potent fertilizers and pesticides, the encroachment of deforestation, and the surge in fires employed to facilitate the expansion of agribusiness operations (Assad et al., 2012). Consequently, there is a pressing need for a conscientious approach toward the proper disposal of agro-industrial waste and by-products, warranting their effective inclusion in the sustainable development agenda of agribusiness.

Addressing these challenges requires a paradigm shift in agribusiness practices, where innovative, systemic solutions are devised to holistically consider the interconnected realms of energy, food, and the environment (Galvão, 2017). Recognizing the integral relationship between these elements, strategic initiatives must be implemented to not only mitigate the adverse environmental impacts but also promote regenerative practices. By integrating waste management into the broader sustainability framework, agribusiness can foster a more resilient and ecologically responsible model that aligns with long-term environmental preservation goals. This necessitates a collaborative effort among stakeholders to adopt forward-thinking practices, ensuring that the growth of the agribusiness sector occurs harmoniously with the preservation of our vital ecosystems.

As an example of the advancement of deforestation, it is possible to mention the Municipality of Sorriso, located in Mato Grosso, where, in the space of 30 years, 353.485 hectares were deforested, totaling 38% of its natural vegetation. During the period from 1991 to 1995, which was the most critical, 128,880 hectares of natural vegetation were suppressed, representing 36.46% of the total area deforested in three decades. In the period from 1996 to 2000, as well as the 5 years that followed, they corresponded, respectively, to 20.79% and 16.06% of total deforestation between 1985 and 2015. The advancement of deforestation so quickly is explained given the voluminous financing received by local producers in the 1990s and early 2000s, directed to the development of agricultural activities. During this same period, the largest number of hot spots were documented in the Municipality of Sorriso, as a result of the massive burning of degraded areas (Azevedo & Pasquis, 2007), where vast areas of native vegetation were exchanged for annual monocultures (Fearnside, 2001).

In 2004, the Municipality of Sorriso already had a considerable amount of environmental liabilities and a total of 73.4% of its territorial extension was deforested, having a direct relationship with the granting of credit. The stability provided by the Real Plan in 1994 coincided with the considerable increase in deforestation and, similarly, in 2004 this intensified with the rise in the prices of agricultural commodities (Ferreira & Coelho, 2015), making it possible to clearly correlate the increase of areas for agricultural use with the deforested area (Camargo, 2017). Therefore, the development of tools that make agribusiness activities sustainable is essential. The Municipality of Sorriso is just one example among many in Brazil that have had a considerable increase in deforested areas over the years for the implementation and maintenance of agribusiness.

The expansion of conventional agroindustry goes in the opposite direction to the sustainability of natural resources, since the activities resulting from the sector cause, among other problems, soil degradation and contamination resulting from intensive management and indiscriminate use of fertilizers and pesticides (Gomes, 2019). The use of fires in open areas, widespread in the Cerrado and Amazon Forest, also generates the loss of aggregates of organic matter and clay in the soils which, when managed intensively for a long period, still have their structure altered, contributing to degradation, compaction, and erosion (Primavesi, 2002). Furthermore, zoning, as an urban planning tool, limited the advancement of sugar cane plantations in the North region but did not eliminate the impacts already caused in the cities where this monoculture expanded. Across the country, the area allocated to sugarcane cultivation has increased consistently since 2001, growing from 5.2 to 8.7 million hectares in 2009 (IPEA, 2011).

Thus, with soil that is conducive to polyculture and other agribusiness activities, Brazil must use instruments that make the sector favorable to go hand in hand with the preservation of natural resources. The adoption of sustainable management practices, with a focus on methods that emphasize crop diversification and rotation, as well as the use of alternative inputs to control pests that do not harm the soil or waterways, are ways of controlling the environmental impacts of agribusiness (Perfecto & Vandermeer, 2010). Furthermore, when carried out properly, these techniques can also benefit degraded areas with low productivity, making them fertile again. Ideally, however, there should be no more deforestation of new areas for cultivation, by making use of the land already used by the agro-industry (Sambuichi et al., 2012). To this end, it is essential that tools are incorporated into the flow of the sector in order to keep agribusiness profitable without the environment deteriorating as has been seen in recent years.

4 The Environmental Impacts of the Sector

Agribusiness is one of the most important sectors of the Brazilian economy, contributing significantly to GDP and job creation. According to data from CEPEA (2023), Brazilian agribusiness revenue in the first half of 2023 represented 50% of all total national exports (3.79% more than the same period in 2022). However, its impact on natural resources is a controversial topic, as such economic benefits tend to bring with it overwhelming damage to the environment.

During the 1980s, there was a significant increase in land occupation, resulting from the granting of credit and investments in works such as road construction, resulting in 18 million hectares of forest devastated. However, it was only after 1994 that deforestation and the agricultural sector were actually associated, due to Brazil's great prominence in the international market, as a supplier of soy and meat (Mello & Artaxo, 2017). It is not by chance that this connection occurred during the period of redemocratization in Brazil, with the demands of environmental non-governmental organizations (NGOs) in line with the defense of human rights, also on the international scene. However, deforestation in the late 1990s remained accelerated, being increasingly associated with the meat market and soybean farming, related to each other in both production and marketing.

Brazil is one of the largest producers of agricultural commodities in the world, such as soybeans, corn, coffee, cotton, and beef, among others, and, to meet this growing global demand, Brazilian agricultural production has been constantly increasing, which can lead to problems such as deforestation, soil erosion, water contamination, and greenhouse gas emissions. It is also important to highlight that the agricultural sector is the main user of land and water globally (FAO, 2021).

Indeed, the conversion of forested areas into cultivated land constitutes a profound and impactful alteration to the original ecosystem. This transformation induces substantial changes in soil properties at physical, chemical, biological, and morphological levels, signifying a comprehensive disruption of natural mechanisms (Lima et al., 2011; Luizão et al., 2006). Such alterations can give rise to considerable environmental impacts, as the intricate balance and interactions within the ecosystem undergo a significant shift. The effects are pervasive, affecting not only the immediate soil composition but also influencing broader ecological dynamics, making it imperative to comprehend and address the multifaceted consequences of this land use change.

Brazil has been seen worldwide as one of the main producers of agrofuels, such as sugar cane, soybeans, and palm oil, among others, serving as raw materials for the production of ethanol and biodiesel. As an example, it is possible to mention the 2021/2022 harvest, which resulted in 578.7 million tons of sugar cane, with the prospect of an increase of 5.4% for the 2022/2023 harvest, according to an estimate from April 2023 (CONAB, 2023). However, the expansion of production is accompanied by environmental impacts, mainly in terms of water consumption, application of pesticides and fertilizers, emission of methane gas, deforestation, and burning of native vegetation to expand agribusiness (Assad et al., 2012).

Furthermore, it is crucial to underscore that investigations into land degradation have predominantly centered on biophysical aspects. This emphasis implies that, over the years, the perspectives and insights of land users themselves have been largely overlooked by both policymakers and scientists (Reed et al., 2007; Forsyth, 1996; Anderson et al., 2011). However, there is a growing recognition among scholars of the necessity to incorporate the viewpoints of local users. This shift in perspective is motivated by the realization that climate change, with its direct and indirect ramifications, significantly influences the social fabric of entire communities, impacting not only the users but also the broader ecosystem. Acknowledging and integrating the experiential knowledge of local communities is paramount for devising comprehensive strategies that address the complex interplay between human activities, environmental conditions, and the challenges posed by climate change. This inclusive approach ensures a more holistic understanding of land degradation dynamics and facilitates the development of sustainable and community-centric solutions.

Therefore, when comparing land degradation, climate change, and poverty, it becomes evident that individuals dependent on land and residing in disadvantaged agricultural regions are directly confronted with the interplay between poverty and environmental degradation. This interaction can lead to the expansion of degraded lands as a consequence of the climate change impacting these regions (Barbier and Hochard, 2018). This is because climate change can lead to soil degradation and a reduction in agricultural production, negatively affecting the income and food security of poor rural populations. In turn, poverty can lead to unsustainable land use practices, accelerating environmental degradation and worsening the impacts of climate change. Hence the importance of considering the interaction between poverty and the environment when formulating policies and strategies for adapting and mitigating climate change. The emerging relationships between land degradation, climate change, and poverty are shown in Fig. 2.

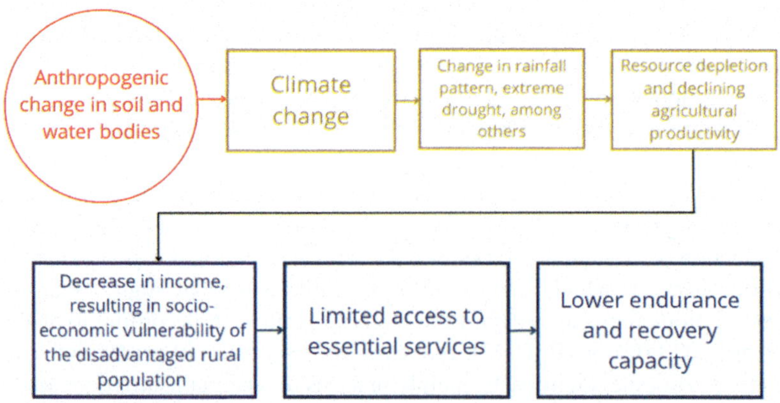

Fig. 2 Schematic representation of the relationship between resource depletion and the socio-economic condition of the rural population (Source: The Author 2024)

It is also important to contextualize the rise of Brazilian livestock farming on the world stage, given that at the end of the 1950s, poultry farming for food production amounted to less than 10 thousand tons, while pig farming already produced almost 500 thousand tons of carcasses. However, the pork industry remained stagnant until the 1980s, while the poultry industry continued to increase (de Miranda, 2020), starting the 1990s with exponentially growing chicken meat production, taking Brazil to the position of the largest exporter in the world (USDA, 2023).

In 2022, according to Brazilian Animal Protein Association data, Brazil assumed the position of the second largest chicken producer in the world, contributing to food security and moving a significant economic chain, with the generation of approximately 4 million direct and indirect jobs, in addition to around of 100 thousand families working on farms and 500 thousand people in agro-industries (Governo Federal, 2022). But all this progress and economic projection brings with it an overload: with approximately 21% of the occupied territory, Brazilian livestock farming represents a sector in real need of sustainable actions (EMBRAPA, 2018).

Soil erosion and compaction caused by cattle trampling change the geometry of slopes and remove soil particles, providing sediment for water courses, among other things. It is essential to highlight that degenerated pastures exhibit high rates of soil and water loss, in addition to low productive capacity, making them more susceptible to attacks by pests and invasive plants. Furthermore, the deterioration of more extensive cultivated and pasture areas only increases the demand for new land, since the cost of deforestation to incorporate new agricultural frontiers is usually much lower than that of recovering degraded and unproductive areas (Gomes, 2019).

5 Efforts to Make Agribusiness Sustainable

Despite the challenges, it is possible to find examples of efforts to make agribusiness sustainable and reduce the carbon footprint. Producers have been adopting agroecological practices, such as agroforestry and low-carbon agriculture, aiming to integrate agricultural production with the conservation of natural resources. With an extensive list of environmental impacts, especially related to deforestation, soil erosion, environmental contamination, and GHG emissions, agribusiness with a focus on sustainability seeks to make agricultural production less harmful, in order to guarantee the conservation of natural resources over the long term (Alves et al., 2015).

However, the transition to sustainable agribusiness can sometimes put a strain on the budget. This occurs because, in addition to the lack of effective awareness about the environmental liabilities of agribusiness, the amount available in the budget is often incompatible with what is necessary to implement actions that result in significant benefits. With this in mind, some tools were created with a focus on agribusiness, such as Green Certificate of Agribusiness Receivables (CRA), which is suitable for investors, being a great financing alternative for rural producers. Because it is versatile and fast, this financial instrument can help producers finance the implementation of technological innovations to recover degraded areas, regenerate native forests, and implement low-carbon agriculture, among others (Mattos Filho, 2019).

Several critical areas of agriculture and related sectors must receive immediate attention from decision-makers in reducing the carbon footprint, using green financing for actions such as: (i) efficiency in water use, with the restoration of soil erosion and its nutrients through the preservation of water in the soil, also reducing losses through runoff; (ii) use of technologies such as minimum cultivation; (iii) use of alternative energy for the operation of agricultural machinery and post-harvest processing; (iv) training small farmers to adopt technologies that encourage the conscious use of natural resources (so that they remain renewable), use diversified crop rotations, as well as implement livestock and agricultural integration; and (v) raising awareness among communities aimed at reducing the use of chemical pesticides and herbicides, through the implementation of integrated management of pests and weeds, among others that do not harm the environment (Das & Cungu, 2023).

Within the framework of green bonds, the Green CRA stands out for its distinctive feature of channeling the funds raised exclusively toward financing or refinancing projects that unequivocally contribute to environmental well-being. This strategic allocation of resources not only aligns with environmental sustainability goals but also brings discernible advantages to rural producers, fostering a harmonious synergy between ecological considerations and the agricultural sector (Mattos Filho, 2019). Table 1 provides illustrative examples, showcasing the tangible positive impacts that result from the targeted allocation of funds through the Green CRA, elucidating its role in driving sustainable practices and delivering tangible benefits to rural stakeholders.

During the 15th Conference of the Parties (COP-15), the Brazilian government voluntarily committed to reduce around 37% of projected GHG emissions by 2020 (approximately one billion tons of CO_2-eq). To achieve this target, several measures were proposed, including (i) reducing the rate of deforestation in the Amazon by 80% and in the Cerrado by 40%; (ii) intensively recovering pastures degraded by agriculture; (iii) actively promoting Integrated Crop-livestock (ICL); (iv) increasing sustainable soil management technologies such as Direct Seeding System (DSS) and Biological Nitrogen Fixation (BNF); and (v) increasing energy efficiency, the use of biofuels, and the supply of alternative energy sources, among others (MAPA, 2012).

Consequently, in 2010, the Sectoral Plan for Mitigation and Adaptation to Climate Change for the Consolidation of a Low Carbon Economy in Agriculture (ABC Plan) was drawn up, with the aim of reducing greenhouse gas emissions from the Brazilian agricultural sector (Silva & Vieira Filho, 2020), promoting actions through seven programs, referring to six technological processes and climate adaptation, as shown in Fig. 3.

During the preparation of this plan, the original agriculture commitments, signed at COP-15, changed to the adoption of actions such as the recovery of 15 million hectares of degraded pastures through adequate management and fertilization, the expansion of the adoption of Integrated Crop-Livestock-Forestry (ICLF) and

Table 1 Benefits of the Green CRA

Advantages for rural producers	
Conquest of impact investors	Obtaining the "green" approval is motivated by the growing demand for investments in projects that are sustainable, as funds, family offices, and economic groups direct part of their resources to investments of this type. With the publication of Provisional Measure 897/2019, issues can also be distributed abroad, reaching the consolidated green bonds market at an international level
Reputation gain	As the issuance of a Green CRA requires transparency and informational organization, resource allocation reports are prepared by the agribusiness company to attest to the application of resources in accordance with environmentally friendly actions. By complying with these requirements, the company assures the market that it follows international standards of transparency and organization, demonstrating socio-environmental compliance and corporate governance, which becomes a preponderant factor for investors to invest in better credit opportunities
More attractive rates	The "green" qualification helps to improve the pricing of fees for issuers, since the transparency and corporate governance commitments linked to the "green" bond qualification process demonstrate to the investor compliance with socio-environmental legislation, which may result in a reduction of credit risk
Preference when choosing investments	Investors attentive to climate change and goals related to the Paris Agreement may see Green CRA as a priority option given other conventional investments. By offering a certified security with the same financial conditions as a common debt security, the green characteristic tends to be a categorical factor for decision-making. The commitment made by the issuer to be transparent in "green" issuance becomes a complementary benefit for investors, while ensuring that the resources will be allocated to activities that are truly beneficial to the environment

Source: Based on Mattos Filho (2019)

Agroforestry Systems (AFS) on 4 million hectares, the even wider use of DSS on 8 million hectares, the increased use of BNF on 5.5 million hectares, the promotion of reforestation actions with expansion of the area with Planted Forests relative to the production of wood, fibers, and cellulose, and the expansion of technologies for treating animal waste in energy generation and organic compound production (MAPA, 2012). It is important to remember that ICLF and AFS allow the integration of agricultural production, livestock, and forestry in the same area, making the most of land use, resulting in economic returns through increased productivity in agriculture, among other benefits, (EMBRAPA, [s.n.]; Padovan, 2021).

Brazil has established public policies, such as the Low Carbon Agriculture Plan (Plan ABC), to reduce the carbon footprint and decarbonize agricultural activity. In this way, the increase in profitability in the sector is aligned with the reduction of GHG emissions, stimulating the expansion of production while meeting the commitments of the Paris Agreement. Increasingly, extreme weather events impact communities around the world (as a result of climate change), leading commercial

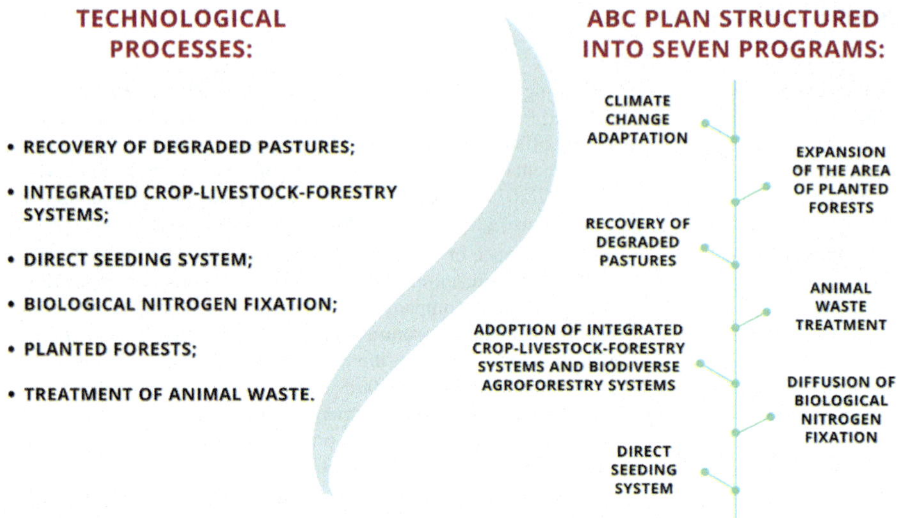

Fig. 3 Actions to reduce GHG emissions from the Brazilian agricultural sector through the ABC Plan. (Source: Based on MAPA 2012)

partners to pressure food, fiber, and energy-producing countries to effectively reduce GHG emissions in their production. Therefore, with BNF, which significantly reduces or eliminates the use of nitrogen fertilizer, it is estimated that Brazil has stopped emitting hundreds of millions of tons of CO_2 equivalent in the last 30 years. The DSS and the Integrated Crop-Livestock-Forestry Systems (ICLSF), which are also part of the ABC Plan, sequester carbon in the soil and equally contribute to reducing the carbon footprint of the agricultural sector (Moretti & Ferreira, 2021).

Regarding adaptation to climate change in the agricultural sector, it is also essential that production systems are structured in such a way as to reduce the vulnerability of producers, rural populations, and ecosystems, aiming to increase the resilience of systems and the dissemination of sustainable use of both biodiversity, as well as water resources. Furthermore, other factors to be considered as they are inherent to climate change are those related to indirect impacts on production costs, commercial product transactions, and energy availability (MAPA, 2012).

The Soy Moratorium is another example of action beneficial to the environment, related to agribusiness. Brazilian Association of Vegetable Oil Industries (ABIOVE) and National Cereal Exporters Association (ANEC) came together to reduce deforestation for soybean cultivation in the Amazon in order to ensure that soybeans, coming from the Amazon biome and negotiated by their signatories, are free from deforestation occurring after July 22, 2008 (ABIOVE, 2022). Thus, a report by Greenpeace (2018) on the Soy Moratorium revealed that, after more than a decade of existence, the search for zero deforestation is not a utopia, pointing to an 85% reduction in deforestation in the 89 monitored municipalities, including the states of Mato Grosso, Pará, Rondônia, Amapá, Maranhão, and Tocantins.

The report also pointed out that less than 2% of the expansion of soybean crops took place in deforested areas after July 2008. Of the 4.5 million hectares cultivated in the 2016/2017 harvest, only 47,365 hectares were not in compliance with the moratorium. The reason is that, with the moratorium, soy buyers began to demand that producers prove the origin of the soy sold, to mitigate the possibility of production from illegally deforested areas continuing in the production chain (Greenpeace, 2018).

6 Guidelines for Issuing Green Bonds

Sustainable financing in the agricultural sector does not differ in relation to the evaluation criteria, but they do differ in issues related to the: (i) application of investment (type of subsidies, as the agri-food sector involves initial subsidies and final subsidies, such as the price of fertilizers and support for exports, respectively); (ii) geographic location (rural and agricultural areas require different approaches from urban areas, as they present different levels of carbon footprint, for example); and (iii) the natural resources available (forests, lakes, sunlight). Such financing mechanisms seek not only to reduce climate footprints but to guarantee economic returns (Das & Cungu, 2023).

A study conducted between 2013 and 2017 by Gianfrate and Peri (2019) indicated a statistically significant advantage for issuers when a bond is labeled green. This advantage is quantifiable, with an average of 18 basis points of lower annual interest paid to investors (which represents 0.18% of the bond's value), being observed by both companies and non-corporate entities, such as municipalities and government agencies. Therefore, this study suggests that, even considering the extra costs necessary to obtain a green certification for issuance, green bonds continue to be advantageous for issuers. Therefore, green bonds not only benefit society but also reduce debt financing costs for issuers.

The ICMA Voluntary Guidelines for Green Bond Issuance relate the Green Bond Principles (GBP) with the Social Bond Principles (SBP) as well as the Sustainable Bond Guidelines (SBG) and the Principles for Sustainability-Linked Bonds (SLBP), since they are published under the governance of the Principles, denoting that the environment is directly linked to the social. The principles aim to cooperate to advance environmental and social sustainability, through the compilation of voluntary milestones with a declared mission and vision. Through universal guidelines and advice, which aim to promote transparency and dissemination of information, they reinforce market completeness by presenting best practices for issuing debt securities, meeting social and/or environmental objectives. Equally important, they cooperate to clarify, among financial market participants, the importance of environmental and social impacts, ultimately aiming to bring more capital to support sustainable development (ICMA, 2021).

Speaking specifically, the GBP are voluntary guidelines that provide issuers with guidance on what needs to be fundamentally considered during the process of issuing a reliable green bond. In addition, they support investors by providing

Table 2 The main components for alignment with the GBP

Main components	Description for a clear process and disclosure for issuers
Use of proceeds	The first step and basis of a green bond is the application of resources to eligible Green Projects. These eligible Green Projects must inform the environmental benefits to be evaluated and, when possible, quantified by the issuer. All of this must be described in the legal title documentation
Process for project evaluation and selection	The issuer must communicate to investors: (i) what are the objectives relating to the environmental sustainable development of eligible Green Projects; (ii) what process is used to determine which eligible category of Green Projects it falls into; (iii) additional data on how perceived socio-environmental risks relating to the project are identified and managed
Management of proceeds	All liquid resources must be made available in a subaccount or tracked by the issuer (if otherwise), as well as attested in an internal protocol process relating to the issuer's lending and investment procedures. The balance of resources must be checked periodically throughout the useful life of the green bond and the issuer must inform investors of the temporary allocation for the balance of unused liquid resources
Reporting	Issuers must provide and maintain all available information up to date regarding the use of resources, with annual updates being mandatory until the total allocation of capital. This annual report must contain a list of the projects for which green bond resources were allocated and the amounts allocated, as well as a brief description of the projects and the expected result

Source: Based on ICMA (2021)

essential information for assessing the environmental impact of their green bond investments, as well as helping financial coordinators by providing processes that will facilitate transactions that seek to preserve the integrity of the market. As such, the GBP highlights how transparency, accuracy, and integrity are essential for the data that will be exposed by issuers to consider the key components and additional recommendations for stakeholders (ICMA, 2021), as shown in Table 2. Furthermore, as the GBP strongly insists on transparency throughout the process, there are still two additional recommendations which are the green bond frameworks and the External Reviews.

Concerning the use of resources, there are eligible categories of Green Projects, but they are not limited to these and are not specifically listed, as can be seen in Table 3. This means that other categories, which are not included, can be considered when analyzing the framework.

In this context, there is a growing awareness of climate risk on the part of companies and investors who, by adopting tools that benefit the environment, consider social and environmental impacts in their decision-making. From 2011 onwards, annual green bond issuance began to increase consistently, with significant growth from 2015 onwards. In 2020, the Paris Agreement was sealed, and green emissions reached a total of 269 billion dollars, surprisingly surpassing past years,

Table 3 Eligible categories of Green Projects in line with the GBP

The most commonly used types of projects supported or expected to be supported by the green bond market	
Renewable energy	Considered clean and inexhaustible, this category includes production, transmission, equipment, and products related to renewable energy
Energy efficiency	Generating and harnessing energy from renewable sources, with the aim of reducing the use of energy from non-renewable and GHG-emitting sources. Examples include resilient developments with energy storage systems and smart grids
Pollution prevention and control	In addition to reducing emissions, good examples of pollution control are appropriate practices for disposing of solid waste (preventing the creation of open-air garbage dumps), as well as reducing waste production, among others
Conservation of terrestrial and aquatic biodiversity	The protection of coastal environments, marine environments, and watersheds
Clean transportation	Reducing GHG emissions through the use of electric, hybrid, and non-motorized means of transport, among others, as well as the use of public means of transport
Sustainable water and sewage management	It considers the protection of water resources, including sustainability in clean and/or drinking water infrastructure, the treatment of domestic and industrial effluents, and functional and sustainable urban drainage systems aimed at mitigating flooding, among others
Climate change adaptation	As climate change is global in scope and attributable directly or indirectly to human activity, this topic includes efforts to make infrastructures more resilient to these impacts, as well as the creation, strengthening, and dissemination of information support systems that include climate observation and warning systems
Green buildings	Aimed at environmental preservation and the use of renewable natural resources, they are validated by compliance with regional, national, or internationally recognized environmental performance standards or certifications
Sustainable management of natural resources and land use	Through (i) sustainable agriculture using agricultural inputs that do not have a negative impact on the environment, such as biological protection or drip irrigation; (ii) ecologically sustainable animal husbandry; (iii) forestry that considers afforestation or reforestation essential, as well as the preservation and restoration of forests, among others
Products, technologies, and production processes adapted to the circular economy	With the strategic concept based on reducing the use of new raw materials, prioritizing durable, recyclable, and renewable inputs, this topic considers not only products but also circular services and/or certified eco-efficient products

Source: Based on ICMA (2021)

with an average annual growth of 60% compared to 2015 (de Assis et al., 2022a, b). Investments focused on the energy sector, which received the largest share of 2020 investments, followed by low-carbon buildings, low-carbon transportation, water infrastructure, and land use (Makpotche et al., 2023). As a result, the market—and institutional shareholders interested in impact investments—consistently perceive the benefits of green finance, both in terms of contributing to the long-term development of smart cities and boosting inclusive economic growth (Ozili, 2022).

It is also important to consider the performance of green bonds in the pandemic year of 2020 where, while suspending activities and facing the insecurity of many entities, the growth in emissions was a positive surprise for the market and for those benefiting from these emissions (Monteiro et al., 2023). Institutions focused on helping their clients get back on their feet and expanded programs aimed at environmental causes, making the record promotion of green bonds that year emblematic (Monteiro et al., 2023). Also in 2020, 25 new issuers entered the green bond market, launching their first issues, 17 of which came from Brazil alone. In a year marked by uncertainty around the world, green bond issuance rebounded in the second half of 2020, reaching a record 269.5 billion dollars by the end of December (surpassing 2019s 266.5 billion dollars). Although the first quarter was strong, green bond issuance in the second quarter of 2020 was affected by the COVID-19 pandemic, but recovered in a record third quarter, according to the Climate Bonds Initiative (2021).

7 Final Considerations

Agribusiness is present in all sectors of society, whether providing inputs as raw materials for transformation or providing final-stage products ready for consumption, making its role and relevance for the continuous maintenance of all living beings on the planet clear. As a major producer of commodities, Brazil has considerable importance for the internal and external production chain, as its production surpasses other countries in quantity, quality, and variety. However, the economic advance of the 1990s also revealed the depletion of natural resources and the aggressive deforestation of territories to expand agricultural and livestock frontiers.

With a continental territorial extension, Brazil has ample land for various branches of agribusiness such as agriculture, livestock, and forestry, among others. However, the carbon footprint is still a real problem and many companies do not feel obliged to mitigate GHG emissions, as such process changes are considered expensive and the information is not clearly disclosed. When the green agenda is raised, businesspeople see it as an issue that can be postponed and does not require attention, even though the irreversible increase in the planet's temperature continues to worry scientists and its consequences are already being felt.

Green bonds are an important tool for decarbonization, as it is impossible to talk about resilience and mitigating the effects of climate change without including the financial market. It is also impossible to address green issues without social concern for rural populations whose income is in decline, compromised by the consequences of climate change and the depletion of resources resulting from the expansion of borders by large industries. With this, the green seal also becomes an attractive element, increasing the companies' reputation in the eyes of consumers and the market. It is important to note that strict supervision, with defined taxonomies, and accurate disclosure of the use of funds from green finance are essential to maintain the confidence of investors and the general public.

That is why public policies such as the ABC Plan, with guidelines for low-carbon agriculture, become essential when issuing a green bond. In this way, companies reduce their carbon footprint through technologies such as BNF or DSS, financed by these emissions whose use of resources is entirely directed toward the green cause.

As the world faces increasingly urgent environmental challenges, the search for strategies to mitigate agribusiness liabilities is inevitable. It is hoped that this study has provided a comprehensive overview of the need to promote financial instruments dedicated to decarbonization, especially in a country that is recognized as a global breadbasket. Furthermore, due to the need for new materials on the topic, it is advisable that the study be used by other critical sectors of Brazilian agribusiness, as well as in studies of the financial sector.

References

ABIOVE. (2022). *Moratória da soja. Safra 2020/2021.* Monitoramento da soja por imagens de satélites no bioma Amazônia. https://abiove.org.br/wp-content/uploads/2022/10/moratoria20_21.pdf

Almeida, M. & Filkova, M. (2019). *América Latina e Caribe Análise de mercado das finanças verdes 2019.* Climate bonds initiative. https://www.climatebonds.net/files/reports/latam_sotm_19_por_final_01_web.pdf

Alves, F. V., Laura, V. A., & de Almeida, R. G. (2015). *Sistemas agroflorestais: a agropecuária sustentável.* ISBN 978-85-7035-420-4. http://www.infoteca.cnptia.embrapa.br/infoteca/handle/doc/1010949

Anderson, R. G., Canadell, J. G., Randerson, J. T., Jackson, R. B., Hungate, B. A., Baldocchi, D. D., et al. (2011). Biophysical considerations in forestry for climate protection. *Frontiers in Ecology and the Environment, 9*(3), 174–182. https://doi.org/10.1890/090179

Aragão A. & Contini E. (2021). *O agro no brasil e no mundo: uma síntese do período de 2000 a 2020.* EMBRAPA SIRE. https://www.embrapa.br/documents/10180/62618376/O+AGRO+NO+BRASIL+E+NO+MUNDO.pdf/41e20155-5cd9-f4ad-7119-945e147396cb

Assad, E. D., Martins, S. C., & Pinto, H. P. (2012). *Sustentabilidade no agronegócio brasileiro.* Fundação Brasileira para o Desenvolvimento Sustentável. http://www.alice.cnptia.embrapa.br/alice/handle/doc/933999

Azevedo, A. A., & Pasquis, R. (2007). Da abundância do agronegócio à Caixa de Pandora ambiental: a retórica do desenvolvimento (in) sustentável do Mato Grosso (Brasil). *Interações (Campo Grande), 8*, 183–191. https://doi.org/10.1590/S1518-70122007000200005

Barbier, E. B., & Hochard, J. P. (2018). Land degradation and poverty. *Nature Sustainability, 1*(11), 623–631. https://doi.org/10.1038/s41893-018-0155-4

Camargo, K. C. D. M. (2017). *Dinâmica demográfica e transformação econômica recente no Mato Grosso* (Doctoral dissertation, dissertação). Universidade Estadual de Campinas. https://doi.org/10.47749/T/UNICAMP.2017.983910.

CEPEA. (2023). *Desempenho das exportações do agronegócio*. Centro de Estudos Avançados em Economia Aplicada. https://www.cepea.esalq.usp.br/br/indices-de-exportacao-do-agronegocio.aspx

CEPEA & ABIOVE (2023). *Cadeia da soja e do biodiesel: PIB, empregos e comércio exterior – 1° trimestre de 2023*. Centro de Estudos Avançados em Economia Aplicada e Associação Brasileira das Indústrias de Óleos Vegetais. https://www.cepea.esalq.usp.br/upload/kceditor/files/Cepea_Abiove_Cadeia%20da%20Soja%20e%20biodiesel_2T2023.pdf

Climate Bonds Initiative. (2021). *Record $269.5bn green issuance for 2020: Late surge sees pandemic year pip 2019 total by $3bn*. Disponível em: https://www.climatebonds.net/2021/01/record-2695bn-green-issuance-2020-late-surge-sees-pandemic-year-pip-2019-total-3bn

CONAB. (2023). *Acompanhamento da safra brasileira. Cana-de-açucar Safra 2022/2023 – 4° levantamento*. https://www.conab.gov.br/info-agro/safras/cana/boletim-da-safra-de-cana-de-acucar/item/download/47195_94fff688f8d04822096cf788ed0a95d6

Das, P. K. & Cungu, A. (2023). *Green finance as a critical lever for delivering sustainable agrifood systems – A global landscape study*. FAO. https://doi.org/10.4060/cc7402en.

de Assis, T. F., Monteiro, T. G. M., de Abreu, V. H. S., D'Agosto, M. D. A., & Santos, A. S. (2022a). Enabling the green bonds market for sustainable transport projects based on the measure/monitoring, reporting and verification method. In *Carbon footprints of manufacturing and transportation industries* (pp. 1–24). Springer Nature Singapore. https://doi.org/10.1007/978-981-19-7226-3_1

de Assis, T. F., Ricci, L. M., Monteiro, T. G. M., de Abreu, V. H. S., D'Agosto, M. D. A., & Santos, A. S. (2022b). Sustainable transport indicators and mitigation actions applied to the green bond principles. In *Carbon footprints of manufacturing and transportation industries* (pp. 139–169). Springer Nature Singapore. https://doi.org/10.1007/978-981-19-7226-3_6

de Miranda, R. A. (2020). *Breve história da agropecuária brasileira*. Dinâmica da produção agropecuária e da paisagem natural no Brasil, 31. https://www.alice.cnptia.embrapa.br/bitstream/doc/1122598/1/Cap02-BreveHistoriaAgropecBR.pdf

Duarte. (2020). *Destruição da Amazônia pode transformá-la em deserto e desencadear pandemias*. O Eco. https://www.oeco.org.br/reportagens/destruicao-da-amazonia-pode-transforma-la-em-deserto-e-desencadear-pandemias/

EMBRAPA. (2018). *VISÃO 2030*. O Futuro da Agricultura Brasileira. https://www.embrapa.br/visao/o-futuro-da-agricultura-brasileira

EMBRAPA. [s.n.]. *Sistemas de produção integrados – ILPF*. Empresa Brasileira de Pesquisa Agropecuária. https://www.embrapa.br/qualidade-da-carne/carne-bovina/producao-de-carne-bovina/sistemas-de-producao-integrados-ilpf

FAO. (2021). *The state of the world's land and water resources for food and agriculture – Systems at breaking point*. Synthesis report 2021. Rome. https://doi.org/10.4060/cb7654en.

Fearnside, P. M. (2001). Soybean cultivation as a threat to the environment in Brazil. *Environmental Conservation, 28*(1), 23–38. https://doi.org/10.1017/S0376892901000030

Ferreira, M. D. P., & Coelho, A. B. (2015). Desmatamento Recente nos Estados da Amazônia Legal: uma análise da contribuição dos preços agrícolas e das políticas governamentais. *Revista de Economia e Sociologia Rural, 53*, 91–108. https://doi.org/10.1590/1234-56781806-9479005301005

Forsyth, T. (1996). Science, myth and knowledge: testing Himalayan environmental degradation in Thailand. *Geoforum, 27*(3), 375–392. https://doi.org/10.1016/S0016-7185(96)00020-6

Galvão, R. R. D. A. (2017). *O biogás do agronegócio: Transformando o passivo ambiental em ativo energético e aumentando a competitividade do setor*. FGV Energia. https://bibliotecadigital.fgv.br/dspace/bitstream/handle/10438/19250/Coluna%20Opiniao%20Rodrigo%20Regis.pdf

Gianfrate, G., & Peri, M. (2019). The green advantage: Exploring the convenience of issuing green bonds. *Journal of Cleaner Production, 219*, 127–135. https://doi.org/10.1016/j.jclepro.2019.02.022

Gomes, C. S. (2019). Impactos da expansão do agronegócio brasileiro na conservação dos recursos naturais. *Cadernos do Leste, 19*(19). https://doi.org/10.29327/248949.19.19-4

Governo Federal. (2022). *Brasil lidera ranking mundial de exportação de carne de frango*. https://www.gov.br/pt-br/noticias/financas-impostos-e-gestao-publica/2022/09/brasil-lidera-ranking-mundial-de-exportacao-de-carne-de-frango#:~:text=O%20Brasil%20%C3%A9%20l%C3%ADder%20na,US%24%207%2C6%20bilh%C3%B5es

Gozgor, K., & Karakas, M. (2023). How do global financial markets affect the green bond markets? Evidence from different estimation techniques. *Economic Research-Ekonomska Istraživanja, 36*(2), 2177703. https://doi.org/10.1080/1331677X.2023.2177703

Greenpeace. (2018). *Relatório do Monitoramento da Moratória da Soja na Amazônia é divulgado*. https://www.greenpeace.org/brasil/imprensa/relatorio-do-monitoramento-da-moratoria-da-soja-na-amazonia-e-divulgado/

ICMA. (2021). *Princípios para Títulos Verdes*. Diretrizes Voluntárias para Emissão de Títulos Verdes. Junho 2021. https://www.icmagroup.org/assets/documents/Sustainable-finance/Translations/Portuguese-GBP-2021_06.pdf

IPEA. (2011). *Instituto de Pesquisa Econômica Aplicada. Diretoria de Estudos e Políticas Sociais (Disoc)*. Políticas Sociais. Acompanhamento e Análise: n. 19, 2011. https://repositorio.ipea.gov.br/handle/11058/4122

Lima, S. S. D., Leite, L. F. C., Oliveira, F. D. C., & Costa, D. B. D. (2011). Atributos químicos e estoques de carbono e nitrogênio em argissolo vermelho-amarelo sob sistemas agroflorestais e agricultura de corte e queima no norte do Piauí. *Revista Árvore, 35*, 51–60. https://doi.org/10.1590/S0100-67622011000100006

Luizão, F. J., Tapia-Coral, S., Gallardo-Ordinola, J., Silva, G. C., Luizão, R. C. C., Trufillo-Cabrera, L., ... & Freitas, M. S. M. (2006). *Sistemas Agroflorestais: bases para o desenvolvimento sustentável*. Universidade Estadual do Norte Fluminense Darcy Ribeiro, pp. 87–100. https://www.researchgate.net/publication/289534112_

Makpotche, M., Bouslah, K., & M'Zali, B. (2023). Long-run performance following corporate green bond issuance. *Managerial Finance, 50*, 140. https://doi.org/10.1108/MF-12-2022-0588

MAPA. (2012). Plano setorial de mitigação e de adaptação às mudanças climáticas para a consolidação de uma economia de baixa emissão de carbono na agricultura: plano ABC (Agricultura de Baixa Emissão de Carbono)/Ministério da Agricultura, Pecuária e Abastecimento, Ministério do Desenvolvimento Agrário, Coordenação da Casa Civil da Presidência da República. – Brasília. ISBN 978-85-7991-062-0. https://www.gov.br/agricultura/pt-br/assuntos/sustentabilidade/plano-abc/arquivo-publicacoes-plano-abc/download.pdf

MAPA. (2020). *Ministério lança Plano de Investimento para Agricultura Sustentável*. Ministério da Agricultura, Pecuária e Abastecimento. https://www.gov.br/pt-br/noticias/agricultura-e-pecuaria/2020/06/ministerio-lanca-plano-de-investimento-para-agricultura-sustentavel.

Margulis. (2003). *Causas do Desmatamento da Amazônia Brasileira*. ISBN: 85-88192-10-1. https://documents1.worldbank.org/curated/en/867711468743950302/pdf/277150PAPER0Po1az1nia0Brasileira111.pdf

Mattos Filho. (2019). *Desmistificando os Green Bonds: Financiamento do agronegócio*. Laboratório de Inovação Financeira. https://www.gov.br/agricultura/pt-br/assuntos/politica-agricola/financas-verdes/textos-tecnicos-e-apresentacoes/DesmistificandoosGreenBonds_Agronegocio.pdf

Mello, N. G. R. D., & Artaxo, P. (2017). *Evolução do plano de ação para prevenção e controle do desmatamento na Amazônia legal*. Revista do Instituto de Estudos Brasileiros, 108–129. https://doi.org/10.11606/issn.2316-901X.v0i66p108-129

Monteiro, T. G. M., de Abreu, V. H. S., & Santos, A. S. (2023). Uma visão geral sobre os Green Bonds no cenário pandêmico de 2020: An overview of Green Bonds in the 2020 pandemic scenario. *MIX Sustentável, 9*(3), 171–180. https://doi.org/10.29183/2447-3073.MIX2023.v9.n3.171-180

Moretti, C. L., & Ferreira, T. T. (2021). A agricultura de baixo carbono no Brasil. *AgroANALYSIS, 41*(4), 23–24. https://periodicos.fgv.br/agroanalysis/article/view/86640

Ozili, P. K. (2022). Green finance research around the world: A review of literature. *International Journal of Green Economics, 16*(1), 56–75. https://doi.org/10.1504/IJGE.2022.125554

Padovan, M. P. (2021). *Benefícios dos SAFs, mas o que é SAF?* EMBRAPA. Empresa Brasileira de Pesquisa Agropecuária. https://www.embrapa.br/busca-de-noticias/-/noticia/65068763/artigo%2D%2D-beneficios-dos-safs-mas-o-que-e-saf

Perfecto, I., & Vandermeer, J. (2010). The agroecological matrix as alternative to the land-sparing/agriculture intensification model. *Proceedings of the National Academy of Sciences, 107*(13), 5786–5791. https://doi.org/10.1073/pnas.0905455107

Primavesi, A. (2002). *Manejo ecológico do solo: a agricultura em regiões tropicais*. NBL Editora. ISBN 85-213-0004-2.

Reed, M. S., Dougill, A. J., & Taylor, M. J. (2007). Integrating local and scientific knowledge for adaptation to land degradation: Kalahari rangeland management options. *Land Degradation and Development, 18*, 249–268. https://doi.org/10.1002/ldr.777

Sambuichi, R. H. R., Constantino de Oliveira, M. Â., Moreira da Silva, A. P., & Luedemann, G. (2012). *A sustentabilidade ambiental da agropecuária brasileira: impactos, políticas públicas e desafios* (No. 1782). Texto para Discussão. https://www.econstor.eu/handle/10419/91310

Sant'Anna, A. C., Granco, G., Bergtold, J. S., Caldas, M. M., Xia, T., Masi, P., ... & Lorenzani, W. (2016). *Os desafios da expansão da cana-de-açúcar: a percepção dos produtores e arrendatários de terra em Goiás e Mato Grosso do Sul*. GR Santos, Quarenta anos de etanol em larga escala no Brasil: desafios, crises e perspectivas, 113–143. https://portalantigo.ipea.gov.br/agencia/images/stories/PDFs/livros/livros/160315_livro_quarenta_anos_etanol_cap04.pdf

Shikida, P. F. A. (2013). Expansão canavieira no Centro-Oeste: limites e potencialidades. *Revista de Política Agrícola, 22*(2), 122–137. https://seer.sede.embrapa.br/index.php/RPA/article/view/312/260

Silva, F. P. D., & Vieira Filho, J. E. R. (2020). *Avaliação de impacto do programa de agricultura de baixo carbono no Brasil*. https://repositorio.ipea.gov.br/bitstream/11058/10101/1/td_2568.pdf

Sinha, A., Mishra, S., Sharif, A., & Yarovaya, L. (2021). Does green financing help to improve environmental & social responsibility? Designing SDG framework through advanced quantile modelling. *Journal of Environmental Management, 292*, 112751. https://doi.org/10.1016/j.jenvman.2021.112751

USDA. (2023). *Livestock and poultry: World markets and trade*. United States Department of Agriculture. http://www.fas.usda.gov/psdonline/circulars/livestock_poultry.pdf

Index

S. S. Muthu (ed.), *Carbon Footprint Assessments*, Environmental Footprints and Eco-design of Products and Processes, https://doi.org/10.1007/978-3-031-70262-4

The manufacturer's authorised representative in the EU is Springer Nature Customer Service Centre GmbH, Europaplatz 3, 69115 Heidelberg, Germany. If you have any concerns regarding our products, please contact ProductSafety@springernature.com

Printed and bound by CPI Group (UK) Ltd, Croydon, CR0 4YY
30/09/2026
02225563-0012